Westinghouse J40 Axial Turbojet Family

Development History and Technical Profiles

Westinghouse J40 Axial Turbojet Family

Development History and Technical Profiles

Paul J. Christiansen

Bleeg Publishing LLC

Bleeg Publishing LLC
Olney, MD

ISBN: 978-0692358528

First Edition

Front and rear cover illustrations: Insets showing front and rear sections of the J40-WE-6, taken from Westinghouse assembly drawing 43J950, June 13, 1952. *(Drawing not to scale.)*

Cover by Carol Hardy Design (www.carolhardydesign.com)

Dedication

To my wife Leslie, who presented me with two lovely daughters and has supported me over the years in all things aeronautic, no greater proof of which was listening to endless hours of my recounting the information in this book as I pieced together the development story and, even then, volunteering to edit this book.

Author's Notes:

I have left all the data in the U.S. English measurements of the original sources. Both Westinghouse and BuAer appear to have been confused at times over any differences between the development engines and the equivalent production model, since so many changes were in process at all times to both. Insufficient records were found to accurately trace the house engines beyond the very early stages of development. Most early development engines were apparently rebuilt into later dash models in the low powered series and are only occasionally referenced in the photo library as "#4" or "#7" on the photo card in addition to the full model designation. The terminology and even the spelling of words related to the technology was in a state of flux during the period covered by the project and I have incorporated the original language with all of its sometimes awkward phrasing to give the reader a sense of the emerging lexicon of the time period. Where a term or its use is obscure in modern parlance I have provided a translation. The Hagley Museum and Library photo collection of J40 program related images is in the form of cropped prints mounted on photo-library index cards. The original negatives no longer exist. The result is that many of the images used in the book are slightly clipped. Hopefully the images selected still enable the reader to gain an understanding of the construction details of the various components and the core engine itself.

It is fortunate that the project basis for every engine model was the specification, since all changes to it had to be requested and approved along with the various document deliverables required by a specific contract. Enough of the technical data related to these discussions has survived within the BuAer contract records. Considerable effort was required to extract this data following a long period of organization of the surviving records to arrive at a more or less sequential series of events for each model. Any errors in interpreting the material are mine alone.

Acknowledgements:

My thanks go to Jeremy Kinney at the Smithsonian National Air and Space Museum, who arranged for NASM's example of their J40 to be taken from storage and pictures taken of both the engine and its data plate. The production contract number on the latter was the key to opening the door at the National Archives for the detailed research that followed. To the many professionals at the National Archives at both the College Park Archives II and downtown Washington locations go my thanks for their assistance in locating the many containers of documents pertaining to the many contracts involved. As well go my special thanks to the professional archivists at the Hagley Museum and Library, stewards of the Westinghouse Electric Gas Turbine Division photographic library, for allowing me to scan the entire collection of J40 photos. Lastly my thanks go to the Aircraft Engine Historical Society for its support and encouragement over the years as this project grew from seeking a few answers into the book you are holding.

Contents

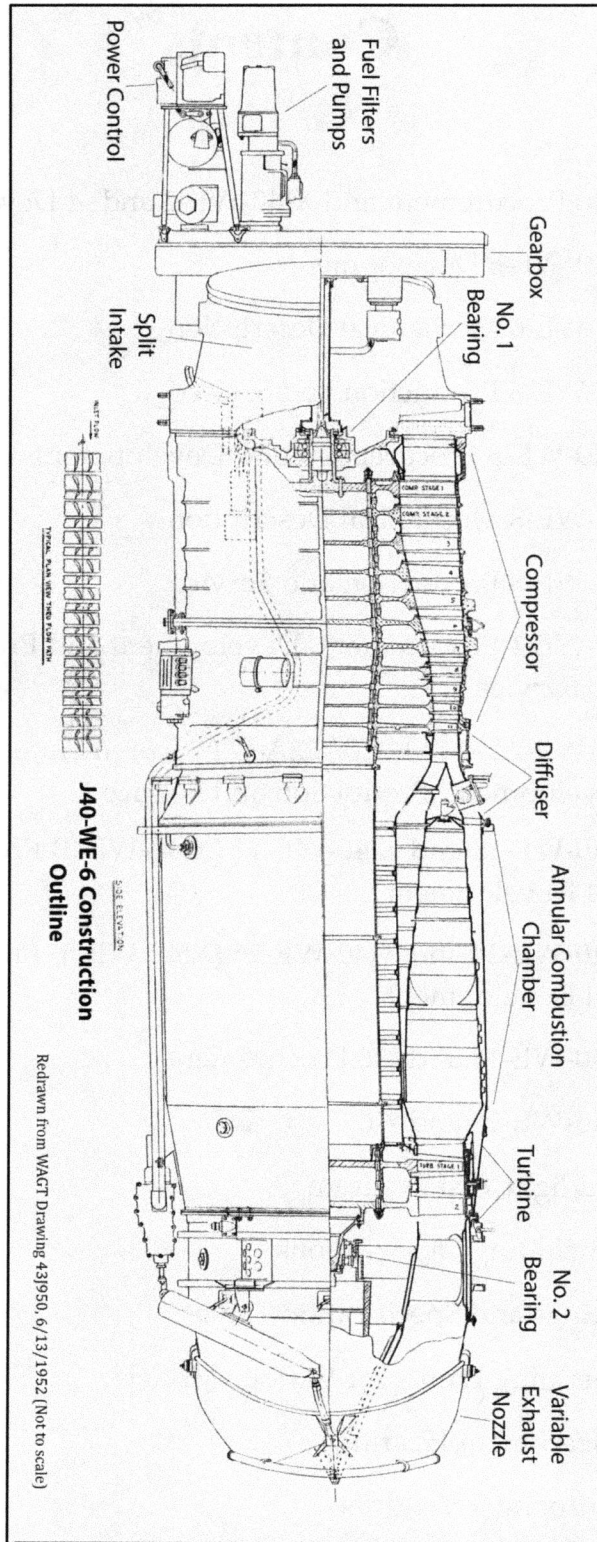

J40-WE-6 Construction outline showing the overall layout of the engine. This general layout was used for all versions of the engine, the A/B versions' nozzles either being the dual eyelid type or an iris multi-petal design. (Compressor elements slightly distorted in original.) *Retraced by the author.*

Introduction

◆——◆

"This is a hazard you take if you are going to have first line aircraft that are the best in the world you have to take some risk like that. Here we were in Korea and the J-40 engine if it had come through it would have been a world beater." – Admiral Russell testimony, 84th U.S. Congress Subcommittee Hearing on Jet Aircraft Procurement

The U.S. Navy began a new turbojet engine procurement in 1947 that ended with the Westinghouse Aviation Gas Turbine (WAGT) Division submission being selected to develop and produce the Navy's next generation of high power engines. The procurement selection process and the reasoning behind the selection were carefully documented by the Bureau of Aeronautics (BuAer) branch of the U.S. Navy as part of their request for funding approval.

No evidence has surfaced that there was much, if any, concern about WAGT's having not only a full production program underway for the J34, but two new versions of the J34 under development contract as well as two versions of another entirely new engine, the J46, to be developed in parallel with the new engine. While it was noted that WAGT had experienced problems in the past transitioning both of their prior engine designs (J30 and J34) into production, the Navy assumed WAGT had learned from these problems.

The Navy could not foresee the rapid change in mission requirements that would be brought about by both advances in aeronautical knowledge and unexpected world events (Korean War, Russia obtaining nuclear weapons) occurring in only a short period of time after the selection was made and development begun. The developing mass of data being collected from the Douglas Skystreaks and Skyrockets, the Bell series of X-1 research aircraft and National Advisory Committee for Aeronautics (NACA) wind tunnel research also began to indicate the need for far greater power

with even better fuel economies than the developing engines had been contracted to deliver.

The explosion of a nuclear weapon by the Soviet Union in 1949 brought sharp focus on the increased need for fast-climbing, supersonic interceptors with increased mission radius. Competition with the Air Force was driving the Navy to develop a long range carrier-based bomber capable of delivering a nuclear weapon. In addition, the outbreak of war in Korea forcefully demonstrated limitations of the then current Navy jets and ramped up the pressure for follow-on equipment with enhanced performance. As these evolving requirements emerged, the Navy contracted with WAGT to deliver more powerful and efficient versions of the J40 in parallel with the original designs. WAGT responded with proposals that indicated the new requirements could be met using the same basic design approach and within the basic installation envelope. By the time the program ended, there had been many different models of the J40 in development or proposed.

As WAGT pressed forward with development, the engineering assumptions and predictions upon which the engines were based proved in many cases to be false. In every area: intake, compressor design and performance, combustion, turbine design and performance, engine management and weight, modifications were required. In addition to the industry's general lack of advanced fluid thermodynamics empirical knowledge, WAGT also lacked the development environment or resources to aggressively attack these challenges and began to address their weaknesses too late to affect the outcome. They used a sequential test approach using a limited number of engines, and assumed the first engines would be basically correct, such an approach having been (more or less) successful in earlier programs. Lacking their own development machine shop facilities, they suffered long delays as subcontractors produced modified parts. As development bogged down, the need for constant re-testing created bottlenecks and the Navy had to supply additional test cells at their own facility. Creation of more realistic project plans as well as additions to their engineering resources and development facilities were late being put into place, primarily due to Westinghouse not wishing to invest their own money in the Gas Turbine Division.

The technical challenges on the testing front also increased, as the military abandoned the standard military engine development, testing and production specifications used in WWII era and moved to a new, greatly enhanced and robust set of "Mil-E" specs. Dramatically stiffer testing requirements in both the hot and cold cycles were added, as well as changes to the 150 Hour Specification Test intended to more closely mirror the actual mission cycles of the engine. All presented new challenges to be met and overcome in order to obtain release to production.

With engine development under contract, the Navy rushed forward with aircraft development for interceptors and bombers. As the need to get the next generation of aircraft in the sky increased, the engines were placed under production contracts before their designs or satisfactory testing could be completed. Early on, the program began to expand across another branch when the Air Force selected one model for what was to become the North American X-10 aircraft. Various later family models were studied in depth for other Air Force aircraft (such as the XB-52) and two Navy programs, code named Skate and Mine Layer. All of these programs placed heavy pressure on WAGT to get basic development completed and working engines through their acceptance tests and into production while complicating everything with the need to coordinate with an increasing pool of airframe manufacturers.

With the higher power later models of the engines seemingly making little progress in development, the Navy was forced to use the original lower power models for the initial testing of the Douglas, McDonnell and Grumman airframes, these engine models being the furthest along and eventually passing (with some caveats) their 150 hr qualification tests. The testing of unique prototype aircraft with incompletely developed engines made for a very complex and unsatisfactory situation for all concerned. Recognizing, albeit a bit late, that the likelihood of receiving the higher power versions actually required by the various aircraft was unlikely to occur, the Navy finally switched to other engine manufacturers for replacements. WAGT's performance throughout left no doubt by the program's end that the Navy had ultimately made the correct decision to abandon the J40 family of engines, even if those decisions were taken too late to avoid embarrassment and strong Congressional criticism.

While the general project outline of the program has been known for some time, the highly complex technical development picture has never been documented in detail. Being a classified project while underway, the many details were hidden from view in the files. The classified security status was removed in 1965, but by then, interest in the J40 engines and their technical details had waned. Neither the Navy nor WAGT had any reason to want the project to be dragged out for review. As a result, the full technical details of most of the models of the J40 have never been published, and the few details in print are often wrong or applied to the wrong model.

It is fortunate that all of the engine models that entered development produced detailed model specifications that became part of the contractual record and these, along with all the change correspondence related to these specifications and the contract deliverables, were retained in the contract files. It has allowed the recreation of the development story in large part, even if some engineering specifics were not preserved.

This volume describes the development, technical challenges and provides a technical description for each model. Some development issues affected more than one model and while every effort has been made to reduce redundancy in the text, it could not be entirely eliminated.

It was not the intention of the author to rehash or question the decisions made by the parties involved in the engine family's development, only to present the history and information as they have emerged from the files. That the project failed to produce viable engines of the desired models within the timeframes they were needed takes nothing away from the good faith efforts of all involved, working within their technical limits and organizational constraints, to satisfy the Navy's requirements. If anything, it provides insight into both the development and the challenges of changing requirements in this transitional period in turbojet development, a time where the vast majority of the turbojet engines designed by industry failed technically or failed to meet the needs of the marketplace. The large research and plant investments made by the traditional aircraft engine industry giants began to allow them to produce successful designs well matched to the military's needs. Westinghouse's late and more modest investments were to prove fatal and by 1965 they had abandoned the aviation gas turbine business.

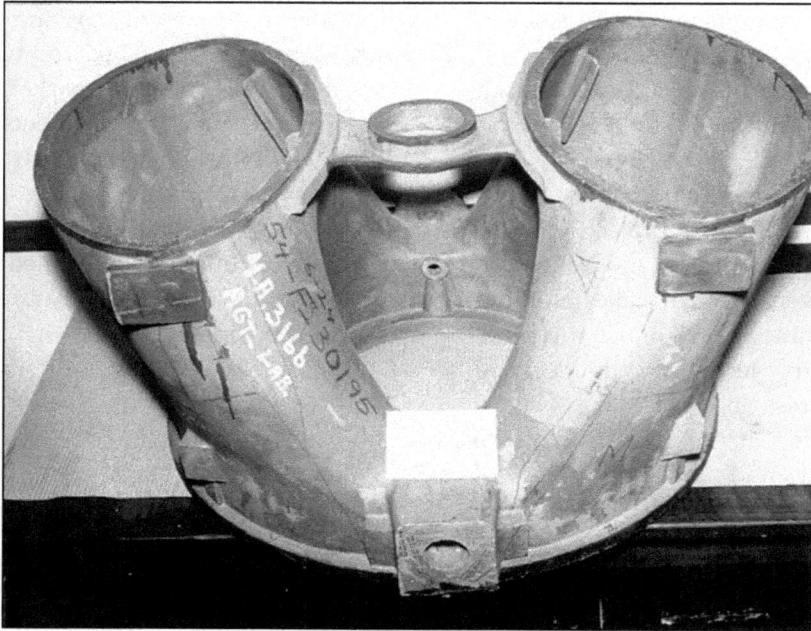

Figure 1 First raw magnesium casting of XJ40-WE-2 intake, WAGT P39237, 7/1/1948. *Courtesy Hagley Museum and Library*

Figure 2 Compressor disc for the XJ40-WE-2 after broaching of the blade slits and before curvic clutch machining, WAGT P39240, 7/1/1948. *Courtesy Hagley Museum and Library*

Chapter 1

XJ40-WE-2 and -4 Procurement and Development

◆————————————————————————————◆

Procurement

After World War II, the U.S. Navy began studies to identify the future needs for turbojet engines, and in October 1945 the VP Design Branch recommended development of a new power plant in the range of 8-12,000 pounds of thrust. The Bureau of Aeronautics (BuAer) included funding in the 1947 budget estimates to initiate development, but recommended detailed studies to identify the most desirable size of engine to be developed. The study was initiated by the Aviation Design Research Branch.

BuAer initiated a request in February 1946 for informal proposals leading to the development of a 7,000 lb thrust engine to be issued to all aircraft engine manufacturers. The proposal was to be broken into two phases; Phase A to cover engine design and delivery of a complete model specification and Phase B to cover full scale engine development through a qualification test. It was anticipated that Phase A would provide sufficient data to permit detailed airplane design studies. Provision was made for scaling the final design up or down from the 7,000 lb thrust design based on results from airplane analyses.

Approval to issue the proposal request was withheld, BuAer being directed to coordinate the proposed project with the Army Air Forces through the Propulsion Sub-Committee of the Research and Development Committee of the Aero Board. The resulting proposal was reviewed at a joint services meeting held on April 4, 1946 and, after reviewing the existing Army-Navy jet engine development programs, it was agreed that development of an engine in the 7,500 lb thrust class should be initiated by the Navy.

On April 12, 1946, BuAer again requested authority to proceed with the proposal request and partial authority was received. BuAer was authorized to invite proposals for the design study phase from Wright, Continental, Allison, and Allis-Chalmers, with the proviso that any work undertaken was not to interfere in any way with the turboprop developments underway. Westinghouse was not included because of this provision, as they were lagging on the 25D turboprop development of the J34 at that time.

In response, Continental declined to bid, Allison submitted a satisfactory proposal, Allis-Chalmers submitted

an unacceptable proposal and Wright requested a delay of a few months in order to submit a proposal for a complete development program, finally submitted on November 12. Wright had previously submitted a technically unattractive proposal for a 6,000 lb engine in February, 1946 (which was funded by the Army Air Forces as the XJ51-W-2) and their new proposal was a scaled up version of the XJ51.

On November 26, 1946, Pratt and Whitney stated that the company desired to stop work on the PT-1 turboprop and would be interested in undertaking development of a 7,500 lb thrust engine. They were instructed to submit a proposal which was forthcoming on February 1, 1947. This proposal appeared to be high in cost and outlined a lengthy development time. As a result, Allison was requested to submit a corresponding proposal, which they did on February 24, 1947.

Westinghouse submitted a letter on March 25, 1947 stating they were interested in submitting a proposal on this development. A prior decision had been made on March 10 to terminate the 25D turboprop development, so Westinghouse was permitted to submit if they could complete the proposal in thirty days. The gating factor at this point was the short time remaining for processing procurements on that year's funds. Meeting the deadline, Westinghouse presented their proposal on April 28, 1947.

General Electric, who was considered technically qualified to bid, was not included in the bidding because they were concentrating all their efforts on development of the TG-190 engine. General Electric had abandoned the TG-100 and TG-110 turboprop engines and had transferred engineering responsibility for the I-40 and TG-180 to Allison. Based on this, BuAer came to the conclusion that little confidence could be placed in General Electric's future ability to deliver an additional completely engineered and developed product.[1]

The design points for the engine to be developed were covered in a U.S. Navy Aviation Design Research memo, ADR Report No. 1012 dated November 1946. Its summary conclusions were:

1. *A turbojet unit developing 12,000 pounds static sea level thrust and possessing Westinghouse 24C operating characteristics results in maximum performance when installed in a fighter similar in configuration to ADR-47 with a combat radius of 300 nautical miles. A further increase in engine size results in a larger airplane of reduced performance.*

2. *Until the engine size (thrust) is reduced to approximately 8,000 pounds static thrust, the reduction in performance associated with reduced engine size is quite small, compared to the reduction in airplane size and weight.* For example as shown in the table to the right:

Performance vs. Thrust Correlations		
Static Thrust (lbs)	12,000	8,000
V_{max}, (mph)	686	670
Gross Weight (lbs)	27,000	19,000

The four proposals available were summarized as to comparative performance and characteristics. All of them proposed development of a satisfactory service engine, after having completed the qualification tests in accordance with the specifications of AN-E-32 (covering engines and engine testing).

Performance Characteristics – 7,500 lb Thrust Turbojet Proposals

	Westinghouse X40-E2	P&W JT-3B	P&W JT-3C	Wright Aero TJ-2	Allison – 505 A	Allison – 505 B	Allison – 505 C	Allison – 505 D
Dry Weight	2,712	2,950	2,710	3,500	2,950	3,200	3,450	3,700
T.O. Spec. Wt.	.362	.394	.331	.467	.394	.427	.460	.494
Basic Dia.(in)	40	50	50	40	40	41	42	43
Length (in)	186	146	146	184	155	160	165	170
T.O. Thrust (lbs)	7,500	7,700	8,200	7,500	7,500	7,500	7,500	7,500
Mil. Thrust (lbs)	7,500	7,700	8,200	7,500	7,500*	7,500*	7,500*	7,500*
Norm. Thrust (lbs)	6,800	6,350	6,950	7,500	----	----	----	----
T.O. SFC	0.98	1.05	0.90	0.95	1.01	0.98	0.94	0.90
Mil. SFC	0.98	1.05	0.90	0.95	----	----	----	----
Norm. SFC	0.93	1.03	0.86	0.95	----	----	----	----
D. P. SFC **	1.06	1.177	1.06	1.023	1.11	1.07	1.03	1.00
Guar. Max. Alt.	50,000 ft	50,000 ft	50,000 ft	50,000 ft	Not Guaranteed			

*   Assumed Value of Military Rating
** Design point SFC at 1525 lbs net thrust at 460 MPH at 35,000 ft. cruise condition.
(a) Dry weights as per AN-E-31, paragraph E-5.
(b) Westinghouse and Wright engine dry weight included a variable area jet nozzle.
(c) The JT-3B was to be the first flight engine and the JT-3C the fully developed and qualified engine.
(d) Allison proposed engines ranged from extreme light weight and small size to best fuel efficiency.

Based on the comparison data, Westinghouse appeared to have the best proposal in the overall performance picture when consideration was given to weight, SFC, and basic diameter. Comparison of the submitted delivery schedules, revealed that here too, Westinghouse was significantly superior to the others. The various delivery milestone phases were defined as:

Phase I – Through one running engine and an available coordinated engine model specification suitable for reference in airplane detail specifications.
Phase II – Through 50 hour flight substantiation test which permitted clearance for flight in prototype aircraft.
Phase III – Through 150 hour qualification test, which permitted establishment of production parts list, release for production, and clearance for unrestricted service flight operation.

Proposed Delivery Schedules (Months)					
Manufacturer	Phase I	Phase II	Cumulative I and II	Phase III	Total
Westinghouse	15	9	24	6	30
Allison	20	14	34	12	46
Wright	22	16	38	10	48
Pratt & Whitney	36-42	6-12	42-54	18	60-72

Proposed Cost of Development (Dollars)[2]					
	Westinghouse	Wright	Allison Target	Ceiling	Pratt & Whitney
Phase I	1,276,799	----	2,900,000	4,060,000	----
Phase II	1,330,411	7,024,527	3,325,000	4,655,000	10,650,000
Subtotal	2,607,210	7,024,527	6,225,000	8,715,000	10,650,000
Phase III	2,054,025	1,600,000	3,675,000	5,145,000	3,950,000
Subtotal	4,661,235	8,624,527	9,900,000	13,860,000	14,600,000
Fixed Fee	326,287	----	----	----	----
Program Total	4,987,522	8,624,527	9,900,000	13,860,000	14,600,000

BuAer noted that while the Westinghouse quotation was on a cost plus fixed fee basis, the actual costs would have to over-run the estimates by approximately 70% before reaching the cost of the next lowest quotation (from Wright). BuAer considered the experimental engine programs underway at the four firms at the time of the proposal analysis and documented the following at that point in time:

Major Experimental Engine Projects with Contractors Concerned[3,4]							
Army and Navy experimental engine projects underway at each contractor being considered. The year following the project was the estimated time when that project was expected to become a minor engineering effort.							
Allison		Pratt & Whitney		Westinghouse		Wright Aero. Corp.	
T-39	1947	2800	1947	J-34	1952	3350	1947
J-33	1948	4360	1951			T-43	1948
T-40	1950	Nene	1951			T-35	1952
		T-34	1954			J-51	1952
						J-34*	1953

*Production engineering only, expected to have a secondary effect on contractor's experimental engineering group

Further review of the various competitors as they related to the proposed program was summarized as follows:

Allison Division, General Motors Corporation – Allison was the leading producer of turbojet engines in the country and was in quantity production of the J33 (I-40) and J35 (TG-180) engines. Allison had accepted engineering and development responsibility for the J33 and was in the process of accepting engineering responsibility for the J35. They had a large engineering and test plant, but experience on axial flow compressor design was limited to that of the Model 500 (XT40-A-2) turboprop then under development.

Wright Aeronautical – Turbojet engine production capacity had not yet been put in place. They were currently designing a 6,000 lb axial-flow compressor (XJ51) and a turboprop (T-35) of centrifugal design for the Army Air Forces.

Pratt & Whitney – The firm was gaining production experience on axial-flow engines through their 19XB production program in support of Westinghouse. Delays were being experienced due to the fact that the 19XB was not production engineered by Westinghouse. They were also developing the T34 turboprop engine for the Navy and were expected to pick up American Nene production when the engine was ready.

Westinghouse - While production capabilities both in engineering and actual production itself were seen as limited at that time, the Navy believed that the liaison between Westinghouse and Pratt & Whitney for the 19XB and other manufacturers for the model 24C (J34) had made Westinghouse management fully aware of the necessity to design an engine for production from the very beginning.

BuAer felt Westinghouse's design engineering ability was unquestionable given that the J34 was the outstanding axial-flow engine operating in the world from the standpoints of weight, size, and performance. The proposed engine was based on fairly conservative cycle temperatures and component efficiencies already obtained in the J34.

While the nearest competitor was seen to be the Rolls-Royce AJ-65 (Avon), it being very attractive on paper, Roll-Royce was known to be experiencing an extremely difficult compressor problem and there was serious doubt the engine could be made to meet its advertised performance.

In a further addendum to the justification analysis added on May 14, 1947, the reviewer pointed out that the Army Air Forces had a significant gap in their engine development programs, lacking a low fuel consumption engine suitable for long range fighter and attack aircraft and were intensely interested in an engine of this size and performance. *(Given budget constraints, perhaps the Navy was hoping the Army Air Forces would help pay for development. This never happened. Auth.)*[5]

One further analysis concerned a typical fighter mission, comparing the relative attractiveness of a given power plant plus fuel weight required for an assigned mission. A quality number for each design was arrived at which weighed the relative importance of engine dry weight versus specific fuel consumption at a cruise design point of 1,525 pounds net thrust at 460 mph at 35,000 ft. It was assumed that for a cruise period of slightly over two hours at design point, the SFC factor would be twice as important as the specific weight factor. This gave a formula of:

$$Q = 1/ (W \times 2 \times \text{fuel consumption at design point})$$
(Where w = specific wt. at design point = dry weight/1,525)

Using this formula, the calculated quality numbers were:

Westinghouse X40E2	2.68
Pratt & Whitney	
JT-3B	2.20
JT-3C*	2.68
Allison 505	
A	2.33
B	2.22
C	2.14
D	2.06
Wright	2.12

However, the design weight did not include an automatically controlled variable area exhaust nozzle.

When the estimated extra weight of a variable area exhaust nozzle was added to the Pratt & Whitney proposal, the quality number dropped slightly, showing the Westinghouse approach to be the most attractive from the performance standpoint. Choosing a weighting factor of 3 instead of 2 resulted in a negligible variation in the relative standings.

The Navy recommended that the Westinghouse proposal, which was attractive from a performance standpoint and was by far the most attractive from a development time and cost standpoint, be used to immediately initiate a development program for a 7,500 lb thrust turbojet.

A Procurement Directive dated May 8, 1947 was approved for a Preliminary Design Study of the XJ40-WE turbojet engine, a detailed engine specification, one XJ40-WE-2 engine prototype, one -2 mock-up, and one XJ40-WE-4 engine for the flight substantiation test, plus requirements and various study deliverables. A Letter of Intent contract was put in place on June 30, 1947 as contract NOa(s) 9212 for a -2, a -4 and an initial -6 development engine and active development was begun. A contract update later added a second -6 qualification engine.

The Westinghouse Proposal Details

The Westinghouse proposal included a preliminary technical description of the engine upon which they had based their calculations. As can be seen from the description, the proposed engine appears to have many of the attributes of the J34 scaled up.

Figure 1 X40E2 Mockup, WAGT photo submitted with the proposal.[6] *Courtesy Hagley Museum and Library*

The enclosed drawing showed an eleven (11) stage compressor single shaft engine running on two bearings. The accessories in the front and the generator are driven off a skew drive off the top of the engine. The double annulus combustion chamber using spark plug ignition leads to a two stage turbine. A movable exhaust nozzle similar to the WWII German engines was planned. The adjacent table is a comparison to the Navy design targets.

Westinghouse claimed the attributes of small diameter (40 inch diameter intake), high performance, aerodynamic cleanness and ease of maintenance in the engine, pointing out that the Navy had expressed a favorable view of such characteristics in the past. They attached Westinghouse report A-469 as Exhibit A to the proposal to explain new features in the design.

Positioning the engine accessories inside the nose fairing, with an external pad for driving airplane accessories either directly or remotely was done to contribute to aerodynamic cleanness. The 7,500 lb thrust rating was chosen to meet the design point and maximum power for take-off specified. It was pointed out that moderate modification in the rating could be made if desired without serious effect to either the SFC or specific weight of the engine.

Steel compressor blades were chosen due to their proven ability to withstand the impact of ice and sand. The combustion chamber followed their established line and had adequate air cooling for long life.

Navy Targets vs. WAGT Proposal		
	Navy	**WAGT**
Mil Thrust, Standard Sea Level, Static Conditions	7,650 lbs	7,500 lbs
Military SFC	Not Specified	.98
Minimum SFC	.826	.81
Design SFC	Not Specified	1.06
Compressor	Axial, 12 Stages	Axial, 11 Stages
Combustion Chamber	Not Specified	Annular, double walled
Air Flow	106 lbs/sec	137 lbs/sec
Shafts	Not Specified	Single
Turbine	3 Stages	2 Stages
Maximum Engine Speed	N/A	7,560 rpm
Length	Not Specified	Cold – 185.75 inches Hot – 186.43 inches
Height	Not Specified	44.5 inches
Width	Not Specified	41.5 inches
Weight (dry)	2,422 lbs	2,680 lbs
Compressor Ratio	6.2 to 1	4.6 to 1
Max Turbine Entry Temp	2,160°R (969°C)	1,290°F (699°C) 1,380°F (749°C) 30 sec
Variable Area Nozzle	Yes	Yes
Provision for Afterburning	Yes	Limited
Fuel	Not Specified	100/130 Grade (AN-F-48)
Max Operating Altitude	50K feet	50K feet

Proposal Scope and Relative Schedule

The WAGT proposal was for development and tooling to construct a small number of experimental engines, including reports and completion of acceptance test of the final engine under the AN-E-30 standards. The design specification would be WAGT-X40E2-2 and the program would consist of:

> **Phase I** – Prototype Engine Design, Construction and Preliminary Evaluation
>
> Item 1 – Manufacture of one (1) Experimental Model X40E Turbojet Engine
>
> Item 2 - Manufacture of One (1) Mockup of Engine X40E
>
> Item 3 – Preliminary Engine Evaluation Report
>
> **Phase II** – Engine Development through Flight Substantiation
> Item 4 – Manufacture of Three (3) Experimental Model X40E2 Turbojet Engines
>
> Item 5 – Flight Substantiation Test Report
>
> **Phase III** – Engine Development through Qualification as a Service Model
> Item 6 – Manufacture of Six (6) Experimental Model X40E2 Engines
>
> Item 7 – Engine Qualification Test Report
> Item 8 – Final Design Drawings and Specifications
> Item 9 – Summary Report and Model Specifications and Engine Qualification Completed

Monthly Progress Reports would be submitted. Engineering services, laboratory services, test services (exclusive of acceptance testing) of 1,315 hours of test stand operation, 50 hours of test stand operation of an GFE X24C engine (J34), and six engines' worth of spare parts would be provided.

> **Engine Testing** was anticipated to be:
>
> Phase I - 75 hrs (X40E) (would become the XJ40E-WE2-2)
> 50 hrs (X24C)
> Phase II – 400 hrs (X40E2)(would become the XJ40E2-WE-4)
> Phase III - 840 hrs (X40E2)(would become the XJ40E2-WE-6)
>
> **Schedule** to be, from contract start:
>
> | Item 1 | 1 Engine | 15 Months |
> |---|---|---|
> | Item 4 | 1 Engine | 18 Months |
> | | 1 Engine | 19 Months |
> | | 1 Engine | 20 Months |
> | Item 6 | 2 Engines | 24 Months |
> | | 2 Engines | 25 Months |
> | | 2 Engines | 26 Months |
> | Item 9 - | Engine Qualification Report | |
> | | | 30 months |

The balance of the proposal covered various terms and conditions typical of all government contracts.

The attached report[7] defined the Westinghouse analyses that led to the design selections. A brief summary of the key design elements provides critical insight into the Westinghouse thinking at the time.

Compressor: The design used the symmetric velocity triangle design previously used in both the 19XB (J30) and 24C (J34) engines due to its characteristic of having the largest stage pressure rise for relative Mach number, resulting in low compressibility losses. Evaluation indicated that increasing the hub/tip diameter ratio from .5 to .6 showed that while it increased the required tip diameter to pass the same flow, the result was lower weight and a higher pressure ratio rise per stage. Based on this approach, a 4:1 pressure ratio would require 10 stages, 4.6:1 would require 11 stages and 6:1 would require 13 stages.

Pressure Ratio	Basic Engine Weight	SFC Ratio at Sea Level Static Military	SFC Ratio at Sea Level Static 60% of Military	SFC Ratio at Design Target (35K Ft, 460 mph, 1,525 lb net thrust)	SFC Average
4:1	2,645	1.048	1.050	1.050	1.049
4.6:1	2,680	1.00	1.00	1.00	1.00
6:1	2,975	.920	.956	.936	.937

The results were analyzed and the above results were used to do a mission effectiveness comparison, which showed that the 4.6:1 design produced a weight disadvantage over the 6:1 design, but was chosen because of *"more practical considerations, principally the more difficult component matching problem and the increased manufacturing cost and complication associated with the increased number of stages. The more difficult problem of matching the compressor and turbine of a 6:1 pressure ratio design could greatly lengthen the development time of the engine, despite our previous compressor experience and the use of a variable area exhaust nozzle."* (This statement would prove to be true when the -10/-12 compressor was designed. Auth.)

Combustion Chamber: This was scaled up dimensionally to operate at the same Mach number as the then current 24C4B chamber with the length increased to provide an even greater margin. Flow distribution tests had shown that a minor twisting of the compressor blades eliminated non-uniformities in the flow distribution leaving the compressor blades and would greatly assist in achieving even combustor temperature distribution. Based on this, the pressure drop usually associated with the use of combustion chamber inlet screens was not anticipated.

Turbine: Two designs were considered. The first was a single stage impulse turbine of 36.3 inch diameter and the second a two stage reaction turbine of 31.8 inch diameter. Basic design was again that of the 19XB and 24C engines. The two stage turbine was expected to show a 2.5 % higher efficiency over the single stage design. The greater heat drop in the nozzles of the single stage turbine permitted operation at approximately 100 degrees higher cycle temperature than the two stage design. This showed the single stage would achieve a 5.4 percent higher thrust per pound of air handled, but the two stage would have a 5.0

percent lower SFC. For the same thrust, the weights of the two designs would be the same. The lower fuel consumption design was therefore selected to be recommended.

Exhaust Nozzle: It was aerodynamically similar to those of the earlier engines, particularly the 19B. The original 24C was a boost engine and lacking a variable exhaust nozzle. The X40E2 design reflected the current 24C4 under development (XJ34-WE-32). It was being designed to be suitable for use in either close coupled or extension tailpipe installations and was to have a hydraulic mechanical actuator driven from the No. 2 bearing oil supplies for the adjustable exhaust nozzle.

Lastly, the report argued that the guaranteed weight of the proposed engine, 258 lbs above the target weight, was offset by the improved fuel consumption it offered of at least 2.9 percent better than the target.

Mechanical Design Features

Magnesium was generally substituted for aluminum in the compressor. It was to be used in the inlet housing, the inlet bearing support, the compressor casing and compressor discs. Stacked, interlocked disc construction was to be used for the compressor rather than a solid forged spindle, providing for simpler manufacturing and greater ease in obtaining acceptable forgings. Steel compressor blades were used as in prior engines. Tests showed these were more resistant to damage from ingestion of snow, ice, sand and small pebbles, allowing the engine to continue operations after such ingestion.

The drive shaft assembly (compressor, shaft and turbine) was supported in two spherically seated (self aligning) anti-friction bearings, the front dual race ball bearings (the foremost a thrust bearing) and the aft end in a dual roller bearing design. These offered easier maintenance and overhaul. Net thrust between compressor and turbine were expected to be equal. The shaft itself was steel of a large diameter thin-walled "torque tube" between the compressor and turbine, offering increased rigidity while reducing shaft deflections below any prior engine, even under the higher maneuvering loads specified by AN-E-30.

Engine mounting was a non-redundant, three point support with two pads in the plane of the No. 1 bearing and one pad above the No. 2 bearing. Maneuvering loads were to be transmitted directly from the rotating elements through the bearing housings and then to the airframe, reducing engine housing loads to a minimum.

Lubrication was to be self-contained, with the 5 gallon capacity oil reservoir (3.5 gal oil/1.5 gal air space) an integral part of each engine. This design confined oil related issues to the engine itself rather than sharing them with the airframe, making it easier to resolve lubrication problems as development proceeded. (Note: The weight for the reservoir and oil was not in the engine weight estimate.) The scavenged areas as well as the oil reservoir were to be maintained at two to five psi above sea level at all altitudes to help control foaming. Use of AAF-3606 oil was recommended to meet the -67°F requirement. The specified high ram air inlet temperature of 196°F would require an oil cooler to be used, which would be an oil to fuel heat exchanger configured with a thermal by-pass valve to ensure oil flow if the oil cooler passages became blocked.

The fuel control system was more complex than the J34 due to the inclusion of the variable area exhaust nozzle. Operation was to have the nozzle open during most of the speed range, only closing down as necessary as the engine approached maximum rated speed. It was to be mechanically scheduled and permit rapid acceleration to rated speed. It included a temperature sensitive override but would permit full military thrust to be obtained at all atmospheric conditions. The report included the following fuel control system schematic:

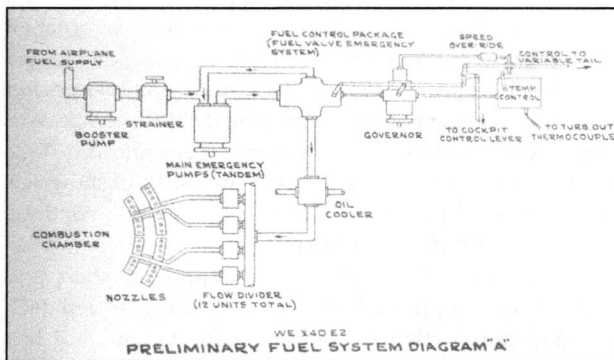

Figure 2 XJ40-WE-2 Preliminary Fuel System. WAGT Drawing EDSK 195613.

The engine power control (throttle) was still under study with two types of governors being considered.

A. Electro-hydraulic Governor

A speed sensing tachometer generator would be mounted on the same shaft as a power supply generator and overspeed switch. It would provide power to operate the governor and temperature control. These controls would be completely independent of the airplane's electrical system. It would maintain the engine speed within plus or minus 70 rpm, with the acceleration rate of the engine controlled by the temperature control. The power level in the cockpit could be provided with trimmer controls to adjust both the idling rpm and maximum rpm on the engine, essentially making all engines respond in the same way to a given pilot.

B. Mechanical-hydraulic Governor

A speed sensing element of the mechanical fly-ball type (self-contained oil actuated) would provide speed control from idle to maximum rpm. At maximum rpm, it would provide approximately isochronous control within a 1% variation. Overspeed protection was provided by a separate sensing element of the mechanical fly-ball type. To prevent over-temperatures during accelerations, an acceleration control was provided, limiting the rate of acceleration in response to the compressor inlet total pressure and compressor discharge pressure. The speed control would provide a control pressure to actuate the main fuel throttle valve in the PD-2 fuel control package. An electric temperature control override was anticipated as well to satisfy AN-E-30. This would measure the average output of a number of thermocouples in the turbine outlet duct and if an over-temp of more than 52°F was detected, it would first open the variable outlet duct and then override the governor. This control would be inoperative if the engine was in the emergency control system mode.

Other features would be an electronic ignition system providing a hot spark up to 40K feet and capable of operating continuously, assuring automatic restarting if the fuel flow was momentarily cut off. Provision for inlet screens in the airplane to guard against foreign object damage (FOD) of the engine was allowed for and it was recommended the screen design be retractable. No provisions for anti-icing were included in the design, but it was noted that a study was underway at NACA, and it was thought that icing of the screens would not be a large problem. Screens were not planned for the X40E2.

Specification WAGT-X40E2-2 was attached to the proposal covering the XJ40-WE-6 proposed, dated April 15, 1947.

The only place a 10 stage compressor was ever mentioned was in the Westinghouse X40E2 analysis discussion of their proposal. It was never mentioned again over the development of any variant of the J40, although compressors with larger numbers of stages were planned or proposed at various points. *(It was not discovered why the extant literature consistently states the J40 was a ten stage engine. – Auth.)*

Figure 3 XJ40-WE-2 Half-section combustion basket, WAGT P39239, 7/1/48. *Courtesy Hagley Museum and Library*

XJ40-WE-2 Development

As contract negotiations began, the Power Plant Division made a vigorous case for strong Navy input and control of the development process, having considered that Westinghouse was trying to independently design the engine with the minimal requirement input from the Navy beyond that which had been given during the month Westinghouse prepared their proposal. A memo concerning this situation and the Power Plant Division's recommendations is included here in part:

1. *"The 7,500 lb thrust turbo-jet engine is considered to be one of the most important new engine developments to support the Navy high performance airplane programs. Although the design point for the engine is set, there are many decisions about the method of reaching that design point that have not been made. It is also important that this engine have optimum installation and accessory features so that it can be used for a wide variety of aircraft types. A number of conferences with prospective contractors on this engine have taken place in the past year. This is not true, however, in the case of Westinghouse Electric Corporation. Westinghouse was not informed of this bureau's requirements until about 1 March 1947. Their subsequent proposal, reference A (summarized above), was prepared in about one month. The evaluation of the various proposals was made immediately after receipt of the Westinghouse proposal. With the full*

recognition that the proposal reflected only the initial thinking on the engine, and no detailed coordination with the Bureau of Aeronautics had been made, the Westinghouse engine was still the most attractive engine submitted from the standpoint of cost, time and general engine characteristics. There are many technical details of this proposal that remained to be negotiated.

2. *The most important single change desired to be made in the Westinghouse proposal is the establishment of an introductory Phase to the contract. This phase is intended to explore thoroughly all the possible engine design details and arrive at the best engine features to assure that the Navy will receive a calculated optimum engine. This is to be accomplished by a "Preliminary Design Study". Upon the completion of the design study the selected engine would then be described in an engine model specification in accordance with the Army-Navy General Specification. This specification would become a part of the contract and would serve as a definite detailed understanding as to what the Navy was buying. Also, to assure optimum airplane installation characteristics, a mock-up of the engine would be built and a Navy Mock-Up Board would pass on its characteristics. Upon the completion of this preliminary phase, the engine characteristics will have been clearly defined; and any subsequent changes could be based on the results of this work. This should prevent future claims for work over and above the contract requirements, difficult amendments to the contract, and misunderstandings as to engine detailed requirements. The configuration and specifications of the 24C engine were not definitely established early in its development contract, NOa(s) 3962, with the result that innumerable difficulties have subsequently arisen in the administration of that contract.*

3. *The necessity for this preliminary work is realized by both the Army and Navy in the general specification AN-E-30 for turbo-jet engines, yet the Westinghouse proposal, reference (a), did not recognize this. Westinghouse proposed to submit the design study during the life of the contract, which would prevent an initial evaluation of the engine. They proposed to deliver the mock-up at about the middle of the contract, a time when the installational (sic) features would have been firmly fixed. This Westinghouse mock-up conception was not planned to be a working medium for the Navy and contractor to evolve an optimum engine; rather, its purpose was to serve as a mock-up engine for use in airplane mock-ups. Finally, Westinghouse would not deliver the engine specification until the completion of the contract. This is totally unacceptable to the Power Plant Division, inasmuch as it repeats the 24C engine mistakes, does not meet the requirements of the AN-E-30, and forces the Navy to write a contract based solely on the "Bid Specification".*

4. *When the Power Plant Division was notified to initiate a procurement directive for this engine with Westinghouse, a conference was held with Westinghouse and the above Navy plan was set forth. The prospective contractor agreed to this plan,*

although he felt that it would require close liaison with the Bureau during the design study to prevent unnecessary delays in the later contract work. Thus, a mutually agreeable technical contract could have been written on a cost plus fixed fee basis.

5. *In the submission of the alternative proposal, reference (b)* (not included), *Westinghouse has taken exception to the original proposal with even greater deletion of details. In this proposal, the contractor has one objective – to build the final engine required in as little time and at the least cost possible. Westinghouse will give what they consider reasonable consideration to the initial design; this, however, will probably not be sufficient. Most certainly, the proposed method of delivery of the mock-up and engine specifications is totally unacceptable. Informally, Westinghouse has indicated that this second proposal, in its technical terms, was calculated to be totally unacceptable to the Bureau of Aeronautics. The purpose of the proposal was to establish the "fixed-price" cost of the engine for comparison with other contractors.*

6. *Enclosure (A) sets forth the technical requirements acceptable to and considered essential by the Power Plant Division........Failure to incorporate these requirements into the contract will produce a repetition of the administrative difficulties encountered in the 24C contract, NOa(s) 3962, and may seriously compromise the eventual service acceptability of the subject engine."* [8]

The net of this effort was contract NOa(s) 9212 Amendment No. 1, dated 3 September 1947, which added various studies, specification reviews and approval and mock-up requirements. Westinghouse began to produce these reports, listed below. Over the entire program, reports of many kinds for various subjects related to both the early engines and later models would continue to be produced.

Report No.	Date Received	Navy Accepted	Title
A-582	10/28/1947		Model XJ40-WE-6 Turbo Jet Engine – Accessory Arrangement Design Study
A-590	10/28/1947	11/18/1947	Lubrication Design Study
A-583	10/14/1947	2/6/1948	Blade Path Arrangement
A-603	12/31/1947	4/14/1948	Selection of Fuel System to Meet Maximum Engine Altitude Performance
A-591	2/4/1948	3/29/1948	Basic Engine Mechanical Arrangement and Mounting Provisions

Report No.	Date Received	Navy Accepted	Title
A-641	2/19/1948	3/29/1948	Features that Provide Ease of Production and Maintenance
A-633	1/26/1948		Starting and Ignition System
A-614	11/28/1947		Studies of Additional Engine Types
A-615	1/27/1948	7/29/1948 (Conditional)	Design Study of Power Control System
A-696	7/13/1948		Supplemental Oil Supply Design Study, dtd 5/12/1948
A-695	7/13/1948		Anti-Icing Design Study for XJ40-WE-6, dtd 5/13/1948
A-721	7/21/1948		Sample Calculations for Use with XJ-WE-6 Performance Curves, dtd 6/18/1948
A-720	7/21/1948		Performance Curves Corrections for Losses of Turbojet Engines, dtd 6/18/1948
A-767	11/22/1948		Sample Calculations For Use with XJ40-WE-6 Performance Curves
A-829	5/3/1949		A Study of the Application of Accessory Quick Disconnect Mountings for XJ409, XJ46, and XJ34-WE-32 Engines
A-840	5/10/1949		Sample Calculations for use with XJ40-WE-6 and XJ40-WE-8 Performance Curves, dtd 4/28/1949
A-749	12/30/1948	6/27/1949	Westinghouse Model XJ40-WE-6 Jet Engine Structural Requirements for Cantilevering an Afterburner and Extension, dtd 9/13/1948
A-928	1/27/1950	3/15/1950	Flight Substantiation Test on Model XJ40-WE-4 Jet Propulsion Engine
A-933	3/16/1950		XJ40 Turbine and Compressor Development, dtd 3/7/1950

Report No.	Date Received	Navy Accepted	Title
A-940	3/20/1950		Design analysis on XJ40-WE-4 and -6 Fuel Pumping System (Interim Technical Report), dtd 3/9/1950
A-938	4/10/1950		Analysis of the Combustor for the XJ40-WE-6 (Interim Technical Report), dtd 3/30/1950
A-942	3/16/1950		Design Analysis of the Engine Starting and Ignition System (Interim Technical Report), dtd 3/9/1950
A-943	3/31/1950		Design Analysis of the Fuel Control System for the XJ40-WE-6 Engine (Interim Technical Report), dtd 3/17/1950

Report No.	Date Received	Navy Accepted	Title
A-810	5/2/1949		Hydraulic Starting System, dtd 3/21/1949
A-954	5/29/1950		Summary Engine Evaluation Report – XJ40-WE-4 Engine, dtd 5/26/1950
A-975	6/7/1950		Design and Tests of High Temperature Materials for the XJ40 Engines (Interim Technical Report), dtd 5/29/1950
A-999	6/25/1950		Stress and Vibration Analysis, dtd 6/19/1950

* Documentation on receipt dates or comments not located for all reports.

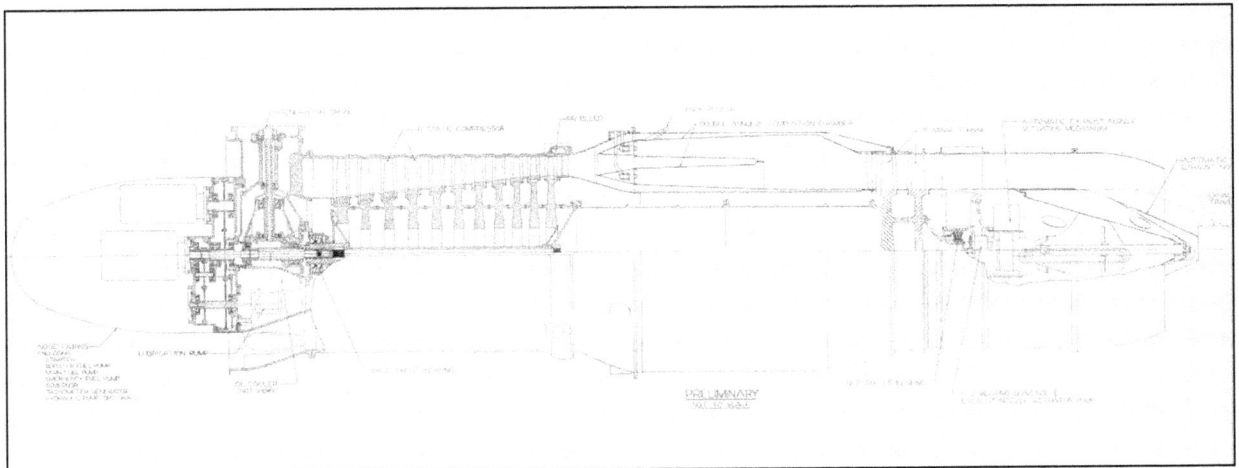

Figure 4 X40E2 Preliminary engine assembly drawing, 43J405, WAGT P36947, abt. 4/15/1947. *Courtesy Hagley Museum and Library* [9]

The September, 1947 Westinghouse activity report indicated that at a meeting on September 17, Navy personnel and Westinghouse had met and agreed to an accessory arrangement with the gearbox positioned inside a split intake. BuAer responded in a memo stating that no formal approvals had been given and would not be forthcoming until a detailed design report had been received and reviewed.[10] This is a good example of Westinghouse trying to move forward rapidly to design the engine in detail and the Navy forcing Westinghouse to pay attention to the formal approvals process. In the end, as in almost all of these situations, Westinghouse proceeded based on the verbal conversations and the official approvals

dragged behind. Only when Westinghouse was trying to gain official test approvals and engine shipment release to the airframe contractors would this situation change.

The referenced memo also confirmed that the Navy intended the engine should accommodate afterburning in the overall design and that eyelid type variable exhaust nozzle designs were being studied to accommodate this eventuality. This is the first time that a requirement for another version of the engine (this would become the -8) was to cause changes in the design of the initial -6 engine, since the eyelid design would replace the moveable cone in the exhaust duct. WAGT was reminded that the initial design conditions for the -6 had stated it should be designed with the view of adding an afterburner later.

The November Monthly Progress report states that the compressor design was 50% complete, that some manufacturing drawings were complete for a small number of blades, and that the stator vanes were done. The initial combustion basket layout was 50% designed with 10% of the detailed drawings complete. It was assumed it would undergo changes before a final design went into the initial -2 engine for testing. The eyelid type variable exhaust nozzle design was 20% complete, with initial calculations showing an eyelid design was less efficient than the moveable inner cone type of control. No final recommendation on the design selection in that area was made at that time. The fuel system and lubrication system design reports had been issued to the Bureau. Mock-up work was proceeding and the engineering and drafting work leading to the production of one XJ40-WE-2 engine was on schedule to be complete September 30, 1948.

A significant BuAer memo was generated on November 20, 1947, the subject of which was the design point of the evolving design. BuAer had reviewed Report A-583 and used this review as a touch-stone to remind WAGT that the design point required by the Navy in the engine configuration for performance was 460 mph, 35,000 ft, 1,870 THP, calculations showing this would produce an engine of 8,650 lb static thrust and .826 SFC at sea level. Static sea level performance was considered to be of secondary importance. The memo requested WAGT state whether either of two blade path considerations in the report met the target or could be expected to meet the target through normal development improvements without major redesign. If so, the cycle configuration would be considered acceptable.

This memo continues, "In the event that it does not appear that either engine D or N can practically achieve the design point goal, it is requested that this bureau be informed as to whether any other of the design study engines will meet this goal. Further, if it appears that an entirely new cycle configuration will be required, it is requested that such a cycle configuration be proposed. Wherever it is considered that a proposed engine will require additional time and money over that set up by the letter of intent to Contract NOa(s) 9212, such additional cost and time should be specified."[11]

WAGT replied to this as part of a comprehensive response to which they attached Report A-614 and referenced A-583. In the first part, they summarized the Navy's request for information on engine types specific to the question of "How does the 'simple turbojet' type of engine compare with the other types of engines (compound engine, ducted fan, turboprop, etc.) for typical combat missions requiring a considerable mission radius". They reported that the differences for the set tasks turned out to

be small, with turboprops and ducted fans having an advantage at lower speeds, but the simple turbojet having an advantage at "practical ranges of miltary thrust to cruising thrust.' At longer cruising ranges, a higher compression becomes advantageous."

The later part of the memo addressed the issue of the BuAer design target for the XJ40 engine and the current design performance. It stated:

It is realized that the 'target' figures given by the Bureau correspond to exceptionally good performance. For instance, the 'target' gas generator curves indicate a thermal efficiency of the power plant of 35.3%. As a matter of interest, this compares with a figure of 33% realizable in the best modern large complex central station power plants. The target cruising efficiency at 35,000 ft, 460 mps corresponds to a fuel consumption of .69 lb/hr/THP. The expected cruising efficiency from the initial XJ40 design under the same conditions is .86 lb/hr/THP. This figure is already some 18% better than what is being obtained with present day turbojets. It is clear, therefore, that considerable improvement and refinement is necessary to reach or approach the target figures.

Fortunately, there appears a number of ways in which improvements can be realized:

1) Refinements in blade path, particularly minimizing the leakage loss around the blades

2) Improved altitude burner performance (efficiency and pressure drop through burner)

3) Adjustable turbine nozzle vanes

4) Increased cycle pressure ratio

Each of these refinements results in improved efficiency. We know of no single factor, which would bridge the gap between the original XJ40 design and the target figure. In particular, designing the engine for a 6.2:1 overall cycle pressure ratio and the same component efficiences gives an SFC of .80, still .11 lb/hr/THP from the target.

The improvements 1 to 4, when applied together, apparently go a long way towards reaching the target fuel consumption.

Some of the refinements discussed above will come about naturally as the engine development progresses. Others will require special effort. Because of the inherent flexibility of the XJ40 design with separate, easily replaceable compressor and turbine discs, the conditions for incorporating improvements are very favorable.

On the basis of the above, the contractor would like to make the following recommendations:

a) That the contractor proceed without delay on the basis of the blade path of engine J of design study A-583 which is substantially the same as the engine described in the proposal and bid spec(ification) with the exception that a single annulus combustion chamber is substituted for the original double annulus combustion chamber.

b) That, if the Bureau so desires, a development program be undertaken on items 1 to 4 listed above in such a manner that, if further evaluation shows incorporation into the engine in order, this can be done with a minimum of cost and without delaying the rest of the engine program.

If the Bureau concurs with these recommendations, the contractor will be glad to make a proposal on such a program.[12]

A BuAer memo then approved the design study A-583 recommended cycle configuration of:

> Cycle Pressure Ratio – 4.6:1
> Maximum Temperature – 1,425°F – 1,500°F
> Burner Pressure Drop – 3.56%
> Single Annulus Combustion Chamber
> Two Stage Reaction Turbine

It also requested WAGT submit a parallel development proposal with cost for a 6:1 compression ratio engine if it could be done with no increase in weight and requested the WAGT-X40E2-2 proposal be updated to reflect the weight decrease from changing from the nose case type accessory gearcase to the hour-glass design, the weight reduction being the primary reason for the change.[13]

Preliminary installation drawings for the -6 were released to the airframe manufacturers in January, 1948, even as the outline of the engine was still being completed. Westinghouse had to prepare a separate transmittal request for each airframe manufacturer, a cumbersome process. The airframe manufacturers involved were: Curtiss Wright Corp., Boeing Aircraft Company, Douglas Aircraft Company, Inc., Lockheed Aircraft Corp., Fairchild Engine & Airplane Corp., Glenn L. Martin Company, Republic Aviation Corp., Northrop Aircraft, Inc., and North American, Inc. Eventually, to ensure BuAer control over these classified documents, Westinghouse would forward all reports and technical data to BuAer and they assumed distribution responsibilities.[14] In cases where urgency was needed, violations of this later process would draw memos from BuAer strongly bringing the violation to WAGT's attention and reminding them of the security process that was to be followed.[15]

XJ40-WE-2 Testing

The engine (serial no. WE003001) was started on the evening of October 28, 1948, run again in the afternoon the next day and then again that evening. The first run was made without an exhaust nozzle (giving an opening of 600 in²) at first up to 3,000 rpm, then up to 6,000 rpm, where vibration was experienced and the engine was shut down. The vibration was traced to an improperly mounted

vibration instrument and corrected. The second run was done with the exhaust nozzle in place (496 in²) up to 6,500 rpm with no issues. The third run was up to 7,000 rpm at first and was then increased to 7,620 rpm (design rpm) when the airflow dropped abruptly and temperatures suddenly increased in the 4 o'clock position (aft looking forward) and the engine was stopped. Total running time was: 1 hr 33 minutes.

> **Initial XJ40-WE-2 Measured Performance:**
> Speed (corrected) 7,260 rpm
> Thrust (corrected) 5,600 pounds static
> Specific Fuel Consumption (corrected)
>   1.06 lbs/hr/lb of thrust
> Air Flow (corrected) 131.2 lbs/sec
> Turbine Inlet Temperature (corrected)
>   1,262°F
> Exhaust Nozzle Area 496 in²
> SFC at 6,800 rpm 1.02

A teardown inspection of the engine revealed compressor spindle axial rubs in the 2nd and 3rd stage discs and radial spindle rubs in the 7-11 stage disc lands. The 5th stage compressor seal metal was burned off in five places. Compressor blade rubbing was visible in stages 4 to 9. The seal strip in turbine stage 2 was broken in one place and warped in several others. The combustion basket was totally collapsed due to a pressure differential. The collapse caused severe hot spots to occur downstream, resulting in damage to the guide vanes, turbine discs, blades and No. 2 bearing support.[16] It was expected the engine would be repaired and testing would restart in two weeks.

Figure 5 XJ40-WE-2 Test Engine. Note fixed exhaust nozzle with onion area control. WAGT P36878, 10/26/48. *Courtesy Hagley Museum and Library*

Figure 6 XJ40-WE-2 With test split intake, WAGT P39879, 10/26/48. *Courtesy Hagley Museum and Library*

On the same day as the first day -2 engine testing began, WAGT sent over the revised specification WAGT-X40E2-2A. BuAer reviewed it and sent it back on November 4 with a list of changes, mostly clarifications and typographical error corrections. Acceptance of the -2, -4 and -6 engines would be based on this modified specification once the changes were made.

The engine was repaired and testing was continued in a variety of configurations. Power control was manual only based on a J34 power control with operator oversight to prevent overspeeds and overtemps.

WAGT reported testing results to date on December 6, 1948 after 11.47 hours of run time, with the last 5.6 being much improved.

XJ40-WE-2 Test Results 12/6/1948								
	Thrust		RPM		SFC		Gas Temp °F	
Rating	Guaranteed	Obtained	Guar.	Obtained	Guar.	Obtained	Guar.	Obtained
Take-off	6,800	7,110	7,260	7,260	1.01	.975	1,240	1,210
Military	6,800	7,110	7,260	7,260	1.01	.975	1,240	1,210
Normal	6,200	6,200	---	---	.98	.945	1,140	870
Cruise(1)	5,000	5,000	---	---	.95	.948	840	793
Cruise(2)	4,000	4,000	---	---	.93	.947	680	750
Idle	450	380	3,000	3,000	---	---	900	755

The test results were obtained using different fixed size exhaust cones. The bell cone had a slight flare on the downstream end of the normal straight cone. Cone sizes used were:

Rating	Exhaust Nozzle Type	Exh. Nozzle Area Sq. In.
Take-off	Cone	432
Military	Cone	432
Normal	Cone	460
Cruise(1)	Cone	497
Cruise(2)	Cone	497
Idle	Bell*	475

* was equivalent to a 500 in² cone nozzle.

Oil consumption had not yet been measured, as oil was blowing out of the breather to the extent that actual internal consumption was not easily measured. It was planned to solve the breather problem by moving it to an idler shaft and centrifuging the oil out of the air. Tests were underway.

XJ40-WE-2 Test Weight	
Configuration Items	lbs
Basic dry weight (minus power control and exhaust nozzle)	3,216
Manual power control and fixed area nozzle	76
Adjusted dry weight	3,292
To dry weight, add weight of variable area exhaust parts, oil-air separator and thermocouple harness	82
Also, breadboard power control with large alternator	191
Total Dry weight became	3,489

Based on the test results, WAGT requested the engine be accepted.[17]

The -2 engine was conditionally accepted by the Essington Bureau of Aeronautics Representative (BAR) for the Navy on December 8, 1948 with the conditions that final delivery weight would be brought down to the guaranteed specification weight and that every effort would be made to improve (reduce) oil consumption. The engine was returned as GFE to WAGT for further testing under the contract.

Construction of the compressor of the -2 differed from later -6 practice in how the stub shaft was attached to the compressor spool. It also had a bevel gear to drive the vertical power take-off accessory pad on the top of the engine. Note: the long rods marked "50" in the drawing below are the compression rods used to hold the compressor stack together. (See figure).

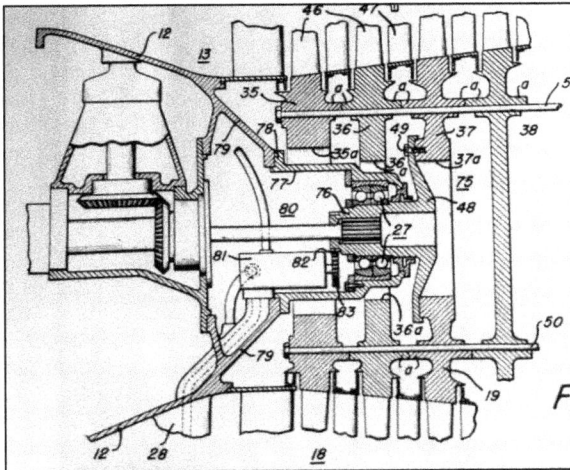

Figure 7 No. 1 Bearing compressor construction of the -2 engine, *WAGT Patent Application Draft (Not Submitted)*

Shortly after the engine was accepted, an additional requirement for starter torque and speed was added, but the engine had already been accepted, so the requirement was only added for the -4 and -6.[18]

The planned control system for the -6 (although the -2 is referenced in the memo's subject line in error, no -2 had adjustable exhaust nozzles of the eyelid type) was found to be acceptable, with the exhaust nozzle eyelids locking in position in the event of a hydraulic failure, but the afterburner version would have to have dual hydraulics and meet additional requirements (covered in the -8 section).[19]

BuAer confirmed in early December that the BAR had the authority to authorize test time on loaned engines to the manufacturer for testing under the referenced contracts. They were to annotate such authorized use in their monthly progress report recap after the fact. It is not known just why this had to be spelled out in detail, as the two J34's on loan had already been used to verify that the modified test cells at Essington were ready for XJ40 testing.[20]

At this point, the first sign of materials shortages in steel became visible. The post WWII consumer market was absorbing almost all of the output of the nation's steel plants. To support defense needs, a "Voluntary Steel Allocation Plan", managed by the Air Force out of Wright-Patterson Air Force Base, was set up to coordinate with the steel industry a guaranteed steel output and availability based on a contract by contract basis. The Navy asked that the J40 contracts in place be added to the list of contracts eligible under that plan.[21] Materials shortages for other reasons would emerge later and affect the J40 production effort.

(NOTE – Further discussion of -6 related items from this point on will be found in the -6 chapters.)

Engine starting came back into focus again with a BuAer memo stating that they did not believe the starter could meet the required 30 second start-up time based on the starter requirements contained in the latest WAGT-X40E2-2A specification. It asked that the latest starter speed-torque requirements be forwarded to the bureau and that any future changes be promptly forwarded to the department. They reminded WAGT that the Navy was responsible to provide the starters.[22]

The -2 was back on the test bed in January, with the compressor being revised and retested. The January progress report (which included items occurring in February) listed a series of problems with the engine. On February 8, while running at 6,490 rpm, the thrust was 4,220 lbs and SFC .913, both better than the specification of 4,000/.947. Back in January, running at 7,070 rpm, an air inlet duct quick disconnect clamp was sucked into the engine, damaging the compressor rotor and diaphragms. After repair, it was returned to test and on February 10, poor fuel distribution against the wall of a bracket burned a 10 x 6 inch hole in the burner basket. The engine was in repair and WAGT had enough preliminary data to produce the preliminary engine evaluation report by March 15. (It had been due in December.)[23]

The starter requirement curves for the -2 were finally forwarded to BuAer on February 23, 1949 as a set of six curves showing torque and horsepower vs. engine speed in a variety of ways. This was in response to an earlier BuAer query, and all the information was then forwarded to AiResearch. AiResearch was requested to notify WAGT of the maximum temperature of the exhaust gas of the starter and *"the possibilities of incorporating a 90° elbow on the exhaust, the elbow to extend no further than 4 inches behind the present exhaust outlet and to be used for a quick disconnect discharge tube."*[24]

February's progress report showed that oil consumption over a two hour period was measured at 1.250 lbs/hr, well over the specification requirement of .7 lbs/hr. Work continued to lower oil consumption. The engine sustained a compressor seal rub on the eleventh stage and was quickly repaired. While the seal was repaired, the basket had to be repaired because of a cracked portion. The gas "scoops" had to be reinforced.

On March 10th, the engine was given a 10% over-speed test (7,260 to 8,000 rpm) and mechanically the engine was satisfactory, with low vibration. Three different exhaust nozzles were used, the largest was 597 in². Air flow at 8,000 rpm was 142 lbs/sec with maximum thrust of 6,360 lbs at turbine inlet temperature of 1,355°F. Compressor efficiency dropped from 81.5% to approximately 75% while operating at 8,000 rpm. The engine had been run a total of 34.38 hours (to March 14, 1949).[25]

Figure 8 XJ40-WE-2 Undressed core prior to testing, WAGT P39607, 9/24/48. *Courtesy Hagley Museum and Library*

On April 7, 1949, a major program progress conference was held, resulting in a BuAer memo summarizing the results.[26]

A. A conference was held at Westinghouse Electric Corporation on 7 April 1949 between Mr. Kroon, Chief Engineer, and the Director of the Power Plant Division, Bureau of Aeronautics, on the progress of the XJ40 turbo-jet engine program.

B. As a result of the conference, it was concluded that the first phase of the J40 development had not attained the development schedule set down two years previous, although to date no major engine difficulty had arisen. It was estimated that the completion of the 50 hour flight evaluation test on the XJ40-WE-4 engine would be about six months late, (December 1949 vs. June 1949) and the 150 hour qualification test on the XJ40-WE-6 would be about four months late (April 1950 vs. December 1949). Westinghouse did admit some tardiness, but not as extensive as estimated. It was believed that this condition would not affect the initial production schedules tentatively set for August 1950 delivery of the J40-WE-6 engine, and therefore should not affect the airplane flight programs. (The effect of the delay upon future models of the engine (the -10 and -12) would not be apparent until late 1949 - Auth.).

C. The lack of any tangible results to date from the design study for the future trend of the J40 engine as brought out at the conference required either that substantial results sufficient for evaluation and initiation of procurement must be forthcoming within six weeks or the future engine program must be considered from another source.

D. The detail summary of the test results program is set forth in Enclosure (A). Additional information was brought forth that indicates that:

1. The XJ40-WE-2 engine was two months behind schedule.

2. The engine test program to date was four months late.

3. The 50 hour test, if successful on the first try, was expected to be complete by the middle of July, one and one half months late. (This assumed that the emergency control need not be submitted to a reliability test, something not acceptable to BuAer – Auth.)

4. Although the first engine was about 500 lbs. overweight a substantial decrease would be obtained in the second engine. (It appears, however, that the problem of meeting the -6 weight guarantees would be difficult).

5. To date no major engine difficulties had occurred. For this reason and because of the extensive internal engine analysis now in progress, Westinghouse believed the total engine test hours to obtain a satisfactory engine will be less than originally planned.

6. The engine fuel consumption, although within the wide limits of the 50 hour engine, had a characteristic of a minimum specific much nearer normal power than originally designed for. To improve this consumption on the present engine design to meet the 150 hour engine requirements at or near normal power would require much attention. If in addition to this the bureau required the original low specific at low power condition, a rematching of the engine components would be required and would result in substantial delay to the program. The bureau's requirements were being investigated further.

7. The compressor operation had been better than expected in most respects. No compressor stall had been experienced to-date, the pressure ratio had been demonstrated above design of 4.6:1 (5:1). The efficiency was still shy about 3% of design. A condition in the 5th stage when determined and corrected may improve the performance.

8. The combustion chamber performance had not obtained design conditions. Two failures had occurred in collapsing walls that should be corrected by added strength of the structure or reduction of the pressure differentials causing the failures.

9. Turbine performance had been good. The present problem was to obtain the optimum match of components through the turbine changes.

10. It appeared that the programs for engine components – namely controls – were later than the engine and would delay the date for the 50 hour test beyond July 1949. Similarly, the component qualification tests might delay the 150 hour test.

11. The future engine design study did not have sufficient data for an evaluation.

12. The "curvic clutch" construction of the compressor and turbine had been completely satisfactory to date. Westinghouse was to be dependent upon Gleason for the machining of these discs until December 1949. (This was considered to be a hazard to adequate rapid engine construction.)

13. The initial plans were to build six test engines and make enough parts for four (4) more engines to support the program through the 150 hour test on the XJ40-WE-6 engine. (This might hazard expeditious prosecution of the development.)

E. It was considered that the problems to be solved for a satisfactory J40-WE-6 engine were, in order of importance:
   1. *Reduce engine weight.*
   2. *Obtain design fuel consumption.*
   3. *Obtain proven engine controls.*
   4. *Accelerate accumulated engine and component test time.*

F. It was considered that a great deal of emphasis by Westinghouse must be placed on the planning of the engine's future. Continuation of the present inadequate progress would result in a program too late for this bureau's application requirements and too late for our fiscal program plans. This particular phase of the conference was completely unsatisfactory and required early and direct action by Westinghouse. Official correspondence to that effect was being prepared for early release.

WAGT stated in Enclosure A their position up until then and summarized here by the author:

A. XJ40-WE-2 #WE003001 had been run 50.78 hours to date.

B. All guaranteed specific fuel consumptions, thrusts, speeds and temperatures for XJ40-WE-2 and XJ40-WE-4 had been met.

C. Total dry weight of engine which met the above guarantees was 3,292 pounds. A fixed area exhaust nozzle and manual power controls were used and were included in this weight. This engine contained several components which were definitely heavier than designed for later engines.

D. Minimum oil consumption measured to date was 2.3 pounds per hour during a two hour run. There was a small oil leak around a faulty seal during this time. Initial large oil losses, due to foam blowing out a vent, had been eliminated by internal baffles in the oil reservoir (gear box) and by centrifuging vented air through a hollow idler gear shaft. Oil consumption was believed now to be within the guaranteed limit but data had not been taken to definitely prove this point.

E. The lubrication system, plus bearings, had been particularly free of troubles. The reservoir venting problem mentioned above was encountered and solved. Minor leaks had been solved as encountered.

F. Fuel pumping, distribution system, and nozzles had been and were now functioning very well. A small leak under each balance valve was discovered during the early engine running but was now fixed.

G. The following components had shown themselves to be absolutely reliable to date:
   1. Inlet duct
   2. Compressor housing
   3. Engine rotor complete
   4. Diffuser
   5. Combustion Chamber outer casing
   6. Rear bearing support
   7. Exhaust nozzle *(handwritten note – "30 hours manual operation but not yoke model")*

H. The following components had shown absolute reliability to date with the exceptions specified:
   1. Gear box – Two oil seals ran hot. – Jets of oil were made to play on these seals and they now ran as they should.
   2. Plastic coating inside compressor was subject to gas erosion at hot end. Coating was removed and the compressor blades lengthened to compensate for removal of coating.
   3. Compressor diaphragms *(stator vanes - Auth.)*
      i. Five forward stages were subject to axial rubs due to not being sufficiently rigid. The tip ends of the vanes were brazed to the outer shroud on the inside surface of the shroud. Also the compressor rotor was shifted forward 1/16 inches. These moves stopped the axial rubs.
      ii. Had had a few radial seal rubs. The seal clearances were opened a few thousands. Also the diaphragm retaining screws increased from three to seven per half. These moves stopped the radial rubs.
   4. Turbine rotor housing. The housing had collapsed due to external gas pressure. It was strengthened and had since been all right.
   5. Turbine housing. One of the 16 segments surrounding the first rotating stage came loose during engine assembly. This was due to insufficient retention of the turbine nozzle diaphragm. The number of turbine nozzle retaining screws was increased from four to eight. This move stopped the trouble.
   6. Combustion chamber liner:
      i. It collapsed once due to external gas pressure. Stiffeners were added to the liner outer wall. No further collapsing had been experienced.
      ii. The inner wall ran hotter than outer wall. A revised liner design had been made that should more adequately cool the inner wall. Also this new liner had a stiffer outer wall which should eliminate need of the stiffeners mentioned above. The new liner would be ready for test about April 30, 1949.
      iii. A general problem existed concerning the bowing of the turbine shaft after engine shut down. This was caused by thermal stratification of air around the shaft causing the top of the shaft to be hotter (and longer) than the under part of the shaft. When the

engine was started with a shaft thusly bowed, excessive vibration resulted. Two solutions had been effective in overcoming this problem. The solution which was intended to be incorporated into the engine was a small electrical shutdown motor, working through a reduction gear, which would be mounted on the gear box. This motor turned the engine rotor at about 1 rpm for as long as needed (estimated 5-6 hours maximum) after engine shut down. A timer would be necessary. The present shut-down motor on XJ40-WE-2 #WE003001 drew one ampere at 12 volts.

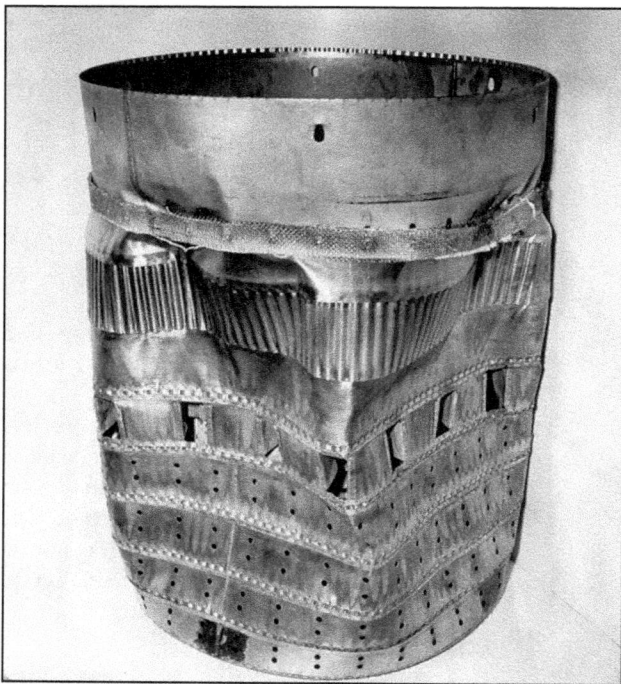

Figure 9 XJ40-WE-2 Collapsed combustion basket liner, WAGT P39893, 11/1/48. *Courtesy Hagley Museum and Library*

I. The engine had been run about 20 hours on AN-F-58 fuel. The general efficiency level of AN-F-58 seemed slightly lower than AN-F-48. No coking problems had yet been encountered.

J. Ignition was accomplished easily, smoothly and consistently on both fuels at around 600 rpm. The engine was self-sustaining at 700 rpm, and could carry itself up in speed from 700 rpm without assistance from the starter.

There were five separate charts presented and included in the attached Appendix A, these showing the proposed revised schedule for all the various J40 models vs.

the aircraft needing them, and test status vs. other past programs. The memo covering the meeting reflects how well the charts went over with the reviewer. It is surprising the true root cause for the delays was not established or apparently discussed. Discussion of detail design and manufacturing issues cannot establish a cause and effect relationship unless the master schedule with its individual tasks is used to determine if the various difficulties were allowed for or not. Only with such an understanding could a realistic determination be made of the probable accuracy of the balance of the tasks and overall chances of meeting the delivery dates, especially when it is considered that an entirely new electronic control system then under development had never been run on the -2 up until the time of the meeting and that the -2 did not have the new eyelid exhaust nozzle design or the developing hourglass gearbox, all three of which would run for the first time together on the -4 engine. No master project plan was presented nor was there any discussion of the cross dependencies of both other J34 and J46 parallel development work. As a result, the meeting failed to make a detailed examination of the likelihood of the new schedules being obtained.

One other key element in this review was the growing concern over lack of progress on the higher power engines (to emerge as the -10 and -12). As it turned out, these were the engines BuAer needed for their airframe programs, not the -6/-8 coming next. Apparently no detailed discussion occurred concerning what WAGT was doing to recover the -10/-12 schedules. There were many red flags but BuAer pressed on with the program. From the distance of time, we can see that all the seeds for a massive program failure were already visible. The lack of published project plans and root cause analysis makes it very difficult for the researcher to determine why decisions were taken as they were.

The March progress report (issued after the April conference) showed the -2 engine had accumulated 67.12 hours of test running. A compressor traverse had been completed over 33 hours of run time and the data were being evaluated. The burner basket showed deterioration and the entire bulkhead of the basket was being replaced. On re-assembly, the engine would have a new burner basket with many new features aimed at improving temperature distribution. The rebuild would have a ceramic coated shaft to improve its ability to slow heat absorption while cooling after shutdown.[27]

The improved burner basket had an additional step in the upstream end, redesigned air scoops, and a new type of stiffening arrangement whereby the ends of the steps were corrugated. These changes were expected to give better turbine inlet temperature distribution, longer life and increased resistance to buckling of the outer wall due to gas loads.

April's progress report still showed 67.12 hours of running completed and the engine was now in overhaul. A better designed basket was installed and the eleventh stage compressor disc was removed because of cracks. Increased airflow was to be gained by opening the compressor inlet guide vanes by eight percent. The pressure gradient pattern at compressor discharge was improved by giving the compressor 10th and 11th stage blades a uniform twist from hub to tip. Delays were incurred when mercury got into the engine and a teardown was needed that revealed both bearings and the number two bearing aft seals had to be replaced. It was expected that testing would resume on May 17th.

May's progress report indicated that (111 hours running to date) progress had been made on the engine basket and that running at full military rpm and temperatures would be attempted. The engine was running with a 30% closed turbine nozzle and both it and the first -4 were showing two percent improvements in fuel consumption. Oil consumption was running at 2.4 lbs/hr, still well above the .7 lbs/hr specification.[28]

June's report indicated WE003001's test time was up to 118 hours and the engine was still experiencing combustion basket and shaft bowing trouble as the main issues.

Progress through July reported engine run time was up to 151.3 hours total. The engine was then torn down and a fatigue crack was discovered in one second stage turbine blade at the root.[29]

The -2 engine was apparently used for component testing during the -4, -6 and -8 development, particularly the power control system, but it is rarely mentioned in the reports from this point on. This very basic engine demonstrated that the planned -6 engine could meet the thrust and fuel consumption targets using the design approach Westinghouse selected. It had value in identifying several key mechanical design issues that would have to be resolved via the -4 engine before an acceptable -6 engine would emerge, among them: continued issues with the burner basket assembly, the unanticipated shaft bowing on shutdown, turbine inefficiency and stress fractures in the turbine blades.

Things still to be tested were the planned yoke eyelid control system, the electronic power control system, magnesium compressor discs, and the planned gearbox and starter system for the -6. The challenges of weight reduction and the control system were looming large as Westinghouse moved on to the -4 as the primary focus.

Only one -2 was constructed, house engine serial number WE003001. It was scrapped at the end of the program.

XJ40-WE-4 Development

The -4 development engines were intended to incorporate all the features of the emerging -6 specification. Developmental testing of the components with improvements would be accomplished prior to settling on a build specification for the -6 for the specification 150 hour test. The December 1948 Monthly Progress Report began to report on the -4, declaring that all bar and sheet stock were on hand and that forgings were expected by the end of February. The blade forgings were being altered to correspond to the redesigned compressor spindle.[30]

It was originally intended that once the -4 engines passed a flight readiness 50 hour test, they could be used in aircraft during flight for test purposes, although not as prime movers of the airframe. In actuality, no -4 was ever taken into the air for flight tests of any kind, being used as house test mules throughout their existence as the -6 and -8 continued to suffer from various problems.

Only a few pictures of an XJ40-WE-4 appear as such in the WAGT photograph collection. They were built up, the second tested briefly enough to pass the 50 hour flight qualification test and then immediately modified into -6 and -8 configurations to support development of those models. It is possible that some of the pictures of early test -6 and -8 engines are really -4's modified to the later configuration(s), but they are not identified uniquely as such.

The performance curves for the -6 were approved on January 7, with one exception. The curve for Ambient Temperature Limits had not been received as yet and full approval of the -2A specification required that curve to be submitted and approved. Also, no performance curves for the -4 had yet been received and BuAer awaited these curves so they could be approved and passed on to the airframe manufacturers intending to use the -4 engine. Apparently no airframe manufacturer ever planned to use the -4 version of the J40.

BuAer also used the memo to remind WAGT that the change noted under the -6 Mock-up Board Item 28 needed to be included in the revised specification (a change to 2A) and also modified to show that the clearance envelope for the generator drive should be in accordance with Drawing AND10343, Generator Installation Clearance (for 10 Inch Mtg. Pad). All generators being developed for the Type XVI drive were in conformance with that envelope.[31] This memo is a good example of BuAer's inclusion of new requirements or topics outside the scope of a memo's subject, a confusing approach. On occasion, such inclusions crossed multiple contracts not related to the J40 contracts.

The January, 1949 (covering December, 1948) activity report indicated that the detailed parts drawings had all been released to the contractors for machining, with construction anticipated beginning in late April.

The February progress report on the -4 indicated that receipt of compressor discs from Gleason Works would be delayed until the middle of May and assembly of the engine would not start until June. The contract required the 50 hour -4 engine to have been completely tested by June 1. [32]

In March, the first XJ40-WE-4 (serial WE003002) had been assembled and sent to test. It was built more as a -2 engine, with the old XJ40-WE-2 compressor spindle that was lighter than the one to be used in the -4 for the 50 hour test. The engine had been run for 18.76 hours, including temperature traversing and checking out the electronic control. Temperature distribution was not good, with a peak temperature of 1,900°F existing between the 2:00 and 5:00 o'clock position with an average turbine inlet temperature of about 1,000°F. The electronic power control performed fairly well, with good stability, acceleration and deceleration performance. Starts were not automatic yet, still being worked on to be in place for the 50 hour test. Results from testing of the control on the J34 test engines were proving helpful in J40 control development. Some work was needed to eliminate carboning(SIC) of the spark plug insulator when AN-F-58 fuel was used.

The fuel filter tests showed it was more than capable of absorbing the required two grams of contaminate per each 1000 gallons per hour pumped.

The gear box had completed two fifty hour tests at full power, with water brakes being used for full power loading. Results were satisfactory. The gear box was being readied for another 50 hour test. (This was the early non-symmetrical design with a wet sump.)

The second -4 engine (serial WE003003) was 40% complete and 25% of the accessories had been received. [33]

The April progress report showed the engine had been through 118 hours of testing. It was identical to the -2 except for minor changes and was being used to test the electronic control. Inlet temperature distribution was poor and because of this, engine thrust was limited to 5,600 lbs. It had been run 107 hours with AN-F-48 fuel and 11 hours with AN-F-58. Poor performance was attributed to the engine basket and cracks in it. It was disassembled and in repair to remediate the cracking and distribution issues. [34]

May's progress report indicated that with 131 hours running to date, progress had been made on the engine basket and that running at full military rpm and temperatures would be attempted. The engine was running with a 30% closed turbine nozzle and was showing 2% improvement in fuel consumption. Oil consumption

was running at 2.4 lbs/hr, still well above the .7 lbs/hr specification. [35]

Figure 10 XJ40-WE-4 Turbine discs with bearing stub assembled to shaft, WAGT P40770, 4/7/49. *Courtesy Hagley Museum and Library*

The electronic control was presenting some problems (unspecified) which Westinghouse was trying to resolve. These were deemed to be "minor".

The first -4 engine (WE003002) was to be rebuilt and used as the 50 hour flight qualification test engine. A second engine (serial WE003003) would be built up to be a back-up engine during test.

June's progress report indicated the first -4 was still configured as a -2 with minor improvements and was still experiencing combustion basket issues and shaft bowing. Test time was up to 195 hours on this engine. The Phase II engine (WE003003, the first real -4 and now intended to be the 50 hour flight qualification test engine) would not be ready until the middle of August even though WAGT had indicated it would be running before the end of July. The reasons for these delays were not explained.

August brought BuAer approval of a significant weight increase for the -4, up from the original estimate of 2,796 pounds to a new limit of 3,185. There were no plans to fly any -4 engine in an aircraft either for test or as a prime power source, so the increase in weight had no practical effect on the -6 directly, but did reflect that the basic design was turning out to be heavier than predicted as development moved forward. [36]

Through July, the first -4 had 195.2 hours of testing. The second -4 was fully assembled with the full complement of accessories, and testing was planned to start on August 19. It was anticipated that the fifty hour qualification test would probably start the week of August 29 if all went well during initial runs, teardown and final run. [37]

Figure 11 XJ40-WE-4 Compressor disc stack assembled to main shaft, WAGT P40779, 4/8/49. *Courtesy Hagley Museum and Library*

WAGT asked for a contract amendment and approval of new delivery dates on three items related to the -4. These were: Item 7 – Delivery of the -4 by September 15, 1949; Item 8 – Report of the 50-hour flight substantiation test delivery by September 15, 1949; and Item 9 – Summary engine evaluation report delivery by October 3, 1949. The changes were based on approval of the engine dry weights proposed and it was noted that the delayed dates would have no effect on the initial production of the J40-WE-6 engine, still planned to begin in July 1950.[38] BuAer accepted the new dates on August 25 but elected not to formally amend the contract.

August was a month of preparation for the completion of the flight substantiation test engine (WE003003) and initial run-in testing (WAGT always referred to this as "green" testing). This engine went through 20 hours of development testing during which the green run was completed with satisfactory performance, with no mechanical defects in evidence. The engine weight was 3,192.5 pounds. However, the teardown revealed problems. The combustion basket had deteriorated and parts had passed through the turbine section causing some damage (repairable). The downstream end of the basket tore loose at the spot welds (about 75% around the circumference). Oil fires in the No. 2 bearing section ruined the bearing. The cause and corrective actions taken regarding these fires were not explained. The combustion basket was further improved and repaired and the bearing replaced. The 50 hour test was slated to start on September 27.[39]

Just prior to this report being received, a final memo arrived that finalized the testing process and established the guidelines to be followed during the 50 hour test. As can be seen, while the engine was to be run through many various cycles, none of them truly reflected a mission cycle. Such testing improvements were far into the future. The WAGT

proposed testing approach outlined below was accepted by the Navy.

50 Hour Flight Substantiation Testing Requirements

I.   **Baseline the Engine**
1. *The engine to be assembled to drawing 43J804-1*
2. *Test to be performed according to the test program outlined.*
3. *Weight of the test engine is 3,182.9 pounds (vs. approved weight of 3,185 lbs.)*
4. *Generator pad of the gear box to be loaded in lieu of the power take-off pad (the same as approved for the -6 engine test).*
5. *The engine to be equipped with a fuel dump valve located in the high pressure fuel line to the flow divider. This valve is needed to stop a small amount of fuel from burning in the combustion chamber after shutdown and causing coking.*
6. *Green testing showed the engine developed maximum thrust at lower temperatures than those in the Model Specification (WAGT-X40E2-2B). The engine to be tested to guaranteed thrust rather than maximum temperature and approval of this approach requested.*
7. *Requested BAR confirmation that the engine as built and operated would be in fulfillment of the contract, provided the 50 hour test was successfully completed.*
8. *AN-F-58 (JP-3) fuel to be used for all testing.*

II.   **Turbine Inlet Temperature Traverse**
After installing in test cell, operate the engine to assure it is mechanically satisfactory.
Perform a Turbine Inlet Temperature Traverse:
    1. At 5,000 pounds thrust
    2. At 6,200 pounds thrust
    3. At 6,800 pounds thrust
Simultaneously with the turbine inlet traverses, take data to determine if the following performance guarantees will be met on a standard static sea level day:

Rating	Thrust	RPM*	SFC
Take Off	6,800	7,260	1.01
Military	6,800	7,260	1.01
Normal	6,200	7,260	.98
Cruise (1)	5,000	6,950	.95
Cruise (2)	4,000	6,750	.93
Idle	450	3,000	--

*Preliminary values subject to revision prior to the qualification test.

III.   **Acceptance Test**
1. *Initial Run – Immediately following running described above, the engine will be subjected to a one-hour thirty minute Initial Run in accordance with the following schedule:*
    i. Make a complete start from zero to idle speed and determine time from initiation of starting cycle to reaching idle speed.
    ii. 5 minutes at idle
    iii. 15 minutes at 5,210 pounds thrust

iv.   30 minutes at 6,200 pounds thrust
v.   15 minutes at 6,800 pounds thrust
vi.   5 minutes at idle followed by rapid acceleration to 6,800 pounds thrust. During this acceleration check:
  1. Operation of speed limiting device
  2. Time required for acceleration
  3. Maximum exhaust gas temperature
  4. Maximum rotor speed
vii.   5 minutes at 6,800 pounds thrust
viii. 5 minutes at idle followed by rapid acceleration up to 6,800 pounds thrust
ix.   5 minutes at 6,800 pounds thrust
x.   5 minutes at idle.

Gas temperatures shall be recorded during all periods.

It being impractical to measure oil consumption during the 30 minute run at 6,200 pounds thrust, the oil consumption for the entire initial run shall be measured and recorded.

2. Inspection – After completion of the initial run the engine will be removed from the test cell and disassembled sufficiently enough to allow an inspection of all working parts. After the inspection the engine will be reassembled.

3. Run-in Prior to Final Run – The engine will be installed in the test cell and operated sufficiently to assure that it and all instrumentation is functioning properly.

4. Final Run – The Final Run will be a complete repetition of the Initial Run described in paragraph III (a) above.

5. Inspection – After the Final Run the engine will be disassembled enough to examine the combustor liner only. If the liner is in good condition the engine will be reassembled.

6. Radio Interference – Radio interference check in accordance with AN-I-27 will be performed on the 4th floor of the "A" Building.

IV.   50 Hour Flight Verification Test

Immediately following the inspection of the liner, the Flight Verification Test will be performed in accordance with the following schedule.

1. Take-Off Run – Five hours of alternate periods of five minutes at take-off thrust and 10 minutes at idle thrust. For all periods at take-off thrust, the power level shall be advanced rapidly to and maintained at the take-off position so that take-off thrust is established by the power control.

2. Military Rated Run – Five hours of alternate periods of thirty minutes at military rated thrust and thirty minutes at idle thrust. For all periods at military rated thrust, the power level shall be advanced rapidly to and maintained at the military position so that military thrust is maintained by the power control.

3. 100% to 80% Normal Rated Run – Ten hours of alternate periods of 2.5 hours each at normal rated thrust and at 80% normal rated thrust.

4. 100% to 60% Normal Rated Run – Ten hours of alternate periods of 2.5 hours each at normal rated thrust and at 60% normal rated thrust.

5. 40% Normal Rated run – Fifteen hours at 40% normal rated thrust.

6. Normal Rated Run – Five hours at normal rated thrust

7. A minimum of 25 starts shall be made on the engine.

8. During the 50 hour run the generator pad shall be loaded by means of a water brake to 2500 inch-pounds torque at 4,000 pounds thrust and above.

9. Radio interference check in accordance with AN-I-27 will be performed on the 4th floor of the "A" building prior to engine disassembly after the 50 hour run.

V.   Inspection

Following the radio interference test the engine will be disassembled and inspected to determine its mechanical condition after the Flight Verification Test.

The 50 Hour Test Run Report

Using the first -4 engine (WE003002), WAGT attempted to run it through an unofficial 50 hour test after 21.84 hours of testing had already been accomplished on the engine. The findings report was dated November 7, 1949 and the run occurred in October, but actual dates involved have not been identified. No explanation was found as to why the second -4 was not used for the test as intended.

This preliminary attempt was designed to:

1. Bring oil consumption within Model Specification limits,

2. Make control adjustments and pass the 50 hour test if possible.

While the first two objectives were met, the run was stopped after 25 hours when it was discovered a shim in the vicinity of the No. 1 engine bearing and combustion liner was failing. Teardown was done and WAGT reported the following:

1. The shim near #1 bearing had begun to fail.

2. The combustor liner had cracked near the downstream end of the inner wall.

3. The compressor inlet guide vane diaphragm and the turbine second stage diaphragm had cracks in the outer shrouds due to poor welding.

4. Cracks in certain other components were discovered and attributed to vibration due to shaft bowing at starting.

Corrective actions taken were reported as:

1. The shim near the #1 bearing was eliminated by a partial redesign of the #1 bearing housing.

2. The combustor liner was redesigned in the cracked area.

3. Intensive welding tests were run to discover the reason for the poor welds in the compressor inlet

guide vane diaphragm and in the turbine second stage nozzle, and also to discover a better method of welding. The tests produced an improved welding method.

4. All parts cracked due to vibration were either repaired or replaced. The source of the cracking was the startup vibration due to shaft bowing. A new type of shaft composed of low thermal expansion material (63.8% Iron, 36.2% Nickel with less than two thirds the modulus of elasticity of steel) was installed in the engine. This new version also incorporated baffles around the shaft to impede the convection that was the cause of shaft bowing.

The reassembled engine was reported ready for a second attempt at the 50 Hour Flight Verification Test on November 8.[40] However, a memo from the BAR a week later indicated the electronic control was troublesome and the engine was being assembled with manual controls. These controls would be used for the checkout and green runs. If the electronic control was not ready for the 50 hour test, WAGT was to advise the BAR of the planned action. At that point, the green run was satisfactorily completed and disassembly was underway. A start of the 50 hour test later that week was expected.[41]

October's progress report indicated the 50 hour test -4 was going to be ready to restart the test on or about November 21, having completed the green run satisfactorily with no issues and having been torn down and rebuilt.[42] This was not to be, however, as the next documentation shows the second try at the test started sometime in the second week of December and was stopped after 11.5 hours of running. The engine had been stopped briefly to change over the fuel tank supply. A quick inspection showed damage to the turbine, with serious nicks and dents to the first stage nozzles and blades and minor nicks and dents to the 2nd stage. Further inspection showed no upstream missing or damaged parts. WAGT attributed the damage to ingestion of a foreign object(s) somehow. They began repairs using Zyglo scans and blending out damage in the nozzles and blades. This would allow them to reassemble the engine with all the original parts and resume the test at the stop point, with restart aimed at December 19.[43] The test was resumed and stopped at the 16.5 hour total time mark due to high oil consumption. Oil was found to be leaking past the No. 1 bearing seal due to the seal rub ring turning and allowing axial movement. It was stopped with a 3/64 in pin to the compressor rotor shaft and the engine reassembled. The delay would require a 16.5 hour penalty run after the 50 hour test was complete. (The -6 oil seal was redesigned due to this problem.[44]) The 50-hour flight substantiation test was completed on January 3, 1950. It

went to radio interference test check-out and then to teardown. The last 33.5 hours of testing were satisfactory except the No. 1 bearing seal gave slight trouble again in the last hour when oil consumption exceeded the .7 lbs per hour guarantee. Thrust and specific fuel consumption were within or better than the Model Specification guarantees. The spark plugs, electronic control and second stage rotating turbine blades had to be further qualified.[45] Initial teardown inspection of the engine found that a fuel nozzle had been incorrectly assembled and caused a 4x6 inch hole to be burned in the combustion basket. Improved nozzle assembly procedures were added to the additional qualification list.[46]

The complete teardown inspection report identified the following problems, all else being acceptable:

1. The combustion basket had a 4x6 in hole burned through due to faulty assembly of a fuel nozzle.
2. The second stage turbine nozzle showed minor cracks in the outer shroud. The first stage turbine nozzle showed minor cracks in both the outer and inner shrouds.
3. Four second stage turbine blades were cracked. (New blade designs were in hand for both first and second stages.)
4. The radio interference tests both prior to and after 50 hours showed the D.C. power leads to be above the Model Specification and AN-I-27 limits.
5. Cracks were found in the exhaust nozzle struts.
6. The No. 2 bearing seal had a 2 inch piece broken off in the Babbitt insert. This was attributed to poor bonding.

Focus on development shifted to the -6 engine at this point, the basic components of the -4 demonstrating the ability of the design to meet requirements for thrust and SFC, albeit with some reliability improvements needed. Little detail was reported on the problems being experienced with the electronic control at this time. Specifics will be taken up in the -6 development section. In summary, the BAR recommended that the penalties be verified on the -6 engine instead of re-running the 50 hour test, noting the entire J-40 program would be delayed if the 50 hour test was re-run for qualification.[47]

Westinghouse submitted report A-928 (not located) on the 50 hr test and on January 27 requested the engine serial number WE003002 be accepted by the Navy and the engine returned to WAGT as GFE for further use in the engine program. The Navy conditionally accepted the test results on March 15 with the following items requiring correction and/or penalty runs: power regulator, No. 1 bearing seal, spark plugs, main oil pump "U" seal packing, second stage turbine blades and exhaust collector strut (prevention of cracks). A positive position lock was required to prevent incorrect alignment of the fuel nozzles. The improved parts could be verified on the 150-hour qualification test of the XJ40-WE-6.

Shaft bowing correction was covered in a memo in this period. The 50-hour test engine did not measure the 16 inch copper plated shaft during the test, but another -4 was measured and showed the copper plating significantly reduced bowing. Effective reduction incorporated three measures:

1. Plate the interior of the turbine shaft with copper plating with a thickness of about .015 inches. This improved rapid conduction of heat from the hot portions of the shaft to the cooler portions.

2. Place two baffles near the downstream end of the turbine shaft: one inside the shaft which rotated inside the shaft and the other stationary baffle placed outside the shaft. The purpose was obstruction of convection heat from the hot first stage turbine disc to the top element of the shaft.

3. The inside-top half and outside-top half of the turbine rotor housing were painted with black sooty paint. This absorbed heat readily and radiated it out of the engine more quickly than the shiny bottom half.

The effectiveness of the measures was indirectly considered to have been proven, as during the 50-hour test bowing was not a problem (not surprising, as the test cycles did not involve engine stopping, cool down to ambient temperatures and restart), the engine having incorporated all three measures. The -6 engine would utilize all three measures and in addition contain vibration dampers located at each of the two main engine bearings.[48]

In May, WAGT reported in a long memo that the fuel consumption figures, which were using a -6% correction factor based on measured differences between AN-F-48 type (petrol 91/96 octane) and AN-F-58 (JP-3), were possibly in error. The factor had been used in reporting the fuel consumption during the 50 hour test. A corrected consumption table was submitted which showed that even with the -6% correction removed, the engine still met the guarantees. Westinghouse explained that all of the continuous testing that had been done to initially validate the difference and amount of the correction and had no explanation as to why fuel consumption measurements between the two fuels on the -6 testing suddenly showed no difference. A more accurate fuel measurement table was placed on order from the Fisher and Porter Company and would become available in June. Until then, no correction factor would be applied to fuel consumption measurements in test until it could be verified it was indeed needed and the amount verified.[49]

The -4 engines were returned after acceptance to WAGT as GFE and used primarily to develop and test the

engine power control system, compressors and afterburners for the -6, -8 and -22 during the remainder of the J40 programs. There were two constructed, house serials WE003002 and WE003003. Both engines were scrapped at the end of the program.

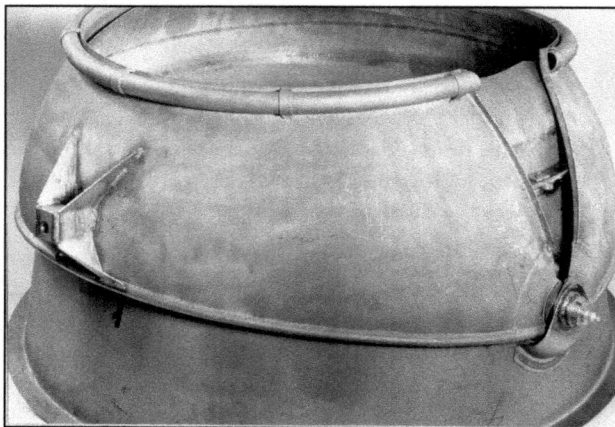

Figure 12 XJ40-WE-4 Initial eyelid exhaust nozzle, WAGT P40771, 4/7/49. *Courtesy Hagley Museum and Library*

Figure 13 XJ40-WE-4 Combustion chamber, inner and outer liner combustion basket assembly, WAGT P40772, 4/7/49. *Courtesy Hagley Museum and Library*

Before leaving the -2 and -4 engines and moving on to -6 development, it is important for the reader to reflect that the later, more powerful -10 and -12 versions of the J40 had already been placed under contract. WAGT presented to BuAer a chart showing the expected position they would be in for engine choice for the coming models. It is telling that not a single surviving document surfaced that talked to the vastly more complex and demanding tasks WAGT was undertaking and the associated risk to BuAer if WAGT failed to deliver. It was a robust growth program that set high expectations and goes a large way to understanding

BuAer inclinations to keep the program in place even as problems in the future grew in severity. Even at the end of the program, Admiral Russell's comment during the Congressional Hearing (see Introduction) shows how hard it was to give up the dream, even as the dream had turned to ashes.

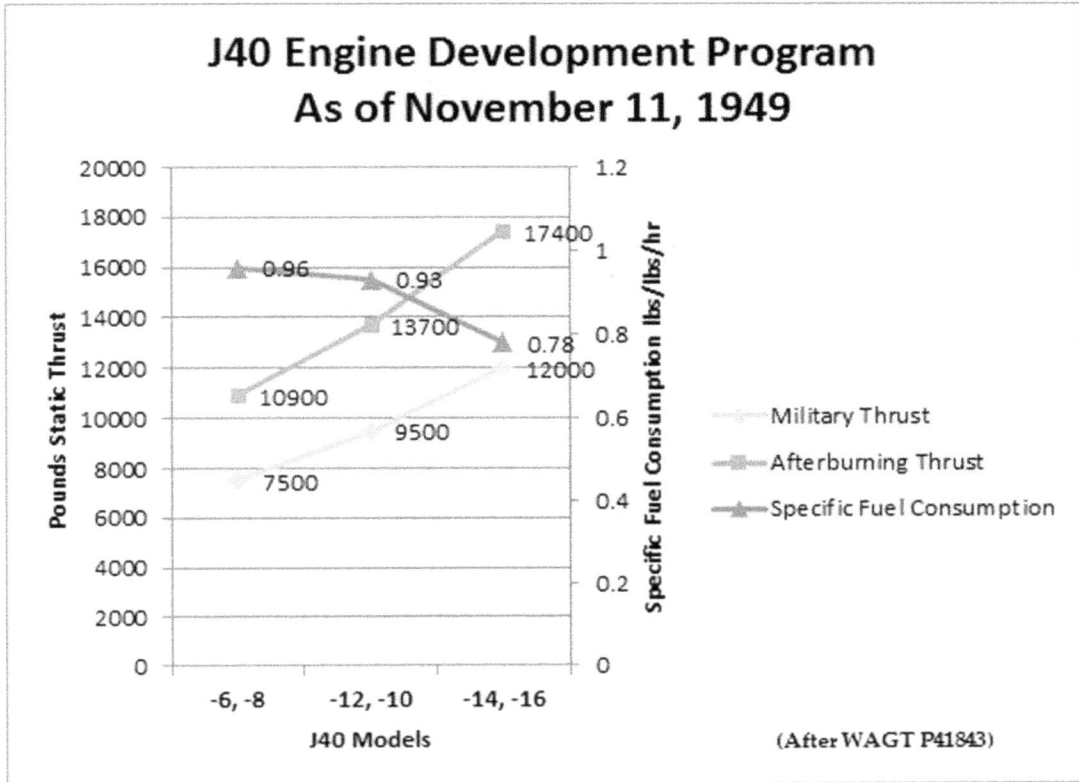

J40 Engine Development Program
As of November 11, 1949

(After WAGT P41843)

Figure 14 XJ40-WE-2 in final assembly with compressor exposed, WAGT P39842, 10/22/1948. *Courtesy Hagley Museum and Library*

Figure 15 XJ40-WE-2 stator vane assemblies in upper half compressor casing, WAGT P39894, 11/1/1948. *Courtesy Hagley Museum and Library*

Chapter 1 - XJ40-WE-2 and -4 Procurement and Development Citations

[1] BuAer Confidential Memorandum, May 6, 1947, United States Navy, Bureau of Aeronautics Contract Correspondence, Box 303 **[B]**, Volume 1 **[V]**, Record Group 72.3.2 **[RG]**, National Archives at College Park **[NACP]**, College Park, MD **B303V1RG7.3.2NACP**

[2] BuAer Confidential Memo, May 6, 1947, Enclosure C, **B303V1RG7.3.2NACP**

[3] BuAer Confidential Memo, May 6, 1947, Enclosure D, **B303V1RG7.3.2NACP**

[4] BuAer Confidential Addendum of May 14, 1947 expands on the analysis of the BuAer Memo of May 6, 1947 on this subject, expanding the analysis to all jet engines known to be under development. **B303V1RG7.3.2NACP**

[5] BuAer Confidential Memo, May 14, 1947, "Addendum to Power Plant Division Memorandum to Chief of the Bureau of Aeronautics dated May 6, 1947" **B303V1RG7.3.2NACP**

[6] Westinghouse Photo P36993, April 18, 1947, Westinghouse X40E Proposal, 4/25/1947 **B303RG7.3.2NACP**

[7] Attachment "A" to Westinghouse X40E Proposal, WAGT Report A-496, "Model X40-E-2 Turbo-Jet Engine Preliminary Design Considerations", April 18, 1947, signed by R.P. Kroon, Manager of Engineering, **B303V1RG7.3.2NACP**

[8] BuAer Power Plant Division Memo, "Proposed Development of the 7500 lb. Thrust XH40-WE (X40E) Turbo Jet Engine by Westinghouse Electric Corporation; Tentative Contract NOa(s) 9212", 23 June 1947" **B303V3RG7.3.2NACP**

[9] Westinghouse Drawing 43J405 on P36967 submitted with the April 18, 1947 proposal **B303RG7.3.2NACP**

[10] BuAer Memo, Comments on Monthly Progress Report A-578 (September, 1947), November 13, 1947, **B303V3RG7.3.2NACP**

[11] BuAer Memo, November 20, 1947, "Cruising Fuel Consumption Improvement Needed" **B303V3RG7.3.2NACP**

[12] WAGT memo, November 28, 1947, "XJ-40 Design – Contract NOa+(s) 9212, Letter of Intent Contract" **B303V1RG7.3.2NACP**

[13] BuAer memo, February 6, 1948 , Approves Westinghouse Report A-583, **B303V3RG7.3.2NACP**

[14] Power Plant Division memo, July 16, 1948, "Contracts NOa(s) 9212 and NOa(s) 9670 with Westinghouse Electric Corporation, Revisions of preliminary drafts of" **B303V3RG7.3.2NACP**

[15] WAGT memo, October 22, 1948, Reference WG -59000 Contract 9212 **B303V1ARG7.3.2NACP**

[16] WAGT Summary Report, November 1, 1948, "Report of Failure at Teardown of the XJ40-WE-2 WE-003001 Engine" **B304V5RG7.3.2NACP**

[17] WAGT memo, December 6, 1948, "Contract NOa(s) 9212 – XJ40-WE-2 Turbo-Jet Engine, Serial No. WE003001, - Acceptance" **B304V5RG7.3.2NACP**

[18] BuAer memo, November 17, 1948, "Turbo-jet Engine Model XJ40WE, Starter Requirements for" **B304V5RG7.3.2NACP**

[19] BuAer memo, December 2, 1948, "Contract NOa(s) 9212, XJ40-WE-2; and Contract NOa(s), XJ46-WE-2 Engine, Design Study of Control System for" **B303V4RG7.3.2NACP**

[20] BuAer memo, December 3, 1948, "Contracts NOa(s) 9051 and NOa(s) 9212 – Development Testing; Control of" **B304V6RG7.3.2NACP**

[21] BuAer memo, January 13, 1949, "Voluntary Steel Allocation Plan – Contractors and Contracts Eligible to Participate in." **B304V6RG7.3.2NACP**

[22] BuAer memo, January 19, 1949, "Contract NOa(s) 9212 XJ40-WE Turbo-Jet Engine, Starting Requirements" **B304V6RG7.3.2NACP**

[23] BuAer BAR memo, February 15, 1949, "Contract NOa(s) 9212 – Monthly Progress Report A-807 (Month of January, 1949), forwarding of" **B304V6RG7.3.2NACP**

[24] BuAer memo, February 9, 1949, "Contract NOa(s) 9212: XJ40-WE-2 Engine Starting Requirements" **B304V6RG7.3.2NACP**

[25] BuAer BAR memo, March 15, 1949, "Contract NOa(s) 9212 – Progress Report for Month of February 1949 – Forwarding of" **B304V6RG7.3.2NACP**

[26] BuAer memo, April 11, 1949, A.L.Baird, Director of Power Plant Division, "XJ40 Turbo-Jet Engine Program Progress Conference, Report of." **B304V6RG7.3.2NACP**

[27] BuAer BAR memo, April 26, 1949, "Contract NOa(s) 9212 – Monthly Progress Report A-834 (Month of March 1949), Forwarding of." **B304V6RG7.3.2NACP**

[28] BuAer memo, June 22, 1949, "Contract NOa(s) 9212 – Monthly Progress Report A-845(Month of May 1949), Forwarding of." **B304V8RG7.3.2NACP**

[29] BuAer memo, August 19, 1949, "Contract NOa(s) 9212 – Monthly Progress Report A-833 (Month of July, 1949), Forwarding of." **B304V9RG7.3.2NACP**

[30] BuAer BAR memo, January 13, 1949, "Contract NOa(s) 9212 – Monthly Progress Report A-797 (Month of December 1948, forwarding of." **B304V7RG7.3.2NACP**

[31] BuAer memo, February 2, 1949, "Westinghouse Specification WAGT-X40E2-2A Dated October 29, 1948 Covering the Models XJ40-WE-2, -4, and -6 Engines, Approval of Performance Curves for." **B304V6RG7.3.2NACP**

[32] BuAer BAR memo, March 15, 1949, "Contract NOa(s) 9212 – Progress Report A-816 for Month of February 1949 – Forwarding of" **B304V8RG7.3.2NACP**

[33] BuAir BAR memo, April 26, 1949, "Contract NOa(s) 9212 – Monthly Progress Report A-834 (Month of March 1949), Forwarding of." **B304V8RG7.3.2NACP**

[34] BuAer BAR memo, May 17, 1949, "Contract NOa(s) 9212 – Monthly Progress Report A-845 (Month of April 1949) – Forwarding of." **B304V7RG7.3.2NACP**

[35] BuAer memo, June 22, 1949, "Contract NOa(s) 9212 – Monthly Progress Report A-856 (Month of May 1949), Forwarding of." **B304V8RG7.3.2NACP**

[36] BuAer Naval Speed Letter, August 19, 1949, "Contract NOa(s) 9212, Increased Weight for XJ40-WE-4, Approval of." **B304V9RG7.3.2NACP**

[37] BuAer memo, August 19, 1949, "Contract NOa(s) 9212 - Monthly Progress Report A-883 (Month of July, 1949), Forwarding of." **B304V9RG7.3.2NACP**

[38] WAGT memo, July 26, 1949, "Contract NOa(s) 9212 - Our Reference WG59000 Contract Delivery Dates" **B304V8RG7.3.2NACP**

[39] BuAer memo, September 27, 1949, "Contract NOa(s) 9212 – Monthly Progress Report A892 for August 1949 - Forwarding of" **B304V9RG7.3.2NACP**

[40] WAGT memo, November 7, 1949, "Contract NOa(s) 9212 – Phase II, 50 Hour Flight Verification Test on XJ40-WE-4 Type Engine." **B305V10RG7.3.2NACP**

[41] BAR memo, November 14, 1949, "Contract NOa(s) 9212 – Phase II – 50 Hour flight Verification Test on XJ40-WE-4 Type Engine." **B305V10RG7.3.2NACP**

[42] BAR memo, November 17, 1949, "Contract NOa(s) 9212 – Monthly Progress Report A-911for Month of October 1949." **B305V10RG7.3.2NACP**

[43] BAR memo, December 15, 1949, "Contract NOa(s) 9212, Phase II – 50-Hour Flight Verification Test on XJ40-WE-4 Engine – Information on." **B305V10RG7.3.2NACP**

[44] BAR memo, December 22, 1949, C-800, "Contract NOa(s) 9212, Phase II – 50-Hour Flight Verification Test on XJ40-WE-4 Engine – Information on." **B305V10RG7.3.2NACP**

[45] BAR memo, January 5, 1950, C-109, "Contract NOa(s) 9212 – Phase II XJ40-WE-4 Engine – 50-Hour Flight Substantiation Test – Information on." **B305V10RG7.3.2NACP**

[46] BAR memo, January 13, 1950, "Contract NOa(s) 9212-Monthly Progress Report A-924 for Month of December" 1949." **B305V10RG7.3.2NACP**

[47] BAR memo, January 23, 1950, "Contract NOa(s) 9212, Phase II, XJ40-WE-4 – 50-Hour Engine – Teardown Results – Information on." **B305V11RG7.3.2NACP**

[48] WAGT memo, February 1, 1950, "Contract NOa(s) 9212 – Phase II – Shaft Bowing in XJ40-WE-4 Engine." **B305V11RG7.3.2NACP**

[49] WAGT memo, May 18, 1950, "Contract NOa(s) 9212 - Correction to Engine Performance Data." **B305V13RG7.3.2NACP**

Chapter 2

XJ40-WE-6 Development

———◆————————————————————◆———

XJ40-WE-6 Mock-up Board

The mock-up board for the -6 was held at the WAGT location in Essington, PA on February 11-13, 1948. To enable the reader to follow each item to closure, the WAGT responses and recommendations immediately follow the mock-up board item. The items affecting the design of the engine were as follows:[1]

XJ40-WE-6 Mock-up Board Daily Items

Item 1 – Change the hydraulic pump drive AND 20002 Type XII K to 1,000 pounds inches continuous torque.

Item 2 – Change the power take-off and generator drive to 2,500 pounds inches continuous torque total. The combined continuous running torque of these two pads shall not exceed this value.

WAGT Analysis and Recommendation (hereafter WAGT A&R):

The two drive gears were driven through the same idle gear in the gearbox and the changes were evaluated together. All gear and bearing loads in the gear train were recalculated. A new layout and necessary detail drawings were made. Roller bearings were selected in place of the existing ball bearing design (precluding later use of a splined drive from the shaft). Detailed design changes rippled through the entire gear case reflecting the larger sizes of bearings and shafts to take the required loads. The total weight change was estimated at 4.5 pounds and two specification changes needed were to reflect a not-to-exceed 2,500 lb/inches continuous running torque of the combined power take-off and generator drives; and the torque rating of the hydraulic pump drive changed to 1,000 lb/ins with the pad type changed from type AND 20002 III J to AND 20002 III K.

Item 3 – Fire seal provisions. The WAGT report of July 6, 1948 was accepted on August 25, 1948. (Report not located.)

Item 4 – Mount the exhaust nozzle actuating system on the exhaust nozzle.

WAGT A&R: Analysis showed that mounting the actuating system on the nozzle itself would make the nozzle longer by at least 20 inches. It was recommended that the actuating system always be mounted on the rear bearing support with the hydraulic actuating piston mounted on the aft under side of the combustion chamber outer casing.

This recommendation was accepted on 24 April, 1949 with the proviso that the actuating system used on the -6 and -8 be as common as possible, incorporating a quick disconnect.

Item 9 – Included three specific actions:

A. Study an increase in oil tank capacity to accommodate 10 hours of engine operation.

B. Study what is required to adapt the engine oil system to an external oil system which can accommodate more than 10 hours of engine operation.

C. Study oil filling provisions to make them more acceptable. (Linked to Item 10 and Item 42 studies.)

WAGT A&R:

A. Westinghouse guaranteed that the engine oil consumption would be a maximum of 0.7 pounds per hour, making an increase in planned tank capacity unnecessary (1.95 gal/14.04 lbs). Ten hours of operation at maximum oil consumption would leave 7.04 lbs (.98 gal) left in the tank, within the safe point of engine operation.

Action needed: Change the guarantee in the engine specification.

B. The study showed it was possible to attach some type of an external oil supply, employing two openings into the gear box for venting and passage of oil. Two such openings already existed in the gearbox design, one below the oil level and the other just under the starter pad and well above the normal oil level. Study A-696 was produced. This indicated that the external tank should be as close to the engine tank as possible and an oil governor should be used to scavenge the wet sump (i.e. the engine oil tank) back to the external tank any time an excess oil level was detected in the engine tank. This would operate just like a dry sump conventional piston engine configuration but allow a certain level of oil to remain in the "dry" sump.

Action needed: None

C. The study incorporated both Items 10 and 42.

Item 10 – Incorporate adequate provisions to show that the oil filter cap and oil dip stick are properly locked and sealed.

Item 42 – Provide an oil tank filler scupper (drain) and drain line with the customer's connection at the bottom of the accessory case.

WAGT A&R: The recommendation was to combine the oil tank filler neck and the oil tank dip stick into one location and integrally attach the dip stick to the filler cap. The cap would have locking spring hasps aligned so that the cap could not accidentally be turned while it was locked. If the cap was unlocked, the spring position would be a visual indicator to servicing personnel showing the cap was unlocked. It was also suggested that the airframe manufacturer provide another device on the servicing door for oil filling/level checking to prevent the door from being closed completely if the cap was in the unlocked position. This subject was discussed in depth in report A-696.

To prevent over-filling of the oil tank, a standpipe was added that drained off excess oil through the scupper any time the dip stick filler cap was unlocked and removed from its locked position.

Item 14 - Study the need to include anti-icing provisions in the XJ40 design.

WAGT A&R: Report A-695 was produced to respond to this action. It recommended that the XJ40-WE-6 be modified to included anti-icing provisions. Specifically, it recommended protection be provided for the inlet duct and inlet guide vanes by means of a bleed of a mixture of compressor outlet and turbine inlet gases, a change that could easily be incorporated into the current design. The system would automatically regulate itself after being turned on by the pilot, so as to have an approximately constant gas temperature at the anti-icing system inlet. The bled gases, after passing through the anti-icing circuit, would be dumped overboard. The maximum amount of bleed contemplated was 1.1% resulting in changes in the static military RPM engine performance of: Thrust Loss – 1.5%, Fuel Consumption Increase – 1.0%, and Turbine Inlet Temperature Increase – 1.5%. The system was projected to add 60 lbs to the weight of the engine. It was noted that if anti-icing provisions had to be extended to other components deeper into the engine, extensive redesign would be needed to accommodate the capability.

Item 20 – Study and report on the problems of trying

to use impingement and powder engine starters.

WAGT A&R: Report A-633 was submitted on January 26, 1948 which was entitled "XJ40 Gas Turbine Engine Design Study of Starting and Ignition System" and included the relevant data. On February 20, 1949, WAGT asked that BuAer accept the data in the study in response to the mock-up board request.[2]

Item 21 – Study the variable area exhaust nozzle to cover the following points:

A. Eliminate the possibility of asymmetric thrust.

B. Reduce the vulnerability of the engine oil system due to failure of the exhaust nozzle actuator lines.

C. Provide a method of moving the nozzle to its minimum position following failure of the hydraulic supply.

D. Incorporate a quick disconnect method for attaching the variable area exhaust nozzle.

WAGT A&R: The response of April 19, 1949 covered each of the points and all but Item B were approved on May 23 by BuAer.[3] BuAer suggested the exhaust pipe extension information be included in the installation manual for the engine.

A. The design included a positive mechanical linkage in the form of a yoke connecting the two eyelids of the exhaust nozzle. The yoke would be actuated by one hydraulic piston. This would eliminate possibility of asymmetric thrust.

B. WAGT believed minimizing the length of the hydraulic lines satisfied this request. Adding a hydraulic fuse (check valve) in the high pressure line to the actuator might be desirable and that was studied. It was determined a fuse would work if there was a complete rupture of the line, but not if there was a slow leak. It would be a very complex item in and of itself and add a potential source of trouble. The line length would be almost as long and have as many fittings as present lines. As a result, the use of a fuse was not recommended. A May 23 memo from BuAer rejected the shortening of the hydraulic lines to satisfy the requirement. Also, they reminded WAGT that the requirement was a commonality between the -6 and -8 nozzle actuator systems and the entire area needed further development.[4]

C. A hydraulic lock would be provided in the actuating mechanism to hold the exhaust nozzle in the position it was in at the time of failure. This would not return the nozzle to minimum position. WAGT pointed to a memo from BuAer (not located) giving specific approval for use of the hydraulic lock for non-afterburning engines.

D. WAGT noted that this was a new requirement not affecting the reliability or efficient operation of the engine and would require cost and specification revision. It noted a quick disconnect clamp for attaching the variable area exhaust nozzle was used in place of the present bolted flange between the aft end of the rear bearing support assembly and upstream end of the exhaust nozzle. No other location was remotely suitable. Such a clamp would be of limited usefulness even in that location because the rear bearing support assembly contained an inner cone which extended downstream and had its apex approximately at the plane of the exhaust end of the exhaust nozzle and obstructed radial removal from the engine. As a result, WAGT recommended not incorporating a quick disconnect.

E. A discussion of exhaust extensions of up to 12 inches, 12 to 70 inches, and over 70 inches explained what constructions would be needed to accommodate each. No such extensions were being designed or built at that time.

Item 25 – Study the maximum temperatures of the engine components and the heat rejection of the components at their maximum temperatures.

WAGT A&R: A listing was provided by WAGT and approved by BuAer August 25, 1948. However, BuAer noted that the maximum given temperature of the electric and electronic components or accessories was given as 165°F, yet the WAGT supplied curve No. 333314 "XJ40-WE-6 Operating Limits", dated May 12, 1948, showed the maximum ram operating temperature for the engine as 200°F. As supplying refrigerated air for cooling accessories or components was considered undesirable, WAGT was tasked with raising the maximum temperature limits of accessories or components accordingly and submitting comments on this subject.

A later memo suggested ram cooling of the Power Control would bring the temperatures down 30°F or so and that high temperature materials only would be used. If these allowed higher operating temperatures, WAGT would request changes on all of their engines.[5]

Item 28 – Add thermocouples to the bearings and main oil pump discharge line and add a pressure port to the main oil pump discharge outlet for customer pressure readings.

WAGT A&R:
Westinghouse added standard thermocouples at the locations requested and a 7/16 inch 20 NF-3 (AND-10050) port on the rear of the main pump. Normal pressure would be 80 psi but would go as high as 500 psi during cold starting, and provisions would have to be made to protect the gauge from the high cold starting pressure.[6]

BuAer responded that the oil port did not meet the specifications spelled out in AN-E-30 and should be changed to *"a 1 -1/16 -12, N-3 internal straight thread opening in the passage feeding from the oil cleaner. The oil pressure at the connection shall not fluctuate more than plus or minus 5 percent under any operating condition. Clearance shall be provided for mounting the oil transmitter directly on the engine as specified in AND10341 unless otherwise specified in the model specification."*

WAGT replied there was insufficient space for the normal transmitter connection and they were providing the smaller one, which had been approved as an exception to AN-E-30 on prior engines. BuAer rejected this and reminded WAGT that prior waivers did not constitute agreement to waive this requirement on this or future engines.[7]

Item 29 – Provide the specified engine handling fittings.

WAGT A&R: WAGT held a conference with the BAR on April 20, 1949 and received approval for the fittings recommended in the study presented. They were to be incorporated on the mock-up and all relevant drawings and related data. The lugs added two pounds to the engine itself.[8]

Item 32 - Study the possibility of beveling the forward intake duct flange to facilitate quick engine changes.

Item 33 - Provide quick disconnect clamp breaks at the top and bottom of the intake duct flanges.

WAGT A&R: It was determined that the flange could be beveled 20 degrees and provide more than adequate clearance for the quick disconnect band even in the face of adverse tolerance mismatch of airframe to engine due to manufacturing/servicing. The quick disconnect band design had not yet been finalized, but it would have two breaks, at the top and bottom of the bands. These changes would add no weight to the engine nor was an engine specification change needed if the items were included in the design.

Item 34 and Item 45 – The first requested that the fuel system planned for the -6 be examined for adaptability for the addition of an afterburner. The latter requested that WAGT submit a study for increasing the strength of the exhaust flange to permit cantilever attachment of a weight equivalent to 45% of afterburner for a range of moment- arms from an afterburner with minimum diffuser length to an afterburner with a diffuser of 2.5 diameters length.

WAGT A&R: Report A-749 dated Sept. 13, 1948 recommended the extension and afterburner not be cantilevered. The excessive increase in engine weight required and the additional development time necessary to design an engine which would operate successfully while under the influence of afterburner vibration transmitted to it through the afterburner extension were major considerations to the recommendation. The increase in weight would range from 16 to 108 lbs depending on the length of afterburner extension. It was recommended the afterburner be supported separately in the airframe, and the extension provided with a flexible joint for expansion and misalignment and that the afterburner vibration be isolated from the engine. The report demonstrates how the analysis was done and the results. One example given of A/B vibration impact was the J34, which was only guaranteed to run a total of 1.5 hours in A/B vs. the normal 150 hours between overhaul of the non-afterburner version.[9]

Both items were deemed satisfied via BuAer approval on Jun 27, 1949, due to the fact the that -6 would not use an A/B, this requirement being handled by the -8.[10]

Item 38 and Item 39 - Incorporate quick disconnect mountings for accessories (hydraulic-electric governor, fuel booster pump, dual fuel pump, afterburner fuel pump) where practicable and produce a study on the best method to incorporate these provisions. Specifically, address the applicability of the Jack and Heintz design (drawing provided to WAGT). The quick disconnect mounting requirements were listed as:

A. Mounting mechanism must be irreversible and self-locking.

B. Mounting mechanism must be completely operable at a single point of application and provision must be made for indexing, such that the point of application may be placed in a minimum of four locations around the periphery of the pad.

C. The mounting must be an integral unit with no loose parts after removal.[11]

<u>WAGT A&R:</u> Westinghouse produced a study that looked at four types: Victory, Eclipse, Lee and Jack and Heintz, and recommended a "bayonet type" of quick disconnect mounting for the dual fuel pump and the hydraulic-electric governor. The bayonet type did not meet the requirements as stated and BuAer requested they restudy the requirements with the objective of meeting all of them.[12]

BuAer responded on March 21 requesting that the Lee type of quick disconnect be included in the revised detailed study, and particularly compare the types for cost, time required for operation, weight, space and availability.[13]

WAGT transmitted report A-829 "A Study of the Application of Accessory Quick-Disconnect Mountings for XJ40-WE-6, XJ46-WE-2, and XJ34-WE-32" in May and informed BuAer that it would not be feasible to incorporate such mountings on the -6 prior to the verification test. If

incorporated on production engines, it would raise the price of each engine $6-700 or more.

Item 47 – Address the air inlet problem to the current pneumatic type starter location on the hourglass-shaped gearbox. Also, increase flexibility of the gearbox such that a generator or alternator might be driven remotely from either the generator or power take-off pads, and possibly accommodate both starter and generator from the pads at the top of the gearbox.

<u>WAGT A&R:</u> Alleviate the air inlet problem through use of a special type elbow and eliminate remaining interferences between the pneumatic starter and other surrounding accessories. Cease further consideration of alternate proposals to use other gear box drives, a gear box redesign, a starter drive extension, etc.

BuAer responded that the study was not satisfactory and that WAGT should reconsider and resubmit the study.[14] In particular, they wanted more flexibility in using the upper pads of the gearbox for alternate and remote uses. WAGT forwarded their new study on January 3, 1949 along with estimated prices. The study gave BuAer three alternatives: 1) Put the starter where currently designed with a special elbow; 2) Move the starter to a top rear power take-off pad on the accessory gear box, driving through an adapter gear box. This would add 30 pounds to the engine and delay testing the accessory gear box until after the -6 had completed the 150 hour acceptance test; 3) Use the hourglass shaped gearbox designed for the -8 which did not require an adapter gearbox for the starter. This would add 25 pounds to the weight of the engine and could be incorporated into the -6 design by December 1949.[15]

It was decided to use recommendation one (1) and the Navy would order the special elbow from AiResearch as part of the 35 hp. starter based on Westinghouse design. Ultimately, all production -6 engines built used the -8 design gearbox. The original -6 gearbox was a wet sump design lacking provisions for negative "g" conditions.

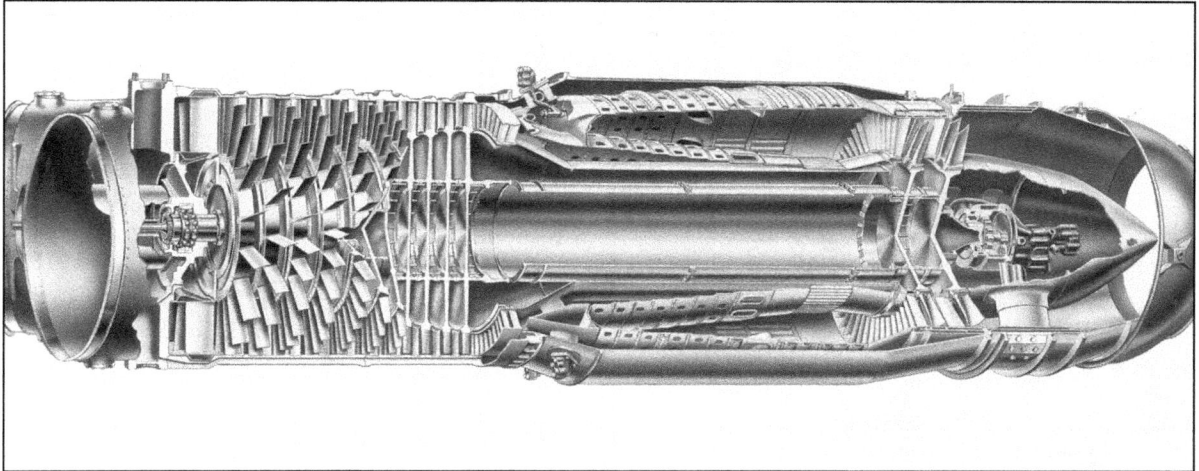

Figure 1 XJ40-WE-6 Aerodynamic cutaway, lacking detail of eyelid exhaust, gearbox and dressing. WAGT P45610, 11/1/1951. *Courtesy Hagley Museum and Library*

After the Mock-up Review Board, design work continued, with WAGT forwarding their first drawings and weight analysis to BuAer on July 22 for the afterburner for the -6. This version of the engine would be developed as the -8 under another contract and will be discussed later. It was apparent that adding the afterburner would change the back of the engine enough to preclude a quick change feature from -6 to -8 in the field. Later, BuAer would require such a capability for the -10/-12 versions of the engine.

The control system report (A-615) was found generally acceptable with additional requirements. The report indicates that an eyelid variable area design was being recommended. The additional requirements were (Westinghouse's replies or comments immediately follow each item[16]):

1. Variable area exhaust nozzle:

A. The nozzle eyelids shall be mechanically interconnected in such a manner as to prevent the occurrence of asymmetrical thrust.

WAGT – Report A-638 described the proposed exhaust nozzle eyelid design and this addressed the interlocking of the nozzles.

B. If the control system utilizes engine oil, any failure of the system shall result in no loss of oil from the engine lubrication system.

WAGT – The design of the system is such that the oil removed from the lubrication system is returned to the lubrication system at the same pressure, therefore, any malfunction of the system will not cause loss of oil from the engine lubrication system. Hydraulic fuses *(check valves)* are used in the oil lines to the actuator cylinder such that damage to these lines will cause no loss of oil from the lubrication system other than the actual oil in the line itself.

C. In the event of failure of the primary control system, the most reliable means that is practicable shall be provided for automatically moving the exhaust nozzle eyelids to the hot day military position.

WAGT – Report A-638 states *"The actuator is designed so that it will be possible to use the variable area exhaust when the main power control system becomes inoperative. Thus, variable area nozzle operation is assured regardless of whether the main or emergency power control is in operation. If the electrical lines to the actuator are damaged, the emergency power control system becomes operative and the exhaust nozzle actuator will regulate mechanically. If the hydraulic lines to the exhaust nozzle regulator are damaged, a suitable hydraulic fuse between the pump outlet and the #2 and #3 bearing oil inlet line will prevent loss of oil. Additional fuses located within the regulator will close at the same time thus providing a hydrostatic lock against movement of the eyelids."* This feature assures that the thrust of the engine will be maintained at the same value it was prior to the hydraulic line failure. In addition the system is so designed that the following emergency provisions may be incorporated at a later date should it be believed that the additional weight and complexity can be justified – duplicated hydraulic lines and check valves (approx. 10 lbs); or a three-way solenoid to allow the pilot to close the exhaust nozzle to a predetermined closed position (approx. 7 lbs).

D. The control system shall be readily adaptable for use with various lengths of exhaust tail pipes.

WAGT – In the redesigned exhaust nozzle control the actuator and rods are mounted on the variable area exhaust nozzle section that is moved rearward and the oil and electrical lines would have to be lengthened as would the emergency control rod. This makes the system adaptable for use with various lengths of exhaust tail pipes with a minimum of changes.

Figure 2 AiResearch compressed air starter for J40-WE-6 and -8, WAGT P40631, 3/4/1949. *Courtesy Hagley Museum and Library*

2. Engine starting:

It was stated that for future turbojet engines, both on the ground and in the air, starting would be accomplished as follows: The pilot would advance the cockpit throttle to the idle position and then actuate a separate starter switch (likely positioned on the throttle quadrant). Automatic starting would commence and end with the engine at idle rpm. Fuel flow initiation and control during starting had to be independent of the actual starting duration, since GFE starters might produce varying results. With the throttle in the IDLE position, fuel flow should commence at the appropriate RPM and at the appropriate volume to accelerate the engine (along with the starter) to the idle RPM without damage to the engine regardless of how long that might take.

WAGT – Proposed a different solution as follows: Start with the throttle in the OFF position and bring the engine to crank speed. Then open the throttle to the IDLE position and let fuel flow to the engine and accelerate the engine to idle speed. The control valve as currently designed had a positive full shut-off valve actuated in the OFF position, eliminating any chance of fuel entering the engine prior to the proper speed being reached and causing an over temperature of the engine. The actual amount of fuel flow would be regulated by the temperature in the engine. Westinghouse pointed out their procedure eliminated the need for a fuel flow initiator control.

3. Service operational requirements:

A. The fuel and control systems shall suffer no detrimental effect when the engine is rotated at 30% military rpm with the power level in the "OFF" position. This is to allow for wind-milling.

B. With the fuel and control system filled with air, it shall be possible to start the engine by restoring fuel flow to

the pump inlet and using normal starting procedure. This covers the condition where the fuel pump has been starved.

C. The fuel and control system shall function satisfactorily when using fuel contaminated to the extent of 2 grams of foreign matter per 1,000 gallons of fuel. For test purposes this foreign matter shall be U.S. Army Standardized Air Cleaner Test Dust having a particle size analysis of:

Particle Size Microns	Percent of Total By Weight
0-5	30 +/- 2
5-10	18 +/- 3
10-20	16 +/- 3
20-40	18 +/- 3
Over 40	9 +/- 3
Through a 200 Mesh Screen	100 Total

If a filter is employed as part of the system it shall be of sufficient capacity to permit a minimum of 10 hours of continuous operation at normal rated fuel flow without being cleaned. The intent is to make fuel systems less sensitive to fuel contamination.

In a comment, BuAer felt the variable area eyelids might not be able to maintain a consistent thrust schedule in practice when manufacturing and service wear occurred with the engine in production and use and asked WAGT to consider this as they continued their design work. Wide variations in engine thrust schedules in service would be unacceptable.

In addition, the 150 psi pressure drop in the fuel nozzles was questioned, noting that carburetors achieved satisfactory results with as little as 5 psi drops. Since no charts of the full fuel flow path in the control were included in the report, the Navy could not understand the need for the high (and very conservative) drop. It noted that the very severe requirements for fuel pumps could be relieved if the fuel control pressure drop could be reduced to its minimal practical value.

The Westinghouse response of October 15 stated they had taken into consideration all of the requirements in the current control design, which they felt would be satisfactory. The pressure drop was based on experience of prior engines and had considered the very high fuel flow of the larger engine. The engine pump was expected to be able to pump at 525 psi but would be reduced if testing later indicated a lower value would meet the performance needs of the engine.

The August, 1948 Monthly Progress Report stated that WAGT anticipated the first operation of the -2 mule engine on or before September 30, but the Bureau Area

Representative (BAR), in his cover letter, stated that based on his observance of the materials on hand and still due from outside sources (sub-contractors) and the test cell prep yet to be done, he didn't expect testing to begin before October 15. Specifically, none of the compressor discs (from Gleason Works, Rochester, NY) had been received yet and were expected September 20. When they arrived they needed to be bladed, assembled and balanced.

As late as October 5, WAGT requested the BuAer developed 35 HP pneumatic starter and auxiliary power unit to supply air to the starter be shipped to Essington as soon as possible. Westinghouse did not have enough air supply for the starter in their shops. Using an AiResearch Model 5500 pneumatic starter as a template, they had discovered two 1/8" removal interferences during the pilot fit, but it was too soon to know if they would still exist after all the actual parts were in place. They could be removed by redesign for the -4. Any such change would require BuAer authorization be made to AiResearch.

This memo also requested BuAer forward any data they had on both pneumatic and solid propellant starter designs in the 15-60 HP range.[17]

On the same day, another memo from WAGT notified BuAer that the current electronic Power Regulator might need to be airframe mounted, at least in the interim, in spite of BuAer's preference that all engine control components be engine mounted. This was the first memo that demonstrated the breadth of engine development programs in which WAGT was engaged for the Navy and how the mass of correspondence being traded would cross program boundaries. WAGT noted the following current issues:

A. Space Requirements – The actual size of the individual components determine the minimum dimension of the entire Power Regulator. If this dimension exceeds the allowable engine envelope the control would have to be located "off" the engine.

B. The Power Regulator would be the same size regardless of the size of the engine with which it is used. On the smaller engines such as the XJ34 and XJ46 there is much less room available than on a larger engine, such as the XJ40.

C. Interchangeability – It is this Contractor's aim wherever possible to use the identical Power Regulator for all sizes of engines in order to reduce spare parts, increase production and reduce costs. A special "tailored" regulator to fit different engines is therefore to be avoided.

D. Ambient Temperatures and Pressures – Ambient temperatures of 165°F appear to be the maximum allowable for electrical apparatus such as the Power Regulator, Generators, etc. Higher temperatures will require heavier, bulkier parts to reduce the heat generated and to better radiate it. Therefore, for the time being, high speed airplanes will require refrigerated air to hold 165°F ambients (SIC).

In addition to temperature, altitude pressures of 50,000 feet and higher present a problem of electrical leakage and flash-over which must be overcome by reduced voltages and improved insulation.

Until the problems of extreme temperature and altitudes are worked out, it may be desirable to locate the Power Regulator temporarily in the refrigerated and pressurized cockpit and to most expeditiously overcome these two problems.

E. Vibration – More vibration is present on the engine than in the airframe, however it is believed that not too much difficulty will be experienced in designing satisfactory mountings for locating the regulator on the engine.[18]

BuAer's handwritten notes on this memo show strong concern for the weight of possible refrigeration equipment in addition to any weight increase from beefing up the Power Regulator components. They indicate that at the XJ34-WE-32 mock-up board, Westinghouse was told that it was the policy of the Bureau of Aeronautics that engine controls, accessories, and components should be mountable on the engine. They referenced two memos of August 25 and September 3, 1948 where WAGT was informed by BuAer that electrical and electronic components and accessories be capable of withstanding the same limiting temperatures that had been identified as operating limits for the basic engines (XJ40 and XJ46).

The Power Regulator would be a source of constant problems as development testing began. Its design and operations issues will be covered later.

With the -2 engine late, BuAer wrote a memo reminding WAGT that acceptance of the engine could not be made without an updated and complete set of specifications. With Westinghouse giving no indication in their progress reports of schedule slippages, BuAer was growing concerned. *"In view of the extensive engine program presently planned around the J40 engine, considerable concern exists as to the reality of the delivery dates set forth by the subject contract, and the corresponding reflection upon other delivery dates set forth in the subsequent proposals. The deferred handling of the engine specification deliveries and the absence of information on the reasons for the initial engine delivery delays do not allay the above apprehension."*[19]

WAGT responded that the specification was finally coordinated with BuAer on 18 and 25 October and would be forthcoming that week. The performance curves would follow the next week. They requested that the XJ40-WE-2 be accepted according to the model specification mentioned. They stated the delay in the -2 was caused by a subcontractor having problems machining the curvic clutch of the compressor discs. Difficulty in balancing the compressor and turbine rotors was experienced due to the new design of the discs. They would be delivering the engine to the test house that week and after preliminary testing, it would be submitted for the acceptance run. Even with delivery of the first engine approximately a month late, they expected to meet the delivery dates for the -4 and -6 engines.[20]

Later in October, it was reported the -2 was not fully assembled yet as the compressor, shaft and turbine had not been balanced. It was expected to begin its testing on October 25.[21] The test cell was being tested using a J34 engine (one of two on loan from the Navy). Days later, Westinghouse sent an approved formal contract back with two suggested changes. These were accepted by the Navy and the LOI contract was formally changed to a cost plus fee contract. One change was to agree that the two J34 engines loaned as GFE to Westinghouse could be returned assembled but not overhauled and the other asked the Navy to confirm that the Navy was still paying to modify two Westinghouse test cells (the latter item was not covered in the contract, but had been previously agreed upon.)

The January, 1949 progress report indicated that 95% of the detailed drawings had been released. The compressor housing was still in the layout stage and the anti-icing features would take considerable re-designing before it would be released.

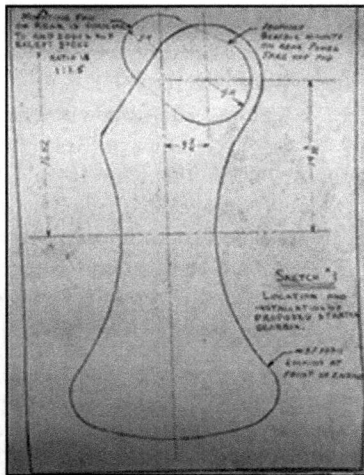

Figure 3 Initial -6 Gearbox Shape with No. 2 starter position option also shown, WAGT drawing attached to a study dated 1/3/1949.[22]

The engine starting debate continued with a WAGT response on February 2. On December 2, 1948 BuAer rejected the WAGT proposed starting process on the grounds that it involved the pilot making a judgment call of when to advance the throttle to IDLE and if confused could damage the engine, particularly if they advanced the throttle first and then accelerated the engine to start RPM.[23]

The WAGT response (below) regarding a pilot's mistake not being able to damage the engine seems to support an argument that turning on the fuel along with the

starter being engaged would not be a problem, although it is not clear that was WAGT's intended point.

"The fuel spray from the spring loaded variable area nozzles is finely atomized when it is injected into the combustion chamber even at very low flows and the velocity of the air passing through the engine will blow the atomized fuel out of the exhaust nozzle. This will prevent fuel from accumulating in the engine and hot, flaming starts will not be obtained. This is particularly true when making a windmill start in flight. The only disadvantage to this is the slight loss of fuel occurring from the time that the power control level is advanced and ignition of the fuel takes place."[24]

They also did not want to change the fuel pump design, arguing that adding the complexity of a pressure regulated check valve to control the start of fuel on startup would mean raising the pressure of the pump and they did not recommend it, advising leaving the pump as it was. They did not know if the pump pressure at 800 rpm was correct as yet, but planned tests to determine if automatic starts could be obtained without damage to the engine.

It was recommended the pump design not be changed because of the risks of decreasing reliability and added weight.

The response to this was that WAGT's current control system for starting had to be changed to be in compliance if it did not meet the requirements.[25] BuAer added that all aircraft being designed were being designed using the same engine starting procedures. Looking ahead to April 6, in a handwritten note BuAer finally accepted Westinghouse's statement that the pilot could simply move the throttle to the IDLE position and press the starter button to start the engine and that this procedure was in compliance with the requirements.[26]

Starting was not the only concern at this time. BuAer discovered that WAGT did not have all of the components required for the emergency fuel system in its design.[27] This was resolved by changing the emergency fuel system test requirements to accommodate the J40 design approach. Exactly what had been out of compliance was not detailed.

Furnishing Westinghouse with an actual AiResearch pneumatic starter and auxiliary power unit had still not been accomplished. BuAer did not anticipate being able to provide either in the short term. They were still discussing changes to the starter with AiResearch. They were also unable to provide data as yet on solid propellant starters, but were planning to visit Victory Engineering Corporation in Newark, NJ in order to obtain information.[28] WAGT had requested that report A-633 that had included a study of problems related to power or impingement starts be accepted.

The February progress report indicated that 95% of the detailed design drawings had been released, with the de-icing provisions still in layout stages and it would be some time before it would be released from manufacture. The preliminary engine evaluation report called for in Item 6 would be ready to submit by March 30. [29]

By March, the progress report showed all layouts of redesigned parts had been completed and detailed drawings very much underway. No new engine parts were released for manufacture during March other than accessories for the fuel system. The weight reduction program was functioning and some weight reduction changes were incorporated during the month. More changes were planned as soon as they could be incorporated. The compressor and bearing housing castings incorporating anti-icing would be released in April. Detailing of the inlet guide vane was started and was 25% complete. The emergency fuel regulator design and layout were completed and 75% of the details drawn, with release expected in April. [30]

The progress conference on April 7, already covered in the -4 development section previously, showed that BuAer was already looking ahead to when the necessary engines of all J40 models would be needed and Westinghouse's chart showed their build plan for Essington. The provision for a production plant is not described in this volume, but work was underway to procure and outfit a production plant in Kansas City and the Ford Motor Company was being considered for additional manufacture capability and capacity. Obviously, getting the engines developed to the point where it passed the acceptance tests was WAGT's first priority. Four separate airframe projects (A3D, A4D, F3H, and F10F) were being designed around the J40. The Navy would soon have the Air Force dependent on the -1 engine for the North American X10 project.

April's monthly progress report showed the gearbox had been through five 50 hour full power runs with no problems. Power loading was done using water brakes. It was projected that an engine for test would be on hand before the schedule date.

BuAer was also looking ahead to the equipment necessary for ground handling, installation and removal. They wanted standardized equipment but if it could not be attained, each airframe manufacturer was authorized to design and furnish equipment to accomplish ground handling for their airframe. [31]

Boeing asked for a conference on June 1 and 2 to discuss the MX-839 (XB-52). Later memos asked for permission for WAGT to give Boeing data on the engine, which was approved.

WAGT requested two more test cells be made available at the Navy test center for development testing for all of the engine types now under development contract. These were to be used for development testing only and not for production testing.

Specification WAGT-X40E2-2A change items were finally approved and WAGT was requested to update the

specification with the agreed items and resubmit it for final approval. It should be noted that changes to bring the specification up to -2B level were already in negotiation. These potentially confusing, overlapping discussions occurred because the 150 hour engine acceptance test had to be approved on the basis of the specification to which it was built. It was entirely possible that multiple 150 hour tests would have to be run to each of the specification levels as they emerged.

Further changes to the intended engine specification WAGT-X40E2-2B, dated May 18, 1949, were being negotiated and the Navy response was issued with many needed changes and corrections on June 9. This specification still covered the -2, -4 and -6 engine models. Concerns were the definition of AN-F-58 fuel aromatic content limits, power control radiated interference shielding, and the emergency power control specification still not being satisfactory but accepted temporarily. New requirements were: 1) To specify that a three-toothed starter jaw adapter be furnished on the Type XII-F starter drive. It would permit mounting a starter incorporating a jaw designed to mate with the proposed AN4140 jaw adapter. (This was being driven by BuAer's adoption of a policy to procure only starters with jaw adapter drives); 2) Add a specification for the Flange Adapter in accordance with AN4139 and request a spline adapter in accordance with AN4052 be furnished on the Type XVI generator drive. This was added to accommodate the initial generator installations; and 3) Increase the volume of the fuel filter to accommodate 50 grams of contamination per 1000 gallons of fuel. It would take until August for WAGT to respond. WAGT included in their response other design issues as well.

WAGT amplified the testing cycles to be followed with considerably more detail and responded to the various requirements changes as follows:

A. Emergency Power Control – prior correspondence (not located) amplified the emergency operation. A supplementary failure analysis was being conducted and would be available by the week of August 29.

B. Fuel Filter – WAGT argued it was not practical (not explained – Auth.) to increase the filter capacity to handle 50 grams/1,000 gals of fuel. In addition, it was felt evidence was lacking to prove the fuel system components could handle particles larger than that passed through a 10 micron filter and still perform satisfactorily. The existing filter could accommodate up to 3 grams/ 1,000 gals without being increased in size. Further study would be needed if the dirt content of fuel were to be larger than the previously specified amount.

C. Oil Pressure Adapter – there was no space available on the -6 gearbox to allow a 1 1/16 inch 12-N3 straight thread opening for the AND10341 adapter. The necessary weight increase was not desirable given that the oil pressure line had to be reduced in size again to accept the line to the remotely located transmitter. The planned -8 gearbox under development appeared to be able to provide the necessary clearance, but a weight study would have to be done to determine the effect on aircraft installations.

D. The -6 gearbox presented difficulty in changing to a three-toothed starter jaw adaptor on the type XIIF drive for the gearbox. "Snapspring" grooves were required in the spline to retain the adapter and these were very difficult to work. The -8 gearbox would provide the three-toothed starter jaw adapter. Also, it was going to be difficult to procure the flange adapter in accordance with AN4139 and a spline adaptor in accordance with AN4052 for the Type XVI generator drive in time for the qualification test. Again, they would be supplied on the -8 gearbox. The parts added about 4.0 pounds to the engine weight.

E. The tachometer generator drive adaptor weighed approximately 0.8 pounds.

The May 1949 progress report indicated WAGT anticipated having the first -6 engine on test in late September or early October. This engine would be for development testing only and the second engine constructed was slated to be used for the 150 hour qualification test. [32]

Having addressed by this time all of the mockup board's installation related items, WAGT was asked to confirm if the -6 installation drawings could be submitted for approval and the mock-up delivered prior to the end of June so that the airframe manufacturers could verify their installation plans. [33] They responded that the installation drawings would be available on or before May 31 (eventually transmitted July 1) and the mock-up on or before June 30. [34]

AiResearch notified BuAer that the starter interference had been eliminated by redesign by WAGT of Westinghouse components and that AiResearch would not be making any changes to the standard 35 hp starter part number 55000 unless BuAer requested them to do so. [35]

BuAer finally dropped their requirement for the -6 to have an extended exhaust pipe (or A/B) cantilevered off the back end of the engine. They concluded they would never put an A/B on the -6, having contracted for a specific A/B version of the J40, the -8. They approved report A-749 and closed mock-up board items 34 and 45. [36]

Accessory quick disconnects were back on the table in July, with the Navy instructing WAGT to expand the study to the J34 and giving a suggested approach to doing the study and coming to a conclusion. The intent was to use the findings to decide on standard fittings to be used across many engines. [37]

Trying to negotiate a weight change to the specification, WAGT wrote a letter to BuAer on June 15 asking the specification weight of the -6 be increased to 2,973 lbs (and that of the -8 to 3,543 lbs) and offering to change the guaranteed maximum altitude of operation for both engines from 50,000 to 60,000 feet. They would also guarantee a compressor bleed air of 10% of the engine air flow up to 20,000 feet. The Navy requested more details on this and WAGT responded that the engine would allow a ram pressure ratio of 1.89 (up from 1.70) on a standard day on a continuous basis and 1.89 (up from 1.70) on a hot day on a limited basis. [38] (Later the Navy would realize that the current specification of engine pressure limits based on a standard day might mean the engine could not tolerate operation at temperatures below standard day at or below sea level due to the increased inlet pressures.) The memo included detailed changes in the engine in an attempt to justify the increase.

BuAer rejected the request for a weight increase over that already approved (for the -6 2,680 to 2,796 lbs). They would only accept requests for weight increases based on:

A. Ability of the presently proposed airplanes to assimilate the weight change without serious compromise of performance.

B. Improved performance over and above the original engine contracted for, adequate to compensate for the weight increase.

C. Improved engine endurance and life that resulted from the weight change.

The improved altitude guarantee was dismissed as simply another data point on existing performance curves. Also, the bleed air requirement was set by AN-E-30 and any performance above that required was of no value to BuAer. Since the engine had not been put through an acceptance test (at this stage, not even run yet), it was impossible to determine if the engine would offer significant performance and durability above existing specification to warrant considering the increased weight. At that point, WAGT's own weight improvement program had only identified a 48 lb reduction that could be incorporated.

The airframe manufacturers were asked to comment on the impact of the requested weight increase on their airframes. In anticipation of them accepting some ability to absorb a limited weight increase, WAGT was requested to provide:

A. An engine parts original design weight estimate.

B. An engine parts design weight estimate at the time of the release of the detailed drawings.

C. An engine parts actual weight.

D. Where parts have been weighed but do not correspond to final design, they should be so marked and the expected actual weight indicated.

E. Where parts' weights are different than original

estimate, indicate whether weight can be brought back to design condition or to a more acceptable weight, and if so, what weight would result.

F. What effect on the engine development program will result from the insistence that the original design weight objectives of the engine be met?

G. Since all the design issues had been decided, BuAer assumed the original weight contingency of 5 to 10 percent remained in the engine. (There is no documentation on just what weight contingency WAGT used in their original estimate.) BuAer requested WAGT state the original contingency and a new contingency be substituted thereafter, post mock-up board. BuAer assumed this would result in a weight reduction of 100-200 lbs. (How they arrived at this amount was not explained and would appear to indicate that BuAer was under the impression that the "contingency" factor could be used to determine a firm upper limit on weight regardless of development findings.)

In closing, BuAer pointed out that the impression given was that WAGT did not give engine weight the same level of importance as thrust or fuel consumption in the airplane. The requested increase of 177 lbs would not be offset by sufficiently improved fuel consumption and the other improvements would be minor. Accepting these new weights would not be in the best interests of the bureau and would seriously compromise the contractual objectives.[39]

With weight a continuing problem, BuAer took the opportunity in July to point out to WAGT that various deliverables on their contracts were *"unexecuted"*. They noted that WAGT had not asked for schedule extensions formally although BuAer had learned from various sources that these items would be late. BuAer interpreted the situation as Westinghouse unawareness of contractual obligations. The lateness of the J34-WE-32 and -38 in passing their tests when a production program was already planned was given as an example and reminded WAGT that the J40 program was also intended to support early airplane programs. BuAer emphasized the importance of establishing and accomplishing realistic engine programs (schedules) in the circumstance. They insisted on tightened program schedule focus through:

a. Delivery be effected on or before the agreed upon dates or mutual agreement reached on an alternate date.

b. Progress Reports should be reviewed with the intention of presenting a report that actually showed the progress or lack thereof, together with substantiating reasons in the latter cases.[40]

WAGT answered (in August) that they were fully aware of the importance of meeting agreement dates and were reviewing their system to see what might need to be changed. They confirmed they had already given BuAer all estimated date changes and would comply with the requirements.

June's progress report anticipated a -6 engine being ready for test in late Fall. Determination of whether this engine would be the 150 type test engine would hinge on the successful completion of the 50 hour test of the second -4 (at that point not even in test).

WAGT responded to BuAer's requested weight breakdowns for each of the engine models currently under development. BuAer had asked for a current actual weight as well, but this was not included in the chart.[41] The entire chart is presented as a matter of interest.

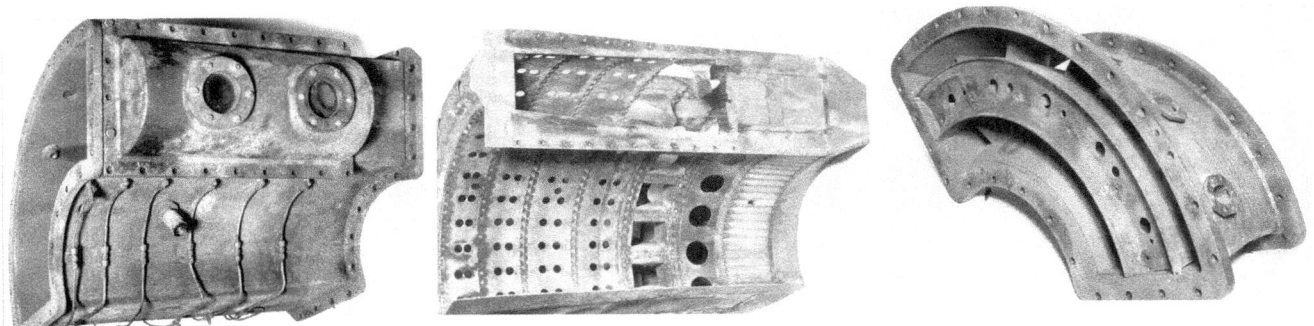

Figure 4 Quarter section combustion chamber test device components, WAGT P41588, P41589, P41590, 10/3/1948. *Courtesy Hagley Museum and Library*

Westinghouse Turbojet Engine Dry Weights - Estimated (lbs)						
Engine Components	J34-WE-32	J34-WE-38	J40-WE-6	X40E8A (J40-WE-10)	X40E7A (J40-WE-12)	J46-WE-2
Engine	1,890.0	1,440.0	2,975	3,968	3,598	1,886
Compressor	457.0	457.0	1,130	1,566	1,566	498
Combustion Chamber including diffuser	254.6	254.6	368	395	395	273
Turbine 1st Stage Disc, Blades, Nozzle	95.4	95.4	241	301	301	128
2nd Stage Disc, Blades, Nozzle	76.7	76.7	254	289	289	108
Shaft, Turbine Housing, Bolts, etc.	60.3	60.3	375*	409*	409*	62
Exhaust System Nozzle	----	86.0	62	----	82	----
Inlet Assembly	57.0	57.0	108	111	111	200**
Accessory Drive and Attached Engine Components	419.0	353.0	459	529	465	355
Afterburner, exhaust collector & nozzle	270.0	----	----	568	----	287
*Included turbine bearing **Included Integral Accessory Drive and oil reservoir						

J40-WE-6 Revised SFC's			
Rating	Thrust	Current[42]	Proposed[43]
Military	7,500	0.98	0.96
Normal	6,800	0.93	0.91
#1 Cruise	6,120	0.825	0.875
#2 Cruise	5,100	0.81	0.85
Idle	450	----	4.0

August 1949 brought another WAGT request to change the SFC guarantees (above) to reflect the current development status of the engine. They argued that the engine was producing SFC's slightly better than current guarantee at high thrusts and slightly worse at cruise powers. These, they said, were better characteristics from aircraft performance point of view than would result from modifying the engine to achieve lower cruise SFC to guarantee, which would raise the high thrust SFC to above guarantee. They presented a new set of SFC's and asked that the specification guarantees be changed to these new values.

These changed guarantees were accepted and incorporated in the -2C version of the specification for the -6.

A continuous effort was being expended to define the type test specifications in detail. Prior communication had clarified the various throttle settings and cycle times in detail. Another area needing attention was the power loading of the accessory gearbox during the test. Although the gearbox itself had to undergo its own type test, during the 150 hour engine type test, it had to be loaded to reflect maximum power extraction to ensure the engine results incorporated this worst case situation. WAGT stated that, because there was not sufficient space to load all the pads, they would load the alternator pad at its maximum of 2,500 pound-inch rating, this being the worst case scenario in any case, applying the maximum loading for the entire transmission to one pad only.[44] The request and approach were approved by BuAer on August 19.[45]

July's progress report did not show any changes from the June report, with most of the focus on getting the -4 qualification test engine through its upcoming first runs and test. [46]

WAGT finally responded in August to BuAer's push for better design to reduce risk of hydraulic fluid loss in the event of damage to the hydraulic lines or the nozzle actuator mechanism. This was Item 21 from the Mock-up Board, issue B, which was still not resolved. The analysis of alternative approaches and WAGT's recommendations is in the table below. These items may be the missing fail-safe elements in the control system BuAer had referenced earlier.

Mock-up Board Item 21 – "Provide Reduced Risk of Hydraulic Failure of Exhaust Nozzle and Provide for Emergency Operation"				
Option	Description	Pros	Cons	Other Comments
Dual hydraulic lines and high pressure elements.	Has duplicate lines and incorporates a hydraulic fuse.	Allows for full backup if one line fails (i.e. full emergency operation).	• Hydraulic fuse hard to develop, wouldn't detect small leaks. • Twice as likely to leak.	• Needed a year of development. • Added approx. 25 lbs to the engine weight.
Additional high pressure element	Duplicates the actuator.	Backs up element if it fails.	No leak or line failure protection.	• Needed a year of development. • Added approx. 40 lbs to the engine weight. • Possible installation envelope size increase.
Full separate emergency system	Duplicated the entire hydraulic system to the nozzle.	Full emergency operation	Installation disadvantages	• Development less than a year. • Weight increase of approx. 20 lbs per engine and 40 pounds on the airframe.
Use of fuel system as emergency backup.	Add fuel line piping to oil cooler to bypass oil cooler.	Decreased hydraulic vulnerability	Increased fuel system vulnerability	• Development of about a year. • Would add approx. 40 lbs per engine.
Use existing system as designed.	No delays in development or weight increases.	• Minimum of components. • Components proven low risk of failure.	Some hydraulic vulnerability remains, although minimal.	Recommended action: use system as currently designed.

Further documentation regarding the final closure of this item was not located, but all J40's constructed appear to have the original nozzle actuator/hydraulic system design, indicating that under time pressures to get a certified engine, BuAer elected to wait and see if the current design gave problems during testing. This wait-and-see approach was applied to the -8 version as well. [47]

The increasingly imminent need for "production" engines for flight test of the emerging airframes under development finally forced BuAer's hand regarding weight of the engines. They accepted the new weight of 2,973 pounds for the -6 "reluctantly". As for the -8, emerging competitive engines had considerably reduced the margin of attractiveness of the -8 engine, particularly if weight was increased to the 3,543 pounds requested. It was expected that weight reduction programs would reduce the increase

before the expected November 1950 qualification of that engine. The concurrent weight increase request for the -8 was not approved.

BuAer continued to push WAGT for an improved weight control process, even outlining the expected elements of such a program.

1. Be able to furnish, at any time, an up-to-date resumé on the weight situation for any engine, correct to within 72 hours, without significant extra effort on the part of the contractor's engineering department.
2. Report current engine weight situations in future monthly progress reports for all experimental engines up until they pass their 150-hour qualification tests, including:
   a. An estimate based on preliminary layouts, or

b. Calculated weight based on firmed details, or

c. Actual measured weight.

d. Current overall weight contingency situation.

e. Assembly, sub-assembly or component breakdowns agreed between WAGT and the bureau.

In passing, the memo noted that the current definitions of the control system, lubrication, ignition system, compressor sub-assemblies, exhaust system nozzles and afterburners were not sufficiently detailed to compile agreed boundaries for weight tracking and control and asked that WAGT study the entire weight control requirement and provide BuAer with a detailed description of the new weight control system Westinghouse had recently established.[48]

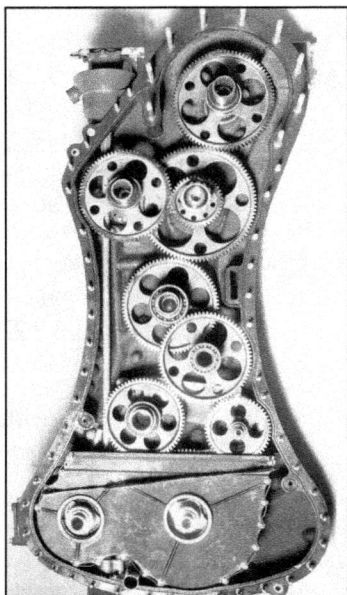

Figure 5 XJ40-WE-6 Early gearbox internal gears, baffles and wet sump gear case shape with filler and dipstick on upper left, WAGT P41952, 12/12/49. *Courtesy Hagley Museum and Library*

WAGT proposed to BuAer in the latter part of July that the -8 hourglass gearbox be used on all -6 production engines rather than the asymmetrical one currently being developed for the -6. BuAer accepted the recommendation, agreeing the new gearbox would be more practical and useful. The original gearbox would be tested as part of the -6 qualification test, as the -8 gearbox would not be ready. The -8 gearbox, as used for the -6, would not include components necessary only for the afterburner, but would be capable of accepting such parts without modification or major rework. A weight increase (between 5 and 44 pounds) was undesirable and WAGT was requested to

minimize the increase and inform BuAer of the exact amount as soon as possible. BuAer accepted the fact that a few of the "first" airplanes might need modification to accept the new gearbox, but having the multiple generator pads, improved (30 seconds vs. 10 seconds) negative g flight oil support, and a common starter motor were worth the changes.[49]

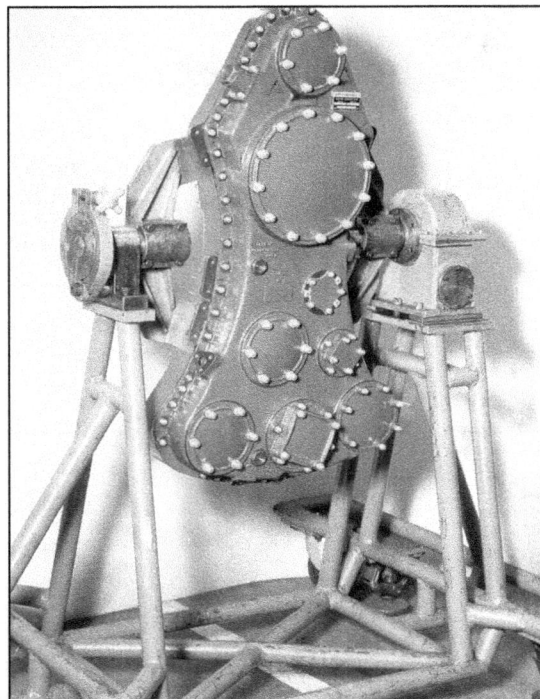

Figure 6 Final J40-WE-6 Gearbox shape, dry sump, same as that of the -8, WAGT P43786, 1/9/1951. *Courtesy Hagley Museum and Library*

The BAR's outlook at this time was that completion of the 150 hour qualification test was likely by the end of the year if no other significant problems were encountered in completing the -4 50 hour test.[50] Only a few weeks later the BAR qualification test date expectation moved out into early in the new year (1951), as the full set of parts for assembly were not in evidence yet. WAGT failed to include a weight breakdown in their monthly status report per the new requirement.[51] It now appeared that the very first -6 assembled would be used for the 150 hour test in an attempt to recover part of the schedule delays and get the engine cleared for production.

Detail design installation drawings had by now been sent to the airframe manufacturers and all engine changes that affected the shape of the engine and its accessories required a survey be done to attain approvals before a change could be incorporated. The very first survey involved the adding of the fuel dump valve to the fuel system. This valve released residual fuel pressure left in the lines after engine shut down to stop fuel from leaking out of the injectors, which caused burning and coking of the injectors. A very small amount (500 cc) of fuel was released to the aircraft drains (in Navy parlance "scuppers") after each false start or shutdown from running.[52] This requirement to survey all proposed such changes marks the start of treating the engines basically as production engines rather than development prototypes. It created delays in the engine decision making and placed an additional burden on the airframe manufacturers to continually review such proposed changes and comment on them as development moved forward.

The No. 2 bearing assembly jack pads were slightly relocated to provide more clearance for oil lines on the right (starboard – although the cover memo references the "standard") side of the engine. WAGT was careful to contain dressing changes within the original installation boundaries, but future changes could and some did cause headaches for the airframe developers, who were wrapping the airframe quite tightly around the mock-up provided.[53]

Another change in design was reflected in the draft Specification WAGT-X40E2-2D, the addition of five support lugs on the engine in place of the welded lugs then part of the design. The -6 added five support lugs on the engine gear box for support of gearbox mounted accessories with large overhung moment arms. This added three pounds to the weight of the engine. An Air Force requirement on the -8 (this applied to the Air Force version of the -8 intended for the X-10 program) and the desire to maintain interchangeability between the -6 and -8 meant that the -6 had to be changed as well. These added three pounds to the weight of the engine, but WAGT was instructed to leave the lugs off the -6 (and the -8 as well), as they were not needed for the Navy's projected -6 (or -8) uses. It was later determined that the Air Force actually requested the lugs on the -6 for the B-52 installation, not the -8. This convoluted discussion resulted in yet another stern memo from BuAer concerning WAGT's apparent lack of focus on weight control. The net result was that the lugs were to be left off both the Navy -6 and -8 to save the 1.5 pounds per engine, the other 1.5 pounds was not recoverable as it was necessary strengthening incorporated in the walls of the gearbox for the lugs.[54] The Navy refused to accept the weight increase and WAGT removed the lugs on early -6 engines by grinding them off and also changed the gearbox case casting to eliminate the strengthening on the Navy version of the gearbox for the -6 and -8.[55]

WAGT finally produced a memo on their weight control process to explain the weight determination and inspection procedures. It specified that a "small" contingency factor was added to the estimated weight during the proposal stage, but this was not explained

further. It acknowledged that weight changes would be reported both as soon as they were known and also in the monthly status report.[56] At this time, further revisions to the engine specification (WAGT-X40E2-2E) that would cover the production version of the J40-WE-6 included a weight change breakdown of increases over the XJ40-WE-6:[57]

J40-WE-6 vs XJ40-WE-6 Weight Changes		
Item Description	Weight Change lbs	Source
WAGT-X40E2-2D (XJ40-WE-6) weight	2,972	Base Weight
Symmetrical gearbox	+44	Provided improved function and commonality of -8 gearbox
Re-faired tail	+7	Improved gas flow on exit
Flowmeter provision	+6	Airframe manufacturers requested support, particularly for the B52 program originally
5 Point engine mount	+3	Better support of different airframe requirements
5 Support lugs on Gearbox	+3	Required for -8 (Air Force Requirement)
Oil filler strainer and scupper – Alternate left or right hand oil filler	+3.5	Fleet operational requirement
Seal attachment provisions on fire wall	+1	Redesign to satisfy mock-up board requirement
Relocate Junction Box	+1.5	Better connectivity to airframe and assist in providing additional space for other items.
Remove tach adapter gearbox	-0.8	Deleted requirements
Add dipstick for side oil filter	+1.0	Fleet operational requirement
Preflight check system	+2.0	Fleet operational requirement
Oil tank drain valve extension	+3.0	Fleet operational requirement
New J40-WE-6 weight	3,046	

BuAer noted that some of the changes above applied only to the XJ40-WE-6 and not to the J40-WE-6 and the total weights finally agreed upon were: XJ40-WE-6 – 2,981 lbs, J40-WE-6 – 3,038 lbs. Confusingly, to lock these new weights in for the XJ40-WE-6 engine, specification WAGT-X40E2-2F was issued, leap-frogging the WAGT-X40E2-2E written for the J40-WE-6.

The approach planned by WAGT to control shaft bowing on the -6 test engine was the same as used on the -4 engine with one addition. The teardown of the -4 after the Flight Substantiation Test showed no engine conditions that could be related to unsatisfactory shaft bowing, so they recommended the same approach be used for the -6. These control mechanisms were:

1. Copper plating the interior of the turbine shaft to a thickness of 0.15 inch to improve conductivity of heat from the hot sections to the cooler sections.
2. Placing two baffles near the downstream end of the turbine shaft: one inside the shaft which rotates with the shaft and one stationary baffle placed outside. These restricted heat from the first stage turbine disc to the top element of the turbine shaft.
3. Painting the top inside half and top outside half of the turbine rotor housing with black sooty paint to reject more heat from the top than the bottom, equalizing the temperatures.
4. Additionally on the -6, vibration dampers would be used on both main bearing housings to absorb any residual vibration.[58]

At this point, first running of the planned test engine was anticipated at the end of February, 1950. There was a full court press underway to complete the many required reports and analyses and get them approved prior to the actual test beginning. The fuel pumping system design was approved based on report A-940 of March 9, 1950. It was the system used on the -4 and was essentially unchanged for the -6, but a later memo from the BAR revealed that many small issues and defects (not detailed) remained before the system could be used operationally.[59] As late as January 1, Westinghouse was still awaiting delivery of the AiResearch starter and air compressor for the engine.

The testing outlook stated that the first engine itself would be completely assembled and ready for its 150 hour test in February. There was apparently nothing from the 50 hour test of the -4 indicating it would experience any other than minor problems.[60] Three AiResearch starters and one compressor unit were on order and scheduled to be delivered on or about February 15 to the Navy and they would be forwarded on arrival to Westinghouse. January's report indicated that -6 initial testing was scheduled to start February 20.[61]

J40-WE-6 Testing Begins

The first -6 engine (serial WE030001) was started in the first week of March 1950 and stopped after 2 hours of running due to vibration. This was found to be caused by turbine shaft spigot deformation. This was corrected and testing restarted on March 16. After another hour, the compressor case cracked due to blade rubbing. The engine was overhauled and rebalanced, and running continued for another 10 hours while turbine temperature traversing tests were conducted. Compressor stalling was experienced during this period, as well as high oil consumption of 1.5 lbs per hour. The testing was discontinued to allow the compressor blades to be "twisted" to new angles to stop compressor stalling and to address the oil consumption issue. As of the 18th of April, the plan was to go into a 150 hour test attempt if the green testing and other minor testing were satisfactory.[62]

The combustor analysis report A-942 was submitted on April 10, 1950 with a request for approval.

Specification WAGT-X40E-2E covering the production version of the -6 and -8 was under negotiation at this time. The main issues were still related to defining which components belonged in each weight group and the total weight of the production engines. The production -6 was to have a guaranteed maximum dry weight of 2,981 pounds.[63] The initial bid weight had been 2,680 pounds, so this reflected an 11.2% increase.

Navy review of report A-942 resulted in a comment that a separate ignition circuit be included for each spark plug to increase reliability. A single breaker-motor and radio interference filter would be sufficient to meet requirements otherwise. It was also noted that a Scintilla TLN-2 ignition system was under test as part of the J34-WE-32 test program and results would be forwarded as they became available. Such changes in design were proposed constantly by the Navy and here is an excellent example of a late change that shows such things were being proposed for other parallel engine programs (the J34-WE-32 in this case) and testing done on them. Here we see BuAer wanting to review the results from the Westinghouse J34 tests and then forward them back to the Westinghouse J40 development team. It should be noted that Westinghouse was under constant pressure to move faster but if they had to shortcut the process internally and act on the J34 results prior to Navy communication to do so, they almost certainly would have been reprimanded by BuAer.

The centrifugal dual fuel pump intended for the production -6 passed its qualification test and Westinghouse requested the test be considered qualified for the -8 as well. The Navy approved this request on May 2.[64] (The XJ40-WE-6 with non-symmetrical gearbox used a vane type pump.)

All was not well and progress continued to be slow. Westinghouse finally had to send a memo on May 18, explaining they understood the urgency to get engines tested and available for the airframe test programs, but listed the following reasons for the delays:

1. Meeting specification performance at the lowest cruise design point.
2. Casting difficulties involving weight and quality on compressor housing and inlet duct.
3. Decrease from anticipated production flow of curvic clutch components.
4. Fabrication difficulties involving other engine components.

The cover letter from the BAR on the 18th added that there had been a work stoppage for 5 days in May. As a result, they gave new estimated delivery dates for some items, the main one of interest being the new qualification test completion date of June 30 with the qualification report date of July 15.

The April progress report was forwarded by the BAR on May 15 and his generalization of the status of the program was not encouraging. Only 17 total hours of running had been accomplished as of the 10th. The longest continuous run was only 50 minutes. An insulation blanket was damaged. This was the blanket covering the turbine shaft to minimize shaft bowing, its inclusion showing that bowing was still a problem. The blanket approach was being abandoned temporarily. The engine had been reassembled for test to restart on the 15th, changes in the compressor and turbine being incorporated to bring it into compliance with the guaranteed performance. If the initial test runs indicated it was meeting the guarantees, a green test would be run. A second engine was going to be built in June.

The implications of these delays from the various development and production issues meant the likelihood of the engine passing a qualification test in June 1950 in time for shipments of production -6 engines in July was now considered doubtful.[65]

High temperature materials test results report A-975 was accepted in June. The BAR reported *"The work on the development of the high temperature materials has been reflected in the XJ40-WE-2, -4 and -6 engines to-date and results have been good. Materials developed have illustrated soundness, workability, and generally all around good qualities, although further work is still necessary."*[66] Exactly what work and research still needed was not specified.

The fuel system dump valve was conditionally approved as long as it did not increase the weight of the engine. Grumman's suggestion that the fuel system drain line and the dump valve drain line be combined was rejected.[67] Along with the surveys of drains for the fuel, Douglas recommended and the Air Force agreed that the installations be checked to determine that fuel could not flow between one drain and another due to pressure differentials. Also, complete drainage should be checked for

nose-up and nose-down installations.[68] WAGT should have paid more attention to the installation issues relating to non-level operating conditions, as will be seen.

Figure 7 J40-WE-6 Compressor spool, multi-section shaft and turbine wheels, WAGT P43868, 1/17/1951. *Courtesy Hagley Museum and Library*

In early June 1950, WAGT reported that their fuel flowmeters using AN-F-58 fuels were now showing inconsistent results. Originally, testing was done to compare the meters using AN-F-48 vs. AN-F-58 fuel and showed with the latter a minus 6% correction factor should be applied to readings taken. Recently, however, the flowmeters were re-compared (this was done on a regular basis) and now showed identical readings between the recently tested fuel types. The reason for this was unknown and Westinghouse was investigating, but all likely sources for the change had proved negative to date. The implication being that all test results reported for fuel consumption when AN-F-58 type fuel was used might be 6% too low. Fisher and Porter calibration equipment was to arrive in July to help determine the problem's source. Results for the -4 engine as reported still met the SFC targets even with the minus 6% correction removed.[69]

May's progress report showed little actual progress toward being ready for the type test. The engine had only completed 20 total hours of running and all efforts were going to improving temperature distribution across the turbine stages and meeting specific fuel guarantees. Some modest progress was seen, but improvements were still needed. The XJ40-WE-4 engine was being re-assembled to provide a baseline to try to duplicate its performance and compare it to the current -6 configuration.[70] WAGT appeared to be at a loss to explain why the performance differences being observed were happening. Given the purpose of the 50 hour test on the -4, namely to prove the basic components of the engine were sound and could meet the performance estimates, the fact that the next engine built did not meet them was not encouraging and caused yet another delay in the schedule.

June showed Westinghouse trying to discover why their J40 engines were exhibiting exhaust nozzle thrust coefficients different from other engine models. They suspected it might be due to the larger size of the J40 which might be creating cell depressions due to the velocity of air passing by the outside of the engine. The small size of the WAGT test cells might be creating this problem. WAGT suggested comparison tests between the AEL test cells and WAGT's for various subtypes of the J40 and the J34 to determine the root cause of the differences being observed. The tests might take a week to conduct and WAGT would provide installation and support at AEL.[71]

In an effort to discover why these discrepancies were occurring, WAGT theorized that the size of the test cells themselves might be to blame. They argued that they were finding that the exhaust nozzle thrust coefficients for the J40 did not agree with those experienced with their other (smaller) engines in the same cells. Since the Navy's Aeronautical Engine Laboratory (AEL) in Philadelphia, PA had much larger test cells, they proposed a testing approach to establish a baseline to determine if cell differences were affecting the test results. They proposed:[72]

1. Test an XJ40-WE-4 in WAGT's cells #9 and #10. Test cell #10 was of different construction, so some differences in performance between the two were expected.
2. Take the same engine and test it at AEL for static performance, measuring thrust and SFC's plus some internal performance readings. This would take approximately one week.
3. Make the same tests as steps 1 and 2 above using one or two -6 engines.
4. Test all WAGT cells with standard ASME nozzles to create pressure surveys.
5. WAGT would transport the engines to and from AEL, provide maintenance and an engineer to aid in testing.
6. AEL would install and run the engines.

The BAR recommended that they assist WAGT as requested, given all the time being spent on trying to clear up various SFC discrepancies.[73]

The multiple development programs now underway (J40-WE-6, -8, -10 and -12 and the multiple model J46/J34 programs) were all having unanticipated delays, creating an expanding crisis for the Navy. The recent news that the first -6 production engines delivery schedule had moved from July 30 to September 30 with only a ten week prior notice finally triggered a strong memo from the Chief of the Bureau of Aeronautics on the entire schedule issue.[74]

Progress seemed to elude the program and Westinghouse had had to inform BuAer in a May 18 memo that the first two production -6 engines would not ship until the end of September instead of the end of July. BuAer responded with a stern letter from Admiral Pride, Chief of the Bureau, reminding Westinghouse of their constant delays. BuAer referenced their previous letter in April of 1949 on the same subject, which stated the Navy would have extreme difficulty in justifying further sponsorship of the J40 project unless WEC could assure an appreciably accelerated rate of effort and accomplishment. The latest delays supported the Navy's feeling that WAGT had not attained the rate of development progress needed for the J40. Such rapid progress was required to meet the needs of the high-priority aircraft development procurement programs planned around the engine. Lack of progress and timely notice of delays as soon as they became apparent was placing the Navy in an extremely embarrassing position both financially and otherwise. Discussions between BuAer and WAGT representatives over the last six months had produced continuous assurances from WAGT that the problems would not affect the production schedule. *"Such continuing poor delivery performance by WEC obviously results in an increasing loss of Westinghouse prestige, both within this bureau and the aircraft industry."* The memo went on to express grave doubts to even the -6 schedule being realistic. A thorough study of WEC's organization and all turbo jet development programs under contract appeared justified, the end goal being the production of *"realistic"* development and production schedules.[75]

W. B. Anderson (Manager, Aviation Gas Turbine Division of WEC) responded immediately and after acknowledging the Admiral's urgent memo, had to tell BuAer in the second paragraph that the -6 production schedule had now slipped to October 30, in part due to a work stoppage in the South Philadelphia plant. Production would shift to the new Kansas City location with Philadelphia now being exclusively development work, producing production parts or engines only if spare capacity became available. An organizational and work schedule review was arranged to occur in the week of August 7. A personal meeting between Mr. D. W R. Morgan, WEC Vice President, and Admiral Pride was requested to discuss the situation.[76]

One wonders where BuAer was all this time, more than a year now having passed since their big review on schedules. Even now, WAGT's response (below) would still not include enough detail to determine if they could actually make their schedules. Even so, the program continued.

Figure 8 XJ40-WE-6, -8 and -1 Fuel nozzle angle alignment jig, WAGT P42038, 1/12/50. *Courtesy Hagley Museum and Library*

While contract management issues are not a focus of this volume, it was obvious that this exchange of memos finally moved BuAer into the micro management of WAGT's work in an effort to get the program onto a more dependable and predictable basis. Awareness of the enormous pressures now being placed on Westinghouse is important to the understanding of the overall development story. The Navy's level of frustration becomes increasingly evident in the tone of many memo exchanges and in particular in some of the pencil comments on the back of BuAer's internal document circulation and tracking sheets. Both parties had an enormous amount at stake and as the recurring, seemingly unsolvable technical problems continued to plague rapid progress, it would seem to underscore every communication from this time forward.

The June/July monthly progress report had no new news except that the first -6 would be back on test August 21 and the second -6, intended to be the 150 hour test engine, was in build and beginning to become visible in the assembly area.[77]

The schedule conference was held (the minutes are discussed below) and WAGT forwarded a letter on August 28 to BuAer and presented them with the following -6 production schedule choices:

A. An XJ40-WE-6 with available design and uncertain specific fuel consumption – completing its qualification test by October 31, 1950.

B. An XJ40-WE-6 engine meeting all contract guarantees – completing its qualification test by March 31, 1951.

The contract items involved and the different schedules for them were:

Proposed New Contract Milestones			
Item No.	Description	Schedule A	Schedule B
10	Qualification Test Engine Testing Complete	10/31/50	3/31/51
11	Qualification Test Report	11/30/50	4/31/51
12	Final Design Drawings and Specifications	11/30/50	4/30/51
13	Final Summary Report	12/31/50	5/31/51

The final paragraph of this proposal reads: *"Since the conference, it is our understanding that the Bureau wishes to defer a decision on performance requirements of the program until we have proposed revised fuel consumption guarantees corresponding to Schedule A. Such guarantees will be submitted as soon as the results of the test cell calibration project at AEL are analyzed."*[78]

The conference referenced covered a full program review of all the BuAer projects underway with WAGT, not just the J40 programs. Many other non-J40 program items were related directly or indirectly to the J40 in that in many cases funds or resources could be redeployed to the J40 program or to improvements in engine field support, which was shortly to become critical.

The minutes of this meeting, presented back to BuAer by WAGT, are summarized:

A. Westinghouse proposed that all Plancor Location 2061 facilities (close to the South Philadelphia plant) except for the Research Laboratory be purchased. The lab was only useful if the power house could be obtained and it was part of Plancor 181. The entire question of funding for test cells and production was discussed and WAGT hoped that funding would be available soon from Westinghouse itself.

B. Again WAGT suggested an interim type test be run on the J40-WE-6 to meet the specification in all respects except for SFC. This was tentatively agreed to pending a formal review, particularly with the Fighter Branch. WAGT pointed out that further development of the -6 to meet all SFC's would dilute the other development efforts *(the -10 and -12 – Auth.)*. WAGT was willing to make financial adjustments to the target price contract if the interim test approach was approved.

C. Control problems were extensively discussed. The electronic control planned to be used on the J40, while advanced in many respects, was demonstrating many weaknesses, such as inability to tolerate high temperature environments in airframes. A "simplified" control had been under consideration but no one could say whether electronic or hydraulic controls would prove best. WAGT recommended a separate development effort be put in place to explore this issue. The requirement for an emergency control system for take-off and low altitude use was a hard requirement on any new control system until it had proven itself in service experience.

D. A discussion on the desirability of running a complete qualification test on the J34-WE-38 vs. the -32 (afterburning model) took place. The net was that the requirement changes for the -38 would delay production needs and create a production issue, as the build rate that would be required could not be met by Westinghouse. Further discussion was deferred until BuAer could decide what to do in the circumstances.

E. A chart of specific development schedule dates was presented by WAGT, who asked if the dates were acceptable to BuAer. It was emphasized that WAGT's dates had to be suitable for the airplane planning program and that it was necessary to reorganize WAGT's such that the proposed schedule dates would be realistic and could be met. WAGT responded that they were *as realistic as could be made on the basis of present knowledge* and that their *"development capacity would be expanded as necessary to accomplish that aim."*

F. Flight testing was considered to be essential and aircraft selected had to be roomy enough to accommodate the necessary instruments to record transient conditions. It was agreed that BuAer would investigate obtaining either an F3D or B-45 and advise WAGT of their findings. (Later, a B-45 was used for air testing of the -22A model of the J40, but the -6 and -8 went into prototype aircraft for first use without ever having flown before. As it happened, the flight test program was very late being initiated, very slow to achieve flight test capability and suffered endless delays during test flying. The program contributed almost nothing to development of the engine and likely diverted resources of more use elsewhere.)

G. Lack of adequate WAGT support in the field for new engines and incorporation of fixes in the field was discussed. Examples used involved the J34, but the implications were relevant to the J40 program, which was only months away from service use of production engines.

H. Defects of currently delivered production engines reported by the airframe companies were brought to WAGT's attention with a request that positive steps be taken to assure such deficiencies were corrected on the engines delivered in the future.

I. BuAer requested that WAGT investigate establishment of either overhaul facilities on the West Coast or make arrangements with another service agency to provide such support. Training of additional BuAer Service Department engineers was going to be needed to support program expansions now underway.

J. BuAer requested that trouble reports come to BuAer from Westinghouse before they received them from the field. (This is confusing and the implication seems to be that the current process had them forwarded to BuAer and they then forwarded them to Westinghouse. Reversing the process would speed up response times to the field.) Preparation of CID's was also discussed, mainly the fact that many were not complete and revisions were often required.

K. A monthly meeting schedule was established to review overall development programs, alternating between WAGT and BuAer sites.[79]

At the conference, WAGT summarized the steps taken, adjustments to their development approach, and challenges being faced. To allow the reader to gain some perspective on the scope of Westinghouse's challenges, the items covered were:

A. All programs had been surveyed. Immediate steps were being taken to accelerate the development program. Some steps were already complete and others were in process.

a. Development programs would be first priority throughout South Philadelphia, with the production load to be carried primarily at Kansas City and at sub-contractors where required.

b. Additional engineers and draftsmen would be added and additional space made available to improve work conditions.

c. The Engineering department would be reorganized for better supervision, planning and coordination.

d. Additional personnel would be made available and an improved procedure established to expedite development of parts within the Division and at outside suppliers.

e. Two new test cells had already been provided and two additional cells suitable for the J40 would be installed. One of the latter would be segregated for performance development work.

f. A facility for testing and trouble-shooting of early production control components was underway and would be complete in October. Additional facilities for endurance testing of control components were in progress and would be available in October.

g. Additional machine tools for the WAGT laboratory shop had been procured and most were installed and available for work.

h. More adequate facilities for the assembly and rework of development engines were being planned.

i. Based on actions taken to date and planned, revised development schedules were established which were believed to be realistic.

B. Technical Problems were reviewed:

a. Afterburner development and automatic controls were listed as the primary source of delay.

i. Development of afterburners had turned out to be more difficult than anticipated. Destructive pulsations (*screech – Auth.*) in the J34 afterburner had finally been eliminated. The J40-WE-8 had been developed up to the 19,000 lbs of fuel/hr against the requirement at sea level to burn 19,750 lbs/hr.

ii. Automatic controls were required due to the advent of variable exhaust nozzles and afterburners. Optimum performance could only be obtained with automatic controls capable of dealing with the rapid operational changes encountered. With the requirement for a completely independent emergency system, the automatic controls had become complex. (Note – this system was basically the same as the -6 control with modifications to handle the A/B activation and pressure feedback into the control.) The electronic control was the primary control, and temperature and speed drift were still being observed even though it worked satisfactorily during the 150 hour qualification test on the J34-WE-38 engine. Initial operation and setting of the controls was an issue, even with better test stand and house engine checks. The -8 engine would incorporate combined primary and emergency controls to eliminate a large number of parts. Ground checks of J34-WE-32 and -38 engines at Chance-Vought had generally been satisfactory.)

C. The J40 program was still trying to meet the guaranteed SFC targets. Instrument problems (not specified) compounded the problems in correcting the situation, giving readings believed to be unreliable.

D. The -8 model was anticipated to complete a 150 hour test by March 1951 although the SFC's attained might not meet the guarantees at that point.

E. The -10 engine (a -12 with A/B) was expected to complete the 150 MIL-E test by March 1952 and the -12 engine was expected to complete the 50 hour test (*might be a typo, with 150 hour intended. – Auth.*) by September 1951.

F. The design studies for the -14 and -16 models were expected to be complete to the time now specified.

(The balance of the minutes dealt with the specific items related uniquely to the J34 and J46 and are not included here.)[80]

The August 1950 Monthly Progress Report (through early Sept) showed that the first -6 (WE030001) was now demonstrating satisfactory radial temperature distribution and compressor stall characteristics. SFC was running 7% above guarantees. The second engine (WE030002) was being assembled for the 150 hour qualification test in the same configuration as the current No. 1 engine. Assembly was scheduled to be complete by September 28 and the engine ready for a green test. The qualification test was to be started the first week of October.[81]

The program continued, with the Navy desperate for even the -6 to get airframe testing underway, even as they began to look at other emerging engines to replace the J40. Emerging engine designs from competitors had caught up from a delivery schedule point of view with the J40 program and BuAer acted to shift their plans away from the J40 wherever possible. That search is outside the scope of this book but the reader should be aware that none of the possible competitive engines had passed tests either, but were being followed closely.

WAGT requested approval in early October to change the qualification test specification to the approved WAGT-X40E2-2E which covered the production version (J40-WE-6) of the engine. The delays in test start made it possible to test the production version of the engine in the 150 hour qualification test instead of the XJ40-WE-6 version. This would allow all but the fuel booster pump to be certified as part of the 150 hour test. Differences between the two versions were:[82]

1. Improved (symmetrical) gearbox
2. Oil reservoir external to gearbox (providing longer negative-g operation)
3. Differently ported main lubrication pump
4. Differently ported oil cooler
5. A new booster fuel pump (but not available for the test)
6. Lengthened exhaust nozzle
7. New exhaust nozzle actuating yoke
8. Incidental changes to piping

The specification approved for the J40-WE-6 engine was requested to be approved for the XJ40-WE-6 engine as well, and that the engine qualification test be performed based on the J40-WE-6 configuration and the qualification of components as described in the specification's Appendix C be omitted (other than the fuel booster pump).[83]

Report A-1028 "AEL and Westinghouse Test Performance Comparisons of the XJ40-WE-4 Engine" was submitted on October 27. The report itself was not located, but the summary cover letter reported that the test showed little variations between test cells at WAGT and those of AEL, although the 2-3% normal test variations between test runs might have been obscuring any real differences between the two cell sizes. Fuel consumption measurements within the normal variation were in agreement. Continued testing to determine test cell instrumentation inaccuracies would continue with an effort to eliminate errors. It was proposed that the readings as taken in the Westinghouse test cells be used for the type test of the -6.[84]

Proposed J40-WE-6 Qualification Test Ratings Objectives		
Rating	**Thrust lbs**	**SFC**
Military	7,500	1.04
Normal	6,800	0.99
90% Normal	6,120	0.96
75% Normal	5,100	0.94
Avg. Turbine Inlet Temp Military	Old – 1,425°F	New – 1,455°F

In October the second -6 production engine, built to production standard, began its shakedown runs with 16 hours completed. A turbine temperature traverse, vibration checks for shaft bowing and accelerations were conducted. The monthly report went on: *"The temperature traverse revealed that heat distribution in the turbine section was not as desired. Inspection of the compressor vanes at teardown on Oct 13 revealed that the angles of these vanes were not to drawing and were a large contributing factor to the heat distribution in the turbine section. This is being corrected. The acceleration checks showed that the engine can be accelerated without detrimental effects. The new vibration dampers appear to be satisfactorily designed as revealed by the vibration checks conducted. Although engine vibration owing to shaft bowing has been considerably lowered (60 mils to 24 mils) with the use of the dampers, another check will be made when the engine is submitted to test again during week of 23 Oct 50. The possibility that the 24 mils vibration figure could be lowered owing to the fact that "O" ring particles may be clogging some of the oil openings and when these openings are cleared up, better performance can be expected on the ring dampers."*[85] The engine was expected to be ready to start the 150 hour test during the week of October 30.

The status report on the qualification engine covered a visit by Mr. Lafaye of BuAer with the BAR and the monthly

conference agenda held on October 26. The issues with the compressor stator vanes were reviewed and it was reported that twisting the vanes to the correct values at all points of the length of the vane did not correct the temperature issues in the turbine section. Additional changes were being made to the combustion basket to allow more cool air to flow to the root of the turbine blades.

WAGT noted that in future all vanes would be checked at all points, not just the mid-points. BuAer noted that twisting the vanes to correct the angles was poor engineering practice, introducing possible cracks and stress factors leading to early failures. It was also noted that the aerodynamic performance of the compressor appeared to be very marginal, particularly after WAGT noted that the compressor vane angles were critical to correct performance, and manufacturing differences could easily account for the performance differences being seen between engines to date. Such wide differences between successive engines made to the same drawings were intolerable to BuAer. The BAR was to study the situation further and report at the next monthly discussion.[86]

November's report showed considerable improvement of the -6 No. 2 due to the combustion basket changes, but a slight hot spot still remained and another modification was done. The outer exhaust collector had cracks due to vibration and a reinforcing band of .049 inch material was added in a one inch band as a stiffener around the exhaust collector in the failure region. The design now included a sliding strut arrangement to absorb expansion differences and reduce stress. A check run was scheduled for December 15.

The fuel booster pump would not be available for the qualification test, the impeller needing correction. The correction was completed and Nash Company had the pump under evaluation, but it would not be available for at least three more weeks.

A bright note was that the gearbox, power regulator, voltage regulator, governor alternator and power scheduler qualification testing were underway and further along than projected. The altitude chamber check-out would be completed by December 15, with low temperature testing of the exhaust nozzle control to proceed as soon as possible thereafter. In spite of these advances, qualification testing of the engine itself was unlikely before the end of December, as after a test run, the engine would have to be torn down and inspected to see if further corrections were necessary.

The Stress and Vibration analysis report (A-999, June 19, 1950) had raised questions on the need to use the critical material of Columbium stabilized steel in the compressor inlet guide vanes, compressor diffuser inner cone, and compressor diffuser struts. It was not used for parts such as the compressor blades or stator vanes. Such critical material had to be further justified. The inlet duct's ability to withstand a pressure differential tending to collapse the duct needed to be verified. Was such a situation considered, as it would normally exist during static operations? WAGT was asked to respond.[87] They did so in May stating that at the time the engine was designed, it was not realized that Columbium was a critical material. It was selected to avoid welding of unstabilized 18-8 alloys on the hot portions of the engine from creating intergranular corrosion. It also gave 15% greater yield strength in the parts. Also, using the same alloy throughout the engine reduced the possibility of mix-ups in part manufacture. The critical alloy parts were being examined concurrently to find which could use other materials in place of Columbium. The intake's maximum pressure differential occurred in the static start condition and was calculated to be 3.5 lbs/in^2. The intakes had been tested to a differential of 20 lbs/in^2 successfully.[88]

Verification of the booster pump came under consideration again, with BuAer noting that although the specification did not specifically state that the booster pump had to be verified by operation on an engine (as opposed to independent testing), it was requested the pump be verified by 50 hours of satisfactory operation on house engines during miscellaneous tests.[89]

November's monthly status meeting minutes were reported concentrating on the 150 hour -6 test engine status. The stator vanes were reported to have been found not twisted to the correct angles except at the very center of the vane. Westinghouse had verified only the very center of the vanes for correct twist angle as specified in the drawings. They found the vanes were not at correct angles at both attachment points. For the test engine, the vanes were simply bent to the correct angles as close to the attachment points as possible. It was recognized this approach had introduced leading and trailing edge joggles and possible early failure points.

Those actions did not correct the poor temperature distribution on the next run and the engine was removed from the test cell. Westinghouse stated the manufacturing tolerances were allowing critical changes between sets of vanes for otherwise identical assemblies. The aerodynamic performance of the XJ40-WE-6 appeared to BuAer to be very marginal. Performance with wide variations between engines would be unacceptable and "intolerable". The Westinghouse practice of bending vanes was also unacceptable and considered very poor practice, leading to questionable durability between engines. The BAR would study the situation further and make sure it was discussed at the next monthly meeting.[90]

The next monthly activity report repeated the early corrective action report on the test XJ40-WE-6, adding that the testing time was seven hours, bringing the total engine time to 23 hours to date. The engine combustion basket was having more air holes added and a few others blocked to bring more cooling air to the root of the turbine blades. The revised engine was due to resume testing on the 17th. If satisfactory results were obtained, the qualification test would be attempted.

WAGT indicated there was a possibility of assembling a third XJ40-WE-6 engine to be ready for testing during the latter part of November, 1950. The BAR thought this unlikely to happen before the middle of December or even later.

The -4 engine sent to AEL for testing (reported as the WE003002 but for some reason, pencil corrected to WE003001 on the BuAer copy) suffered an exhaust collector liner failure in the vicinity of the No. 2 bearing support. At the time, it was undergoing an afterburner run at AEL. (This implies that the first -4 was modified to handle the developing -8 afterburner. Earlier failures had also occurred with an XJ40 (model unknown) modified with an ASME nozzle installed.) The engine was returned to Westinghouse, who attributed the failure to the extreme pressure fluctuations in the exhaust collector during engine power changes. Some changes had been made in the -6 design but the -8 would need reinforcement. No further changes or reinforcement were planned in that area for the production -6.

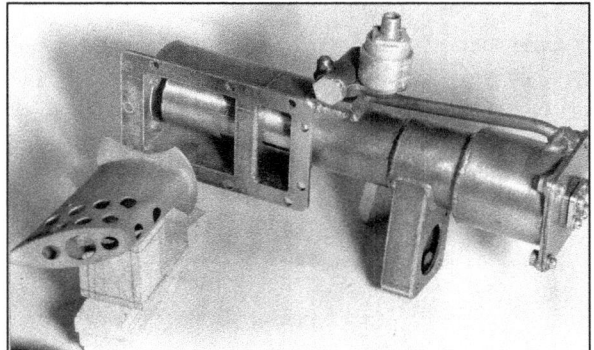

Figure 9   J40-WE-6, -8, -1 and -22 Anti-icing valve assembly, WAGT P43793, 1/10/1951. *Courtesy Hagley Museum and Library*

The second XJ40-WE-6 was now producing SFC's within the revised SFC's requested for the qualification test, however, these had not yet been approved.

The vibration of the engine was now tolerable. Recent tests had produced vibration during start of 16 mils which imposed a 2.5g load on the components holding the rotor. In normal running, the vibration varied between 3 and 5 mils. No vibration limits had yet been established for J40's, but the J34 had an acceptance limit of 3 mils.

The second stage turbine nozzle vanes were being redesigned to eliminate cracks in the trailing edge of the vanes where they were welded to the shroud. The first two production engines were to be accepted with ground vanes, assuming the vanes would be ground the same way, with the redesigned nozzle assembly being installed at first overhaul. Subsequently it was learned that WAGT would produce the first eighteen -6 engines with ground vanes before the "umbrella and vane fix" would start to be used.[91] It was not clear if the BAR was looking for agreement to accept the 16 production -6 engines on order with ground vanes as well.

The airframe manufacturers were already asking for operational instructions, starting with how to properly "depreserve" a J40. The engines would be run and ground tested and then removed as airframe development occurred, possibly many times. The engines needed special flushing between long periods of weeks between runs. Westinghouse sent BuAer instructions covering the -6 and -8 models on November 2. This was the beginning of operational support to the field users and predates the first shipments of the engines, which had not yet passed an acceptance type test.[92]

BuAer's legal branch noted at this point that the tools, dies and fixtures being created as part of development were owned by Westinghouse due to there being no clause relating to ownership of these in the contracts. Such a clause was later added giving BuAer ownership.

In December, BuAer accepted WAGT's recommendations on the acceptance test specifications as described above. They continued to require an independent test of the fuel booster pump. All production engines had to be configured identically to the XJ40-WE-6 test engine unless a specific deviation was authorized. Since the booster pump would not be on the test engine, the fuel pump was required to operate with inlet fuel pressures equivalent to those provided by the booster pump during the test. (A later communiqué asked that at least 50 hours of actual running of the pump on house engines in miscellaneous test be accumulated to verify the pump.[93]) It was assumed the known turbine temperature distribution deficiencies on the XJ40-WE-6 would have been corrected prior to the type test. Also, the accessory drive and gear case loading as previously described was found to be satisfactory. No accessory drives other than those driving Westinghouse supplied components would be loaded. Weights would be installed on the following pads:[94]

Pad	Weight	Overhung Moment
Hydraulic	50 lbs	400 lb/in
Alternator	225 lbs	2,500 lb/in
Generator	75 lbs	625 lb/in

The last paragraph of this memo is most striking. *"In view of the additional delay which would be occasioned by development of the XJ40-WE-6 engine to fully meet all present contractual guarantees; this bureau desires that development be terminated upon successful completion of qualification test at performance levels approved in paragraph 2 above. Action will be initiated at an early date to effect an appropriate downward revision to target price of the subject contract."*[95]

November's monthly activity report shows definite progress on temperature profile correction with the new combustion liner, but the engine still showed one slight hot spot. This was corrected, but the outer exhaust collector revealed cracks due to vibration. A one inch band of .049 inch material was added as a stiffener around the exhaust collector in the area subject to failure. It also now included a sliding strut arrangement. The engine would be ready for a check run on December 15. The booster pump would not be available for the qualification test due to the impeller design needing correction by the Nash Company.

Component qualification tests of the gearbox, power regulator, voltage regulator, governor alternator and power scheduler were underway and ahead of schedule. The altitude chamber check-out was to be complete by December 15 and the low temperature test on the exhaust nozzle control was to proceed as soon as possible thereafter.

The scheduled starting date for the qualification test before the end of December no longer appeared probable. After completion of the December 15 test and check runs, the engine would be inspected to determine if any further correction actions were needed.

The first two -6 production engines were diverted to the experimental and qualification test program. This action made moot the request to accept them as production engines with ungraphited compressor blade roots. In connection with blade root lubrication, MoS2 would not be used since its suitability with regard to salt spray corrosion had not been proven. The material did exhibit promise in vibration damping, galling prevention and inhibiting corrosion resistance. J34 tests were underway on MoS2 lubricated blade roots, using a zinc chromate coating at the edges after assembly to the disc. After the 150 hour test, the disc assemblies would be subjected to salt spray tests.

Wind-milling tests had included collection of air bleed data and the data would be in the final report covering these tests.[96]

J40-WE-6 Production Preliminaries

Even with the -6 not having passed an acceptance test and even some doubt existing that consistent performance and reliability could be obtained on the engines due to guide vane issues in the compressor, the BAR began to report on initial production assembly in a memo of November 17, 1950.

He reported that from production engine three and up, the engines would be identical to the qualification engine. The first two engines would be identical except for:

1. Turbine shaft – would be copper plated. Also it would have the triple layer heat baffle to be used on all other production engines as well.

2. Compressor blade roots – Ungraphited compressor blades to be used, even though the qualification engine would have graphite plated blade roots. The use of ungraphited blades in the J34 had shown that other than occasional minor galling, no difficulties had been experienced. Since the first two engines were intended to be used for ground testing only, the BAR felt this would be acceptable, but asked to be informed of BuAer's decision on their use. (Shortly thereafter, made moot as the first two engines became development engines, not production).

3. Combustion chamber liner – the reinforced combustion chamber liner might not be available for the first two engines based on the current delivery schedule. WAGT had been informed non-reinforced liners were unacceptable unless the engines were designated non-flight engines.[97]

4. Main oil pump "U" seal packing.

5. Second stage turbine blades and exhaust collector strut (cracking).

6. Positive lock on fuel injector assemblies to prevent misalignment on installation.[98]

The originally intended first two production engines were converted to development engines, making the differences issue moot.

The Flight Substantiation Test on the -4 had been completed in November 1949 and the following items were considered not qualified. Qualification of these was moved to the J40-WE-6 150 hour Certification Test agenda: spark plugs, electronic power control, second stage turbine blades, combustion basket and No. 1 bearing seal.[99]

Figure 10 J40-WE-6, -8 and -1 Dual oil pressure pump assembly, WAGT P43799, 1/10/1951. *Courtesy Hagley Museum and Library*

XJ40-WE-6 150 Hour Qualification Test

Engine XJ40-WE-6 No. 2 started a 150 hour qualification test on January 2, 1951. The BAR results report at the 108 hour point in the test cycle was issued on January 8 and is reproduced here:

1. *"The 150-hour qualification test on the subject engine model was started 2 Jan and up to and including the morning of 8 Jan, 108 hours had been completed of the test. The contractor is using an engine designated as the XJ40-WE-2(sic) for the test. The elements comprising this engine had accumulated considerable test hours before actual qualification test started. The compressor spindle and compressor blades (roots graphite) accumulated 69.7 hours before test started. The following elements accumulated 86 hours before test started: front and rear turbine shafts, first and second stage turbine discs, first and second stage turbine blades, second stage turbine nozzles, compressor housing, No. 1 bearing assembly, diffuser, rear*

oil pump and fuel flow divider. The following elements accumulated 40.7 hours before test started: accessory gearbox, dual fuel pump, main oil pump and oil cooler. The following elements accumulated 38 hours before test started: first stage turbine nozzle, primary fuel regulator, emergency fuel regulator, over-speed relay, fuel nozzles, and combustion liner. The no. 2 bearing had accumulated 62 hours before type test started.

Figure 11 J40-WE-6 and -8 Gearbox testing rig, WAGT P43946, 2/1/1951. *Courtesy Hagley Museum and Library*

"The calibration of the engine prior to starting qualification test showed that performance as set forth under reference (a) was met. The engine throughout 108 hours has performed fairly well. Throughout the military periods the turbine inlet temperature was maintained at near peak requirements of 1,425°F. The mechanical performance of the accessories was very good other than the power regulator. Power regulator trouble was encountered early in the test (20 hours). On this investigation a temperature amplifier tube controlling turbine out temperature was replaced owing to a broken filament. When power regulator

trouble was encountered again at 36.5 hours, investigation revealed that the air cooling tube was pulling the power regulator box against the shock mounts, making those mounts ineffective due to engine vibration. This necessitated replacing seven tubes because testing showed these tubes had absorbed too much vibration and deterioration had set in. At 54 hours a thyratron tube controlling the fuel valve had to be replaced. This was evidenced when the speed fell off.

3. "At 36 hours it became necessary to either make a field repair to tail eyelids or replace them (because of cracks). It was decided to replace the eyelids. This element will need a re-design for durability and the contractor has this in mind. The spark plugs had to be cleaned on two occasions (at 20 and 36 hours) because ignition could not be obtained. Also at 20 hours the fuel valve motor had to have the clutch adjusted for torque setting because of incorrect fuel flow. Other than the above trouble the engine has performed satisfactorily.

4. "The fuel rotometers which had showed incorrect fuel flow when checked with the Fischer-Porter instrument before the test, were re-checked after the Fischer-Porter engineers had examined and found nothing wrong. It was concluded that the personnel handling the calibration had taken data which was erroneous. On re-calibration the flow check showed that rotometers were within 1% of the master. This matter then was considered cleared up.

5. "The contractor had been running with a seven thermocouple tail instead of the nine. A hot spot in the engine had caused fuel flow thrust and turbine inlet temperature to increase while turbine outlet temperature remained constant. When the thermocouple in the 4 and 6 o'clock area were unhooked, conditions improved and test got underway. The contractor will investigate further the problem connected with the harness and the Bureau of Aeronautics will be kept informed.

6. "Engine performance shown by calibration data taken before the test started is listed below:

7. "Overall, over the 108 hours, the engine has performed satisfactorily considering that some of the engine elements have been through considerable hours before test began. In conjunction with this, the engine had accumulated approximately 23 hours of military operation before beginning the test. Performance as set forth under reference (a) has been met. Oil consumption has been well within the model specification requirement of .7 lbs per hour. Throttle torque has not been measured. This will be performed after test and will be further checked on engine XJ40-WE-6-3. Radio interference tests have been performed before test and will be performed after test.

8. "This office thus far concludes that the power regulator needs further re-design for durability of tubes. The tail eyelids need re-designing because of cracks and the "coking" of the spark plugs must be cleared up before these items will be acceptable."[100]

XJ40-WE-6 Certification Test 108 Hour Point Performance Data						
RPM Actual	RPM Corr.	Thrust Corr.	Ambient (°F)	SFC Corr.	Turbine Inlet Temp. Actual (°F)	Corr. (°F)
5,660	5,810	2,745	34	1.04		
6,380	6,550	4,080	35	0.927		
6,450	6,610	4,110	34.5	0.910		
6,730	6,900	4,750	34	0.910		
6,900	7,080	5,000	33.5	0.903		
7,200	7,380	5,460	34	0.930	1,005	1,085
7,260	7,460	5,820	34	0.935	1,052	1,133
7,290	7,480	6,100	33	0.935	1,092	1,175
7,290	7,480	6,560	33	0.940	1,150	1,235
7,270	7,460	6,770	33.5	0.943	1,188	1,275
7,260	7,480	7,360	32	0.965	1,270	1,365
7,250	7,440	8,120	32	1.017	1,420	1,520

The reassembled engine ran the balance of the 150 hour test and report A-1093 (*not located*) was submitted on the results. BuAer accepted the engine (XJ40-WE-6-2, Serial WE003003) with exceptions that needed qualification.[101] The engine was returned to Westinghouse as GFE for further work to clear the engine for production. Work still needed was listed as:

1. "Agreement to WAGT's proposed change to braze the sixth stage stator vanes to prevent the minor cracking found during engine teardown.
2. "The fuel booster pump had to pass a 500 hr sea level test and the cold test at the supplier's plant, and WAGT would then subject the pump to the altitude test.
3. "The 7th and 8th stage compressor seals would be resolved as described in CID 14513.
4. "The reassembled engine must successfully complete verification test on the power regulator, the exhaust nozzle gates and other elements. (It was noted that the nozzle actuator, power regulator, governor alternator, voltage regulator and power scheduler had been qualified over 500 hours.)
5. "The TLN-10 Scintilla high energy ignition system was qualified.
6. "The fuel control to be component tested at WAGT and the fuel pumping system at Pesco.
7. "The power regulator must demonstrate repeatability, and reliability on engine testing must be achieved. The power regulator will be the same as the qualification regulator with the difference that the thyratron tubes would be shock mounted."

All of these things appeared to push the final clearing of the production engine to at least the end of February.[102] That did not occur and the production clearing run was not to begin until later in March.

In March the BAR sent the following telegram to BuAer: "94 and ½ hours verification J40-WE-6 power regulator and exhaust nozzle eyelids completed. Power regulator performed satisfactorily throughout. Visual inspection tubes reveals only minor deterioration. No visual cracks exhaust nozzle eyelids. This office considers verification these items satisfactorily completed and has authorized contractor proceed to complete verification engine through 150 hours. This is substantially same engine which ran through qualification. BuAer PP-213 serial 0190 Jan. Nash boost pump ran on above tests. Double ply bellows failed approximately 90 hours. Contractor intends further improve this bellows. Recommend this be considered successful engine verification of pump and final qualification be based lab tests. This office has authorized contractor continue to run this pump through 150 hours on engine prior complete teardown. Request confirm all of above."[103]

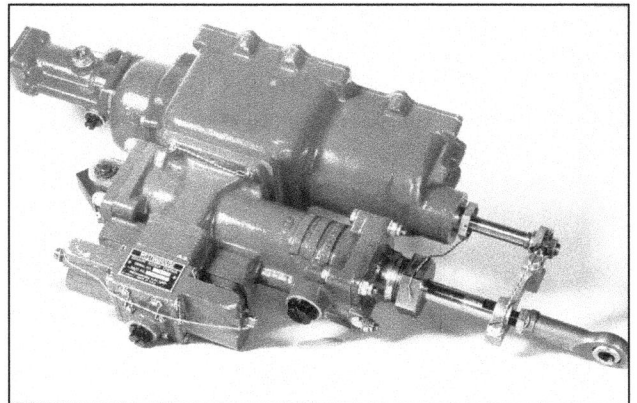

Figure 12 J40-WE-6 Exhaust nozzle actuator hydraulic control assembly, WAGT P44137, 2/28/1951. *Courtesy Hagley Museum and Library*

Another requirement was moving front and center regarding throttle torque, or the amount of effort it took to move the throttle in either direction. The power control assembly on the -6 required about the same amount of torque as the production J34 and since at this point the Navy was only planning to use the -6 in single engine applications at the airframe plants and at Patuxent River test facility, it was agreed that the -6 torque levels were acceptable. But it was clear the -8 version would be a problem, since the addition of the various levers to operate the A/B from the power control unit would add to the torque factor. A long running dialogue over this issue continued right to the end of the program, with the Navy wanting lower torques on the later models.[104] Charts submitted showed the measured torque to be 22 lb/in to move to idle, 15 lb/in between idle and military, and 35 lb/in to move into A/B range.

The end of March showed the Navy accepting the final qualification of the gearbox (however, chafing of the aluminum tubes by safety wiring had to be corrected), the booster pump was satisfactory, and the exhaust nozzle eyelids passed the 94.5 hour verification test satisfactorily. The two thyratron tubes were new at the start of the 94.5 hr. verification test and needed to complete an additional 55.5 hours to reach the 150 hour accumulated mark. The testing schedule of the verification test (various cycles, etc.) did not need to be adhered to, the point being to have the tubes run 150 hours without failure or malfunction.[105]

BuAer was still not satisfied with all the qualification tests as yet. In April, the BAR noted that although WAGT considered the XJ40-WE-6 engine fully qualified with the exception of the power regulator and laboratory testing of the fuel booster pump, BuAer wanted the primary and emergency regulators to complete a successful 500 hour laboratory test to specification AN-E-32. The test was not underway as yet. The 94.5 hour verification test had been completed. The power regulator had also accumulated 80 additional hours of evaluation. The two latest power regulators embodied changes to improve engine "stall" characteristics on jam accelerations (throttle slams) accelerations and temperature control starting. Speed and temperature drives were still being evaluated. The stall characteristics still needed further improvement, which was underway.[106]

Exhaust nozzle control qualification was accepted and approved May 4, 1951. This control was used on the J34 as well as the J40.[107] May's activity report showed the 500 hour primary and emergency fuel control tests had been completed satisfactorily as of June 7. The dual fuel pump was still only partially completed as work on the erosion problem continued.[108] June's report showed the dual fuel pump and booster fuel pump tests were underway. The test had to be stopped because the 50 hp motor drive unit had a bearing failure. Retesting was to get underway 15 August. The BAR added a comment in the cover letter that his opinion was that both Pesco and Nash had to be "pushed" all the way to get the tests started, beginning four to six months ago.[109]

While WAGT had requested the 150-hour test be accepted as satisfying all AN-E-32 requirements as far back as March 15, 1951, BuAer withheld final qualification approval at that time pending completion of all the outstanding items. Additional items were closed out over the next few months as described above and on August 6, the BAR forwarded the test report (A-1164) and all the related materials to BuAer along with WAGT's request for approval, stating final approval of the 150 hour test rested with BuAer.[110]

Cleanup testing and qualification work continued into August, with the dual fuel pump and fuel booster pump testing underway at WAGT. Cold and altitude tests would be conducted at WAGT's plant as well. Following those tests, the pumps would be sent to Nash Engineering for the remaining tests. The fuel boost pump had been upgraded with a Monel shroud to address erosion in the vapor section of the pump. Production -6 engines were experiencing boost pump spline shearing due to misalignment. Redesign changed to a shorter spline length for better alignment and a flexing section instead of a shear section to withstand greater loads.[111]

With the -6 in production, and the -8 under development (under Appendix D of the specification), the WAGT-X40E2-2F specification was sent to BuAer for approval. The Navy responded that they wanted a separate specification to cover the XJ40-WE-8 and -22 engines in view of the change in control and performance of these engines. This would allow them to move toward closing out the NOa(s) 9212 contract for the -6 only and move the -8 into its own specification. They further responded to some minor wording changes in the -6 portion of the -2F submittal, approved the Engine Operating Limits Curve No. 369041 and asked for a Power Control Level Schedule curve for the -6 engine showing power level travel versus temperature, rpm, thrust and throttle actuating torque. The balance of the curves was still under review.[112]

WAGT report A-1101, which covered the Exhaust Nozzle Actuator (used on the J34-WE-32, J34-WE-38, J34-WE-17 and J40-WE-6) low temperature testing, was rejected by BuAer on the grounds that the submitted test did not comply with AN-E-32 and that the modified test actually run by WAGT was not submitted to BuAer or approved as a satisfactory alternate prior to the test being conducted. WAGT argued the actual AN-E-32 test as stated was impractical, as the hydraulic fluid (the engine oil) heated up from -67°F to about 160°F normal engine operating temperature within a few minutes of running. Their tests started after the cold soak at below -67°F with the actuator on the actual engine. The AN-E-32 requirement stated the actuator should remain at -67°F or below during the entire test cycle. The Navy admitted that a modification of the test specification might be necessary, but that WAGT did not get it approved prior to the test. Also, BuAer noted the test results only reflected the start measurements of such things as resistance and emergency schedule torques. They requested this be changed to both before and after measurements. The actual actuator performed satisfactorily during WAGT's testing, noting that some "O" rings had to be replaced due to stretching, which was normal for that type of "O" ring.[113]

September reported every attempt being made to complete the development contract by the end of year, which appeared likely to be attained. The fuel pump shaft seal was replaced during its qualification test due to leakage. The leakage was apparently caused by a foreign "black" substance embedded in the original seal. The engine weight had actually decreased to 3,033 lbs, 16 lbs below the specification. The fuel pump and booster pump were about to be shipped to Pesco and Nash Engineering for their final qualification tests.[114]

October reported the cold and altitude tests on the fuel pumps completed and the dual fuel pump having completed 350 hours on test satisfactorily to date.[115]

November showed closure of outstanding items regarding reports and modified performance curves as both the Navy and WAGT tried to get the development contract closed. Production engines issues will be covered in another chapter.[116] The primary and emergency fuel regulators were qualified and accepted by BuAer on November 20.[117] The interim technical report (A-1125 on the Air Flow Testing of Power Regular) was accepted as partial fulfillment of Item 4 of the contract.

Thyratron tubes were next. Recommendations for accepting the test results were sent in May and further stated that the power regulator of the J40-WE-6 engine had now met all requirements of the engine specification and should be accepted.[118] The fuel pumping system was accepted in its entirety on May 7.[119]

Although the modified re-test of the exhaust nozzle actuator showed it failed at temperatures below -30ºF, BuAer felt that they had to accept the test results so as not to slow down the development of the -8 and -22 engines. The actuator needed redesign and such a redesign was in hand for the -8 and -22 engines. The limited number of -6's to be used and the nature of that use indicated that further development of this part could not be justified. BuAer used this acceptance to point out how dissatisfied they were with such a long lead time test not being completed until over a year after the engines were being used in aircraft. They requested in the future that WAGT keep them notified of all component test schedules.[120]

The wooden -6 mock-up engine was disposed of sometime in 1953. All further development of the -6 ended except for issues related to safe flight or field maintenance. Those items are covered in the production section.

Figure 13 J40-WE-6 #2 Serial WE003003 Certification Test engine, WAGT P43806, 1/10/1951. *Courtesy Hagley Museum and Library*

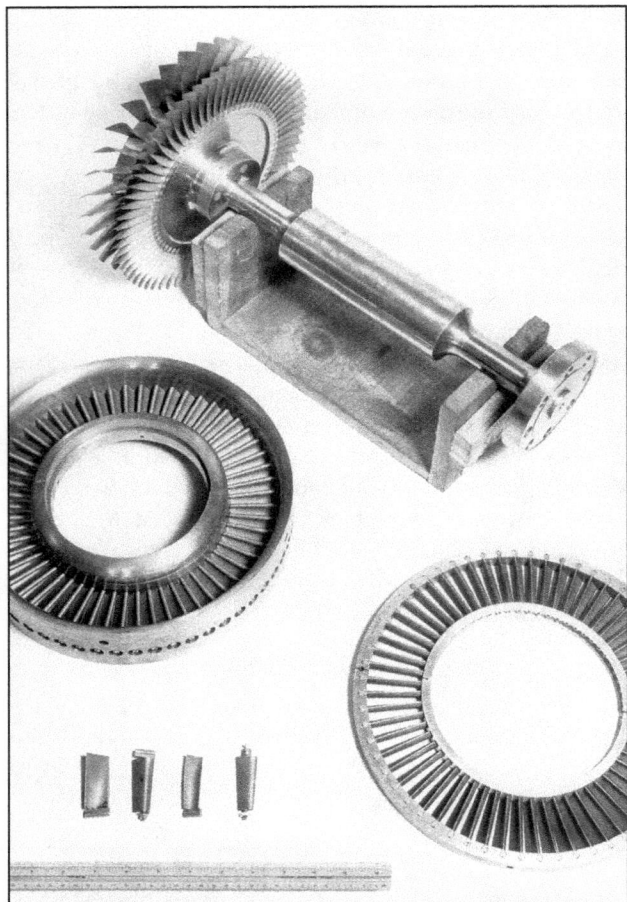

Figure 14 XJ40-WE-6 Turbine test model. Used to confirm aerodynamic performance and test assembly approach. First stage guide nozzles lower right, second stage guide nozzles above to left. WAGT P40768, 4/7/1949. *Courtesy Hagley Museum and Library*

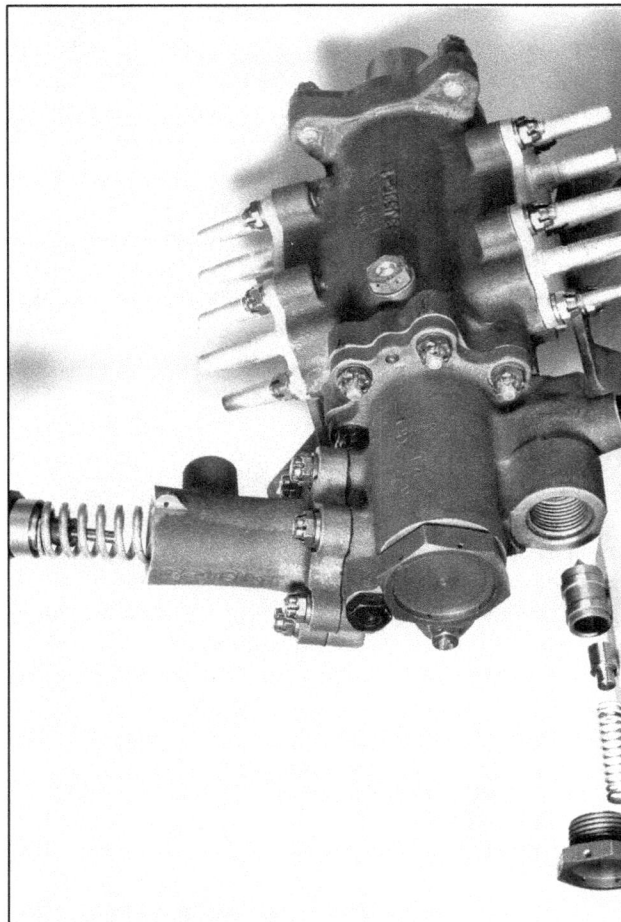

Figure 15 XJ40-WE-6 Initial fuel splitter assembly design, WAGT P41351, 8/13/1949. *Courtesy Hagley Museum and Library*

Chapter 2 - XJ40-WE-6 Development Citations

[1] WAGT Memo, July 13, 1948, Proposed Contract NOa(s) 9212 Changes in the XJ40-WE-6 Recommended by BuAer Mock-up Board **B303V1RG72.3.2NACP**

[2] WAGT memo, February 20, 1949, "Proposed Contract NOa(s) – 9212 Changes in XJ40-WE-6 Engine Recommended by BuAer Mock-Up Board Item No. 20" **B304V6RG72.3.2NACP**

[3] BuAer Memo, May 23, 1949, "Contract NOa(s) 9212 – XJ40-WE-6 Engine Mock-Up Board Decisions, Items 21 and 4. **B304V7RG72.3.2NACP**

[4] Ibid

[5] WAGT memo, January 12, 1949, "Contract NOa(s) 9212, XJ40-WE Engine, Mock-Up Changes Studies; Information on" **B304V6RG72.3.2NACP**

[6] WAGT memo, December 21, 1948, "Proposed Contract NOa(s)-9212 Changes in XJ40-WE-6 Engine Recommended by the BuAer Mock-up Board" **B304V6RG72.3.2NACP**

[7] BuAer memo, April 20, 1949, "Contract NOa(s) 9212 – Item 28 of XJ40-WE-6 Mock-Up Board Report." **B304V6RG72.3.2NACP**

[8] WAGT memo, June 1, 1949, "Contract NOa(s) 9212 Engine Mock-Up Board Report, Item 29." **B304V7RG72.3.2NACP**

[9] WAGT Report A-749, September 13, 1948, "Westinghouse Model XJ40-WE-6 Jet Engine Structural Requirements for Cantilevering an Afterburner and Extension" **B304V7RG72.3.2NACP**

[10] BuAer memo, June 27, 1949, "Contract NOa(s) 9212, Westinghouse XJ40-WE-6 and XJ40-WE-8 Engines, Afterburner Support and Fuel System Provisions for" **B304V8RG72.3.2NACP**

[11] BuAer memo, August 1948, "Contract NOa(s) 9212, XJ40-WE Engine, Mock Up Change Studies" **B304V5RG72.3.2NACP**

[12] Ibid **B304V5RG72.3.2NACP**

[13] BuAer memo, March 21, 1949, "Contract NOa(s) 9212 – XJ40-WE Engine Mock-Up Change Studies" **B304V7RG72.3.2NACP**

[14] BuAer memo, September 16, 1948, "Contract NOa(s)-9212 – XJ40-WE-6 Engine – Location of Starter on" **B304V5RG723NACP**

[15] WAGT study, January 3, 1948, "Study of Item 47" **B304V8RG72.3.2NACP**

[16] WAGT memo, October 15, 1948, "Contract NOa(s) 9212 – XJ40WE-2 Engine Design Study of Control System for" **B304V5RG72.3.2NACP**

[17] WAGT Memo, October 5, 1948, "Contract NOa(s) 9212 – Westinghouse XJ40-WE Engine Starting System" **B304V5RG72.3.2NACP**

[18] WAGT memo, October 5, 1948, "Contract NOa(s) 9212 XJ40-WE-2,4,6; Contract NOa(s) 9670 XJ34WE-32 and XJ46-WE-2. Location of the Electrical Control Components" **B304V5RG72.3.2NACP**

[19] BuAer memo, October 14, 1948, "Contract NOa(s) 9212 – Acceptance of Engines" **B304V5RG72.3.2NACP**

[20] WAGT memo, October 27, 1948, "Contract NOa(s) 9212 – Acceptance of Engines" **Ibid**

[21] BuAer BAR memo, October 19, 1948, cover of the WAGT Sept. 1948 Monthly Activity Report A-757, **B304V5RG72.3.2NACP**

[22] WAGT Study, January 3, 1949, "Mock-up Board Item 47, Revised Study, Sketch #1", **B304V6RG72.3.2NACP**

[23] BuAer memo, February 9, 1949, "Contract NOa(s) 9212 – XJ40-WE-6 Engine, Starting of" **B304V6RG72.3.2NACP**

[24] WAGT memo, February 2, 1949, "Contract NOa(s) 9212 – XJ40-WE-6 Engine, Starting of" **Ibid**

[25] BuAer memo, March 1, 1949, "Contract NOa(s) 9212 – XJ40-WE-6 Engine, Starting of" **B304V7RG72.3.2NACP**

[26] BuAer handwritten note on back of distribution routing sheet, dated on or after March 29, 1948. **B304V1RG72.3.2NACP**

[27] BuAer memo, March 3, 1949, "Contract NOa(s) 9051, 9433, 9791, 9670 and 9212, Model J34-WE-30a, -34, -32; XJ46-WE-2, and XJ40-WE-6, -8 and -10 Engines, Emergency Fuel System, Contractor Furnished Components" **B304V7RG72.3.2NACP**

[28] BuAer memo, March 21, 1949, "Contract NOa(s) 9212, Westinghouse XJ40-WE Engine Starting System, Forwarding of" **B304V7RG723NACP**

[29] BuAer BAR memo, March 15, 1949, "Contract NOa(s) 9212 – Progress Report A-816 for Month of February 1949 – Forwarding of" **B304V7RG72.3.2NACP**

[30] BuAir BAR memo, April 26, 1949, "Contract NOa(s) 9212 – Monthly Progress Report A-834 (Month of March 1949), Forwarding of." **B304V7RG72.3.2NACP**

[31] BuAer, Power Plant Division Memo, May 20, 1949, "J40-WE-6 Turbo-Jet Engine Ground Handling Equipment – Provisions for" **B304V7RG72.3.2NACP**

[32] BuAer memo, June 22, 1949, "Contract NOa(s) 9212 – Monthly Progress Report A-856 (Month of May 1949), Forwarding of." **B304V8RG72.3.2NACP**

[33] BuAer memo, April 25, 1949, "Contract NOa(s) 9212 – XJ40-WE-6 Engine Mock-Up; Delivery of" **B304V7RG72.3.2NACP**

[34] WAGT memo, May 5, 1949, "Contract NOa(s) 9212 – XJ40-WE-6 Engine Mock-Up; Delivery of" **B304V7RG72.3.2NACP**

[35] AiResearch memo, April 27, 1949, "Contract NOa(s) 9212, Westinghouse XJ40-WE Engine Starting System" **B304V7RG72.3.2NACP**

[36] BuAer memo, June 27, 1949, "Contract NOa(s) 9212, Westinghouse XJ40-WE-6 and XJ40-WE-8 Engines,

Afterburner Support and Fuel System Provisions for" **B304V8RG72.3.2NACP**

[37] BuAer memo, July 1, 1949, "Contracts NOa(s) 9212, 9570, and 10114 XJ34-WE-32, XJ46-WE-2 and XJ40-WE-6 and -8 Engines, Engine Accessory Quick Disconnect Mountings for." **B304V8RG72.3.2NACP**

[38] WAGT memo, July 11, 1949, "Contracts NOa(s) 9212 and 10067 – Operating Limits with Revised Weights" **B304V8RG72.3.2NACP**

[39] BuAer, July 14, 1949, "Contract NOa(s) 9212 – Engine Weights – Proposed Revision of." **B304V8RG72.3.2NACP**

[40] BuAer, July 18, 1949, "Contracts NOa(s) 9212 and 9670: Unexecuted Deliveries on contractual Items; Request for Clarification of." **B304V8RG72.3.2NACP**

[41] WAGT enclosure to memo of July 27, 1947, "Contract NOa(s) 9212 and NOa(s) 9670 Engine Research and Development Programs" **B304V1RG72.3.2NACP**

[42] Specification WAGT-X40E2-2, April 15, 1947, **B303V37RG72.3.2NACP**

[43] WAGT memo, August 17, 1949, "Contracts NOa(s) 9212 and 10114 XJ40-WE-6 and -8 Performance" **B304V9RG72.3.2NACP**

[44] WAFT memo, August 4, 1949, "Contract NOa(s), XJ40-WE-6 Engine Qualification Test, Loading of Power Take Off Drive on Qualification Test When Using 43J707-1 Gearbox." **B304V9G72.3.2NACP**

[45] BuAer memo, August 19, 1949, "Contract NOa(s) 9212, Westinghouse Model XJ40-WE-6 Engine Qualification Test – Loading Accessory Drives for." **B304V9RG72.3.2NACP**

[46] BuAer memo, August 19, 1949, "Contract NOa(s) 9212 – Monthly Progress Report A-883 (Month of July, 1949), Forwarding of." **B304V9RG72.3.2NACP**

[47] WAGT memo, July 29, 1949, "Contract NOa(s) 9212 – XJ40-WE-6 Engine Mock-Up Board Decisions Items 21 and 4, and Contract NOa(s) 9670 – XJ34-WE-32 Engine and Contract NOa(s) 10114 – XJ40-WE-8 Engine." **B304V7RG72.3.2NACP**

[48] BuAer memo, November 1, 1949, "Contracts NOa(s)-9212 and -10114: changes in specifications for XJ40-WE-4, -6, and -8 engine weights" **B305V10RG72.3.2NACP**

[49] BuAer memo, October 19, 1949, "Contract NOa(s) 9212 – Revision of gearbox for J40-WE-6 turbo-jet engine; proposal for" **B304V9RG72.3.2NACP**

[50] BAR memo, July 20, 1949, "Contract NOa(s) 9212 – Proposal for Revision of Gear Box for XJ40-WE-6 Engine." **B305V10RG72.3.2NACP**

[51] BAR memo, November 17, 1949, "Contract NOa(s) 9212 – Monthly Progress Report A-911 for Month of October 1949." **B305V10RG72.3.2NACP**

[52] WAGT memo, November 9, 1949, "Contract NOa(s) 9212 and 10114; WEC Turbo-Jet Engines Model J40-WE-6

and J40-WE-8 Fuel System Dump Valve, Survey of." **B305V10RG72.3.2NACP**

[53] WAGT memo, November 28, 1949, "Contract NOa(s) 9212 and 10114; WEC Turbo-Jet Engine Models XJ40-WE-6 and XJ40-WE-8 Relocation of Jack Pads on #2 Bearing Housing, Survey of." **B305V10RG72.3.2NACP**

[54] BuAer memo, February 1, 1950, "Westinghouse XJ40-WE-6/-8 engines – Extra gearbox support lugs." **B305V11RG72.3.2NACP**

[55] WAGT memo, April 20, 1950, "Contract NOa(s) 9212, Westinghouse Specification WAGT-X40E-2D Dated 4 November 1949, covering the models XJ40-WE-6 and XJ40-WE-8 engines." **B305V12RG72.3.2NACP**

[56] WAGT memo, January 1, 1950, "Engine Weight Control System" **B305V10RG72.3.2NACP**

[57] Specification WAGT-X40E2-2E, January 17, 1950, Appendix C **B305V11RG72.3.2NACP**

[58] WAGT Memo, February 1, 1950, "Contract NOa(s) – 9212, Shaft Bowing in XJ40-WE-4 Engine." **B305V11RG72.3.2NACP**

[59] BAR Memo, March 31, 1950, "Contract NOa(s) – 9212, item 4, Interim Technical Report on design analysis of the fuel control system for the XJ40-WE-6; submittal of" **B305V11RG72.3.2NACP**

[60] BAR memo, January 13, 1950, "Contract NOa(s) 9212-Monthly Progress Report A-924 for Month of December 1949." **B305V10RG72.3.2NACP**

[61] BAR memo, February 16, 1950, "Contract NOa(s) 9212 - Monthly Progress Report A-936 for Month of January 1950." **B305V11RG72.3.2NACP**

[62] BAR memo, April 18, 1950, "Contract NOa(s) 9212 Monthly Progress Report A-960 for March 1950." **B305V12RG72.3.2NACP**

[63] WAGT draft specification WAGT-X40E2-2F as of March 30, 1950. **B305V11RG72.3.2NACP**

[64] BuAer memo, May 2, 1950, "Contract NOa(s) 9212, WEC Turbojet Engine Model J40-WE-6, laboratory qualification of J40-WE-6 fuel pumping system" **B305V12RG72.3.2NACP**

[65] BAR memo, May 15, 1950, "Contract NOa(s) 9212, monthly progress report A-971 (month of April); forwarding of." **B305V13RG72.3.2NACP**

[66] BAR memo, June 7, 1950, "Contract NOa(s) 9212, section "D", item 4(h), interim technical report; forwarding of." **B305V13RG72.3.2NACP**

[67] BuAer memo, June 9, 1950, "Contract NOa(s) 9212 and 10114, XJ40-WE-6, and -8, Fuel System Dump Valve; approval of." **B305V13RG72.3.2NACP**

[68] Air Force Air Material Command memo, "June 13, 1950, 'Survey of the Back Pressure Requirements of Main Fuel Regulator Drains on YJ34-WE-32, J34-WE-38, XJ46-WE-2, J40-WE-6 and XJ40-WE-8 Engines" **B305V10RG72.3.2NACP**

[69] WAGT Memo, May 18, 1950, "Contract NOa(s) 9212, Correction of Engine Performance Data." **B305V13RG72.3.2NACP**

[70] BAR memo, June 14, 1950, Contract NOa(s) 9212, monthly progress report A-984 (month of May); forwarding of." **B305V13RG72.3.2NACP**

[71] WAGT memo, June 26, 1950, "Contract NOa(s) – 9212 Our Reference WG-59000 Test Cell Calibration" **B305V13RG72.3.2NACP**

[72] WAGT memo, June 26, 1950, "Contract NOa(s)-9212 Our Reference WG-59000 Test Cell Calibration" **B305V13RG72.3.2NACP**

[73] BAR memo, June 29, 1950, "Contract NOa(s) 9212, test cell calibration." **B305V13RG72.3.2NACP**

[74] BuAer memo, July 17, 1950, No subject, signed by Rear Admiral Pride **B305V13RG72.3.2NACP**

[75] BuAer memo, July 17, 1950, No subject. Hand delivered. **B305V13RG72.3.2NACP**

[76] WAGT memo, July 25, 1950, No subject, references confidential letter Serial 07381 dated 17 July 1950. **B305V13RG72.3.2NACP**

[77] BAR memo, August 16, 1950, "Contract NOa(s) 9212, Item 4, monthly progress report A-1008 (June and July 1950); forwarding of." **B305V13RG72.3.2NACP**

[78] WAGT Memo, August 28, 1950, "Contract NOa(s)-9212 Our Reference WG-59000 Development Delivery Dates" **B305V13RG72.3.2NACP**

[79] WAGT Minutes, September 6, 1950, Minutes of BuAer and WAGT on Development Schedules, Held at WAGT August 11, 1950. **B305V13RG72.3.2NACP**

[80] WAGT submission at program review conference, July 29, 1950, "Westinghouse Development Program", 4 pages **B305V13RG72.3.2NACP**

[81] BAR memo, September 25, 1950, "Contract NOa(s) 9212, Item 4, monthly progress report A-1022 (August 1950); forwarding of." **B305V13RG72.3.2NACP**

[82] WAGT memo, October 6, 1950, "Contract NOa(s) – 9212 Configuration of XJ40-WE-6 Engine, Our Reference WG-59000" **B305V13RG72.3.2NACP**

[83] WAGT memo, October 10, 1950, "Contract NOa(s) 9212, configuration of XJ40-WE-6 engine." **B305V13RG72.3.2NACP**

[84] WAGT memo, October 23, 1950, "Contracts NOa(s) 9212 and 10385 J40-WE-6 Qualification Test – Performance Ratings" **B305V13RG72.3.2NACP**

[85] BAR memo, October 17, 1950, "Contract NOa(s) 9212, item 4, monthly progress report A-1032 (September 1950); forwarding of." **B305V13RG72.3.2NACP**

[86] Lafaye memo to file, November 15, 1950, "Status of XJ40-WE-6 qualification test engine" **B305V13RG72.3.2NACP**

[87] BuAer memo, December 18, 1950, "Contract NOa(s) 9212; comments on stress and vibration analysis report" **B305V13RG72.3.2NACP**

[88] WAGT memo, April 30, 1951, "Contract NOa(s) 9212; Comments on Stress and Vibration Analysis Report." **B305V14RG72.3.2NACP**

[89] BuAer memo, January 5, 1951, "Contract NOa(s) 9212: verification of booster fuel pump" **B305V14RG72.3.2NACP**

[90] BuAer Memo to File, November 15, 1950, Visit by Mr. Lafaye with Lt. Schaefer to Essington, 10 November 1950, "Status of XJ40-WE-6 qualification test engine" **B305V13RG72.3.2NACP**

[91] BAR memo, November 20, 1950, "Contract NOa(s) 9212, item 4, monthly progress report A-1052 (October 1950); forwarding of." **B305V13RG72.3.2NACP**

[92] WAGT memo, November 2, 1950, "Contracts NOa(s) 9212 and 10114 Westinghouse Turbo Jet Engines XJ40WE-6 and XJ40WE-8; Depreservation of" **B823V6RG72.3.2NACP**

[93] BuAer memo, January 5, 1951, "Contract NOa(s) 9212; verification of booster fuel pump" **B305V14RG72.3.2NACP**

[94] WAGT memo, November 3, 1950, "Contract NOa(s) 9212 configuration of XJ40WE-6 engine to be subjected to 150 hour qualification test." **B305V13RG72.3.2NACP**

[95] BuAer Chief memo, December 11, 1950, "Contract NOa(s) 9212, "Configuration and performance requirements of XJ40-WE-6 qualification test engine" **B305V13RG72.3.2NACP**

[96] BAR memo, December 14, 1950, "Contract NOa(s) 9212, Item 4 – Monthly Progress Report A-1070 (November 1950) – Forwarding of." **B305V13RG72.3.2NACP**

[97] BAR memo, November 17, 1950, "Contract NOa(s) 9212, configuration of XJ40-WE-6 engine to be subjected to 150 hour test." **B305V13RG72.3.2NACP**

[98] BuAer memo, March 15, 1950, "Contract NOa(S) – 9212, Model XJ40-WE-4 turbo jet engine, Serial Number WE003002; acceptance of" **B305V11RG72.3.2NACP**

[99] BAR memo, January 13, 1950, "Monthly Progress Report A-920 for Month of December 1949." **B305V10RG72.3.2NACP**

[100] BAR memo, January 8, 1951, "Contract NOa(s) 9212, item 9, XJ40-WE-6; 150-hour qualification test of." **B305V14RG72.3.2NACP**

[101] BuAer Naval Speedletter, January 19, 1951, "Contract NOa(s) 9212; acceptance of qualification test engine" **B305V14RG72.3.2NACP**

[102] BAR memo, February 19, 1951, "Contract NOa(s) 9212, item 4, monthly progress report A-1107 (January 1951); forwarding of." **B305V14RG72.3.2NACP**

[103] BAR twix to BuAer, BEP181, BEF A142, March 15, 1951. **B305V14RG72.3.2NACP**

[104] BuAer memo, March 26, 1951, "Contract NOa(s) 9212, Westinghouse turbo jet engines model J40-WE-6; engine throttle torque." **B305V14RG72.3.2NACP**

[105] BAR Speedletter to BuAer, March 28, 1951, "Contract NOa(s) 9212; XJ40-WE-6 qualification and verification tests" **B305V14RG72.3.2NACP**

[106] BAR memo, April 18, 1951, "Contract NOa(s) 9212, item 4, Monthly Progress Report A-1144 (March 1951); forwarding of." **B305V14RG72.3.2NACP**

[107] BAR memo, May 4, 1951, "Contract s NOa(s) 9670 and 9212, Report of Component Qualification Testing of the Exhaust Nozzle Control for Westinghouse J34-WE-32, J34-WE-38, J34-WE-17 and J40-WE-6 engines; transmittal of." **B306V1RG72.3.2NACP**

[108] BAR memo, June 14, 1951, Contract NOa(s) 9212, item 4, Monthly Progress Report A-1166(May 1951); forwarding of." **B306V1RG72.3.2NACP**

[109] BAR memo, August 14, 1951, Contract NOa(s) 9212, item 4, Monthly Progress Report A-1179 (June 1951); forwarding of." **B306V1RG72.3.2NACP**

[110] BAR memo, August 6, 1951, "Contract NOa(s) 9212. Verification Test Report supplementing "Report of 150-hour Engine Qualification Test" **B306V16RG72.3.2NACP**

[111] BAR memo, September 13, 1951, "Contract NOa(s) 9212, Item 4, Monthly Progress Report A-1214 (August 1951); Forwarding of." **B306V1RG72.3.2NACP**

[112] BuAer memo, September 20, 1951, "Westinghouse Specification WAGT-X40E2-2F dated 23 June 1951 covering the XJ40-WE-6 and J40-WE-6 engines." **B306V15RG72.3.2NACP**

[113] BuAer memo, October 10, 1951 "Contract NOa(s) 9212; J34-WE-32, J34-WE-38, J34-WE-17, and J40-WE-6 exhaust nozzle actuator and control assembly" **B306V1RG72.3.2NACP**

[114] BAR memo, October 15, 1951, "Contract NOa(s) 9212, item 4; monthly progress report A-1231 for September 1951, forwarding of." **B306V1RG72.3.2NACP**

[115] BAR memo, November 7, 1951, "Contract NOa(s) 9212, monthly progress report A-1254 for month of October 1951; forwarding of." **B306V1RG72.3.2NACP**

[116] WAGT memo, November 26, 1951, "Contract NOa(s) 9212, Westinghouse Specification WAGT-X40E2-2F dated 23 June 1951 covering the Model J40-WE-6 engine" **B306V15RG72.3.2NACP**

[117] BuAer memo, November 1951, "Contract NOa(s) 9212, J40-WE-6 engine, primary and emergency fuel regulators, component qualification test of" **B306V1RG72.3.2NACP**

[118] BAR memo, May 6, 1952, "Contract NOa(s) 9212 Verification Test of the Thyratron Tubes in the Power Regulator P/N 61F758-9 for J40-WE-6 Engine." **B306V16RG72.3.2NACP**

[119] BuAer memo, May 7, 1952, "Contracts NOa(s) 9212 and NOa(s) 52-403c. Component Qualification Testing of Fuel Pumping System for XJ40-WE-6 Engine and Dual Pump for XJ40-WE-8 Engine" **B306V16RG72.3.2NACP**

[120] BuAer memo, May 27, 1952, "Contract NOa(s) 9212. Report of Rerun of Low Temperature Phase of Component Qualification Testing of the Exhaust Nozzle Control for Westinghouse J40-WE-6 engines, Transmittal of" **B306V1RG72.3.2NACP**

Chapter 3

J40-WE-6 Mechanical Description

Westinghouse published several descriptions of the J40-WE-6, some focusing on end user education and others detailing many of the construction details of the engine, including assembly and materials used. The -6 was designed to handle the additional loads and pressure requirements of an afterburner producing approximately 40 percent higher thrust over the base military thrust rating. The overall layout of the engine allowed for a fairly simple addition of the afterburner aft of the rear bearing assembly, but the details in making this work in practice will be covered in the -8 related chapters.

The detailed description of the various components and construction techniques of the engine is the basis of all later model discussions, which were broadly similar except in specific areas to be discussed as part of the later models as appropriate. Changes in the basic engine from the -4 model were made because of the weight reduction program and the need to improve performance. These changes will be covered in the related component areas as they are discussed.

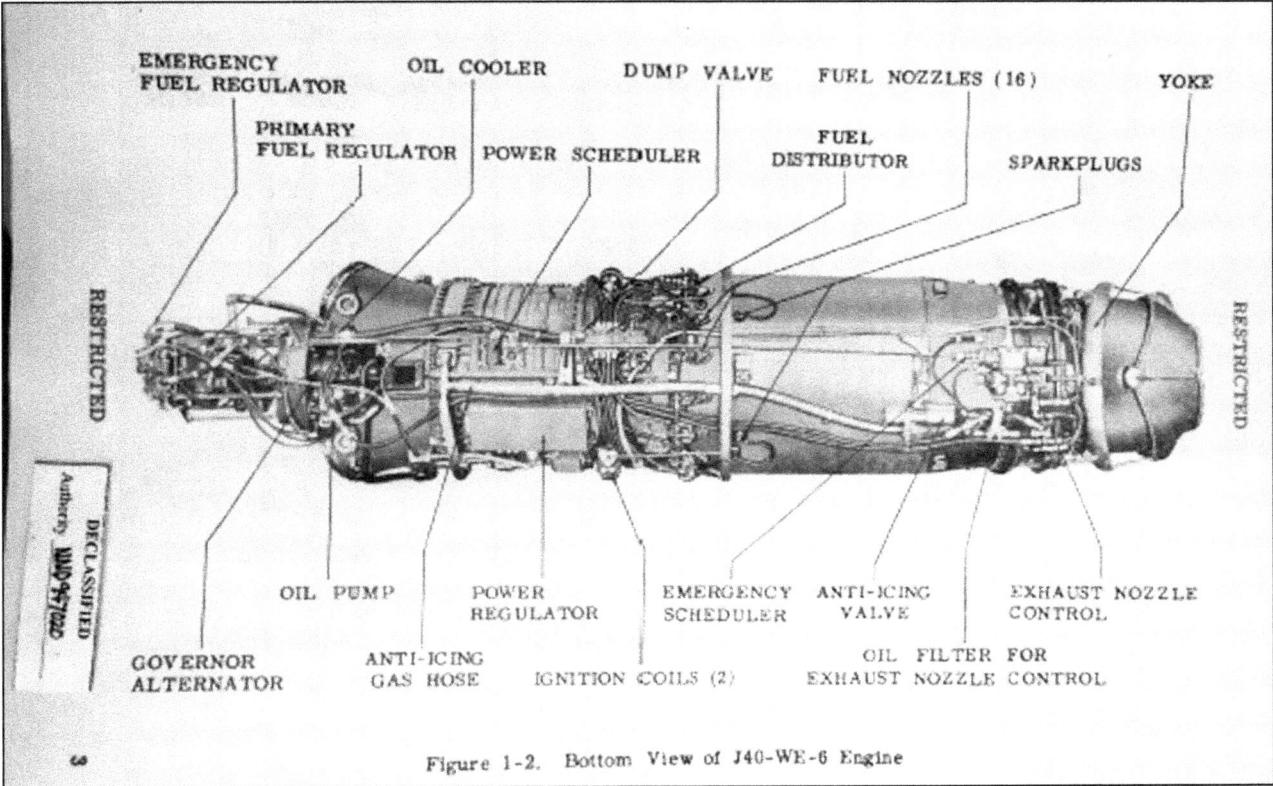

Figure 1-2. Bottom View of J40-WE-6 Engine

Figure 1 WAGT J40-WE-6 Service Instructions, page 3, engine bottom view, 12/15/1951. WAGT "P" number unknown.

Engine Mechanical Design

The basic engine configuration was reflected on engine assembly drawing 43J950-4 for the Qualification Test and as assembled for the Verification Test by 43J950-5.

A. External and Supporting Structure

The external and supporting structure was composed of an inlet duct and No. 1 bearing support, an accessory gearbox, a No. 1 bearing housing, a compressor casing, a diffuser, a combustion chamber outer casing, a No. 2 bearing support, and an exhaust nozzle. Internally the turbine rotor housing and turbine housing were essentially structural members. All these parts were spigot jointed and bolted to form the supporting framework of the engine.

1. Inlet Duct and No.1 Bearing Support. The inlet duct acted both as an air duct and as a structural member containing the front engine mounts. It also served as a mounting platform for the gearbox and oil reservoir. The duct was vertically divided into two elliptically shaped passages positioned outward 20 degrees symmetrically on either side of the vertical axis leaving space for the gear box, accessories, controls and an oil reservoir between them. The gear box was mounted on four pads located at the upstream face of the ovals. The front of the cylindrical oil reservoir was supported by the gear box in the same plane as the gear box flanges and occupied the space between the ovals from this plane aft to the plane of the No. 1 bearing housing.

On the inlet duct just forward of the downstream flange and on the vertical centerline were located two front engine mounts. Only one of the mounts was used in any particular installation, either mount being capable of taking the total engine thrust. The outside of the duct's rear annulus was a spigoted flange for assembly with the compressor casing. The inner annulus was flanged for the attachment of the No. 1 bearing housing. The total thrust of the engine plus a large part of its weight was carried by this housing through the duct to the mounts.

The forward end of each intake oval had separately machined short length oval rings fastened by rivets to the duct allowing the duct work of the engine to be attached to the airframe inlets with "V"-type quick disconnect flanges. The intake duct was a one-piece 3/16"thick magnesium alloy using AN-M-36, Comp. C., Cond. HTA which was a heat treated aluminum-zinc alloy.

2. Gear Box. The gear box had an hour-glass shape dictated by fitting it in between the ovals of the inlet duct. The dimensions were 39 5/8 inches high, 20 7/8 inches wide, and 5 ½ inches thick. The drive for the gear box was

through a splined quill shaft connected to the end of the compressor rotor stub shaft.

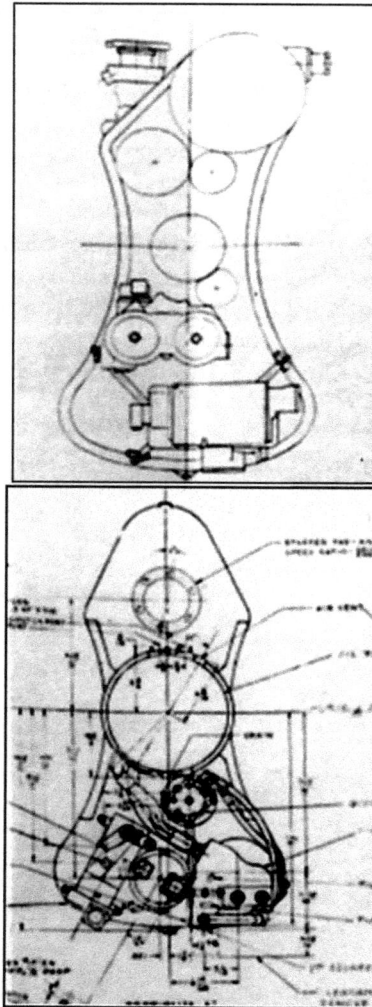

Figure 2 Original vs. final gearbox shape. The original design was a wet sump and the later was a dry sump with the oil tank between the gearbox and the front of the engine, being wrapped around the stub-shaft driving the gearbox. *Images cropped from WAGT engineering drawings.*

On the rear face of the gearbox were mounted the oil pump, oil cooler, overspeed relay, and governor alternator. There was also a pad for power take-off or generator use. On the front face mounted directly to the box were the booster and main dual-fuel pumps. There were pads for a starter, tachometer, hydraulic pump, and generator. A cantilevered mount was fastened to four bosses on the cover. The mount consisted of a plate and four legs which extended outward from ball and socket joints at the bosses. On the fore and aft sides of the plate were mounted the main and emergency fuel regulators.

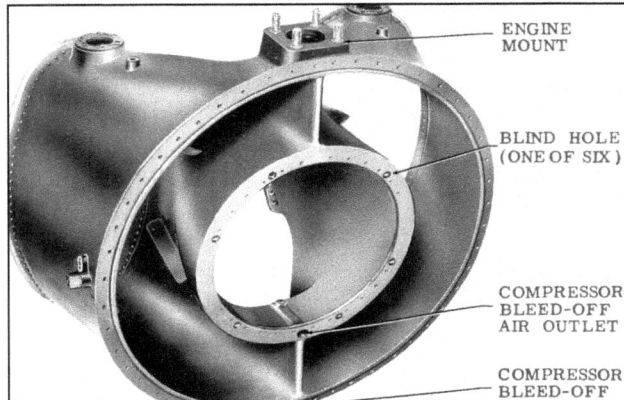

Figure 3 Magnesium intake interface to engine compressor casing, WAGT P43803A, 7/3/1952. *Courtesy Hagley Museum and Library*

Internally, the gear box consisted of a train of straight spur gears all mounted on anti-friction type bearings. Lubrication was provided by numerous oil jets fed from cast-in passageways within the gear box housings. All oil seals were of the rubbing carbon-face type. The gear box was a dry sump being scavenged by three elements of the scavenge pump. The scavenging was arranged to be adequate in all flight attitudes. The cover and base of the gear box were of magnesium alloy castings per specification AN-M-36.

Figure 4 Split gearbox case with gears visible. Layout Shared with the -8 and -1. WAGT P43884, 1/19/1951. *Courtesy Hagley Museum and Library*

3. No. 1 Bearing Housing. The housing was a magnesium alloy casting per AN-M-36 which contained the No. 1 double-row ball thrust bearing, a vibration damper,

rub ring, oil jets, and a seal. The thrust bearing was two counter-bore angular-contact ball bearings placed back to back and mounted in an aligning ring. The bearing was pre-loaded by clamping the two inner races on the compressor stub shaft by the bearing locknut. The outer race was spherically seated to allow for shaft misalignment. Between the O.D. of the outer race and a hardened steel sleeve pressed into the hub of the magnesium housing, there was a floating hardened-steel cylinder which had a wall thickness of .105 inches and was slightly shorter in length than the width of the bearing. The clearance between the floating ring and mating parts was between .0056 and .0085 inches radially. The space was supplied constantly with engine oil under pressure and the action of oil and adjacent metal parts constituted a vibration damper for the rotor. The single oil seal for this housing was located immediately downstream of the bearing and was carried in a cover which was in turn bolted to the magnesium housing. The seal was of the rubbing carbon-face type. Forward of the bearing and bolted internally to the housing was a steel ring (rub ring) .5 inches wide, located concentric with a spacer mounted on the stub shaft. The radial clearance between these parts was .022 inches minimum. The purpose of the rub ring was to support the rotor for a short period of time in case of bearing failure.

The remainder of the No. 1 bearing housing from the hub to the O.D. was conical in shape with its apex pointed downstream. On the rotor side of this cone near the O.D. was a land for fastening an air seal for the space between the housing and the first stage compressor disc. This area was pressurized by compressor discharge air through cored passages along the bottom of the compressor casing and through the inlet duct and bearing housing. The purpose of the pressurized chamber was to relieve the No. 1 bearing of most of the unbalanced thrust loading.

At the forward end of the hub was a spigot for supporting the aft end of the oil reservoir.

4. Compressor Casing. The casing was a four-piece magnesium casting of Westinghouse Specification 9425-1 material. One joint was along the engine horizontal centerline and the other joint was in a plane perpendicular to the engine longitudinal centerline, located opposite the sixth stage of the compressor. The latter joint is not normally parted once assembled. All the compressor diaphragm elements, excluding the inlet guide vane and outlet straightening vane, were generally semi-circular in shape and were fastened within the upper and lower sections of the compressor housing. Each diaphragm half or element was held to the compressor housing by seven .25 inch diameter fillester head screws.

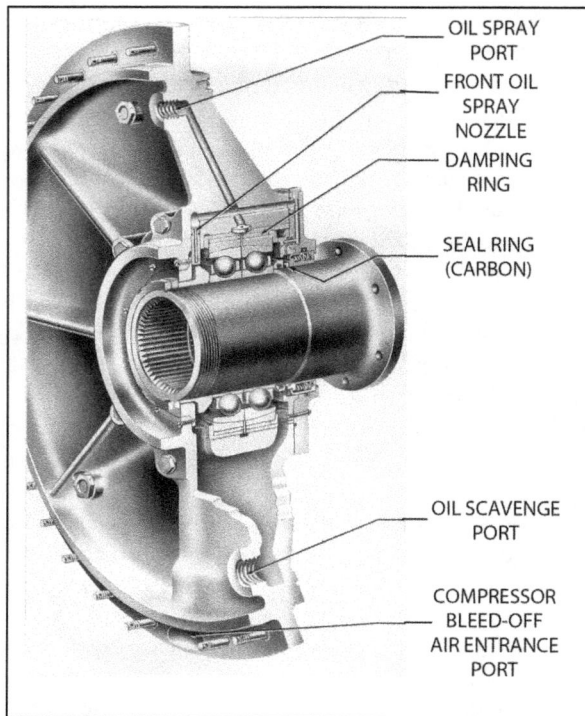

Figure 5 Front (No. 1) bearing assembly cutaway, used on the -6, -8, -1 and -22/-22A engine models, WAGT P47228A, 12/15/1952 (Redrawn). *Courtesy Hagley Museum and Library*

The screw holes in the housing varied in size to permit relative expansion to take place between the steel shrouds of the diaphragms and the magnesium casing. For a similar reason, the torque of the diaphragm retaining screws was somewhat less in the rear half of the compressor than in the front half. This same relative expansion problem also affected the material of the bolts used in the horizontal and vertical joints on the aft half of the compressor. Special "high-expansion" steel (AMS 5624) was used for bolting in those regions.

The forward end of the casting was recessed to receive the hollow outer shroud of the inlet guide vane. The outer shroud was fastened to the casing at three points by large nuts. These nuts were hollowed out in the center to allow hot anti-icing gases to flow through the hollow guide vane.

At the aft end of the casing was another recessed section opposite the outlet straightening vane which was effectively an annulus for the collection of compressor discharge air. This air could be bled off for airframe use through four pots located 45 degrees from the engine horizontal centerline.

Around the exterior of the compressor casing, mostly on the underside, were a number of bosses used for attaching engine accessories. These accessories were: the power regulator, the auxiliary electric control, the power scheduler, the fuel flow distributor, and the ignition coils.

Figure 6 Compressor casing, quarter sections permanently joined, stator vane segments in place, WAGT P43802, 1/10/1951. *Courtesy Hagley Museum and Library*

On the vertical joint of the housing, on the upper half, were located two engine lifting eyes and on the bottom vertical centerline of this same flange was a jack lug for hoisting the engine.

The presence of small circumferential ribs on the external contour of the casting gave the casting stiffness necessary for machining purposes. For strength in areas of high stress, local thickening of the wall was accomplished and longitudinal ribs were added at either end. The rear half of the horizontal flange was thicker than the front portion and used larger diameter bolts than the front portion. This was necessary because the rear half of the compressor was subjected to higher internal pressures and higher temperatures than the front half of the compressor.

5. Diffuser. The diffuser was a series of four short, concentric, sheet metal cones that formed two annular, expanding passages. These passages divided the compressor discharge air into the correct proportions required for later introduction through the outer and inner walls of the combustion chamber liner. The outer cone was the main structural member of the diffuser and was constructed from 18-8, half-hard, stainless steel (MIL-S-5059, Comp. D., 2 B finish, half-hard). The flanges on either end of this outer cone were machined from forgings of the same material. There were sixteen pads equally spaced circumferentially on the outer surface of the outer cone. The pads were for mounting the fuel nozzles. The pads were welded to short sections of streamlined tubing that extended radially into the diffuser and supported the diffuser splitter. The splitter was made by welding together the upstream ends of two of the above cones. The fourth and innermost cone had one flange welded on the upstream

end and another welded on the downstream end. These flanges were for the support of the outlet guide vane at the upstream end, and the turbine rotor housing at the aft end. This innermost cone was fastened to the inner cone of the splitter by eight radial, streamlined struts.

Figure 7 Diffuser end that attached to compressor casing, WAGT P43797, 1/10/1951. *Courtesy Hagley Museum and Library*

The two cones that form the splitter were seam welded together at the upstream end and spot welded to a header at the opposite end. The header helped support and position the fuel nozzles by means of reamed holes and saddle brackets located within the splitter. When the fuel nozzle heads were bolted to the flanges on the outer cone of the diffuser, the fuel tubing had a spring action which applied a constant positive force to the nozzle, forcing it radially inward into the saddles and longitudinally aft against the header bushings in the combustor liner. The combustor liner was fastened to the downstream side of the diffuser header by engaging the upstream end of the liner over "mushroom" headed bolts in the diffuser header. This formed a bayonet type of fastening. After the liner was rotated into the proper position it was locked by the fuel nozzle heads protruding into the upstream end of the liner. The material of the diffuser, other than in the outer cone, was annealed and stabilized 18-8 stainless steel (AMS 5510).

6. Combustion Chamber Outer Casing. This was a sheet metal cylinder with a sheet metal cone welded to the downstream end. There was a flange on the upstream end of the cylinder and on the downstream end of the cone. The forward flange was the largest basic diameter of the engine (40 7/16 inches) and from flange to flange the casing measured 50 inches long. The sheet was .038 inches thick of 18-8 and of half-hardened stabilized stainless steel. The flanges were of 18-8 half-hard forgings.

The conical section was slightly less than one third of the total length of the casing and tapered down to a diameter of 36 3/16 inches at the flange O.D. At either end of the cylindrical section on the bottom vertical centerline were located two fuel drain bosses to which were attached fuel drain valves. Near the front flange and in the bottom half, 33 3/4 degrees from each side of the vertical centerline were two bosses for spark plugs. At the joint of the cylinder and cone and adjacent to the rear drain boss, was a pad for the anti-icing valve. This pad had two holes through it, one rectangular for extraction of compressor discharge air and another stream-lined in shape into which was inserted a tube long enough to permit the extraction of some hot gases from within the combustion chamber liner.

On the conical section of the casing were two circumferential rows of bosses. The front row contained 17 bosses for attaching turbine inlet thermocouples which were used during preliminary running to check temperature distribution. The second row had eight bosses equally spaced. Into these bosses were installed radial retaining pins that restrained the second stage turbine nozzle torque loads.

Also on the cone located at the aft bottom area was a bracket for supporting the exhaust nozzle actuator. The actuator mounted on a swivel joint that was pinned to the bracket.

7. No. 2 Bearing Housing and Rear Support. For clarity, this part was broken down into its main components: Casing, bearing housing and support struts, and exhaust collector. The casing was a cylindrical sheet metal member with a flange on each end and two support rings between the flanges. The support rings had welded between them four strut-end castings equally spaced. These end castings held the No. 2 bearing support struts in place and transmitted their loads from the aft end of the rotor to the support rings. The support rings were designed and had provisions for engine mounts at five locations; two at the horizontal centerline, two approximately 45 degrees above the horizontal, and one on the top vertical centerline. The use of these mounts was restricted to certain combinations and loading directions. Two support brackets for the exhaust nozzle yoke were welded to the aft support rings, part way below the horizontal centerline. All materials in the casing except the strut and castings were heat-treated 12 percent chrome steel (AMS 5504 and 5615). Bosses were provided on the sheet for four pressure rakes located between the support rings, and nine thermocouples located between the support rings, and nine thermocouples located between the support ring flange and the exhaust nozzle mounting flange. Also straddling the front flange and the forward mount ring part way below the horizontal centerline were two jack lugs for hoisting the rear of the engine. A lifting lug was fastened to the top of the forward

flange and front mount ring.

Figure 8 Rear (No. 2) bearing assembly, WAGT P46226A, 12/15/1952 (Redrawn). *Courtesy Hagley Museum and Library*

The No. 2 bearing housing was an all steel cylinder machined from an 18-8 steel forging and welded to it were four pads around the O.D. for attachment of the main support struts. Within the housing were the bearing retainer and oil seal, vibration damper ring, rear roller bearing, and oil jets. The damper ring functioned the same way as that at the No. 1 bearing. A small recessed pocket was welded to the bottom of the housing for oil scavenging purposes. The pocket was baffled to allow efficient oil scavenging in normal flight. The rear of the housing was flanged for attachment of the rear oil pump. The four main support struts were .090 inch sheet metal stampings with solid blocks welded to both ends and threaded to receive three bolts at each end. The material was a high-temperature chrome, nickel, cobalt steel (AMS 5532). The struts were not exactly radial, but were arranged tangential to the No. 2 bearing housing. This mounting allowed free expansion of the struts because when they were heated by

the exhaust gases they caused the bearing housing to rotate slightly without moving off center. This also resulted in a favorable stress pattern on the bearing housing under flight loadings.

The exhaust collector was a sheet metal duct work for guiding the exhaust gases smoothly from the turbine to the exhaust nozzle. It had an outer cylinder, an inner cone, and four struts. The inner cone narrowed down slightly from the front to rear, ending in a flange for the attachment of the exhaust collector end-cone. The inner cone was supported by four tangential struts that in themselves had a unique construction. Each strut was a hollow sheet metal tube of streamlined cross-section. Each end of each strut had two generously filleted collars. The outer ends of the struts were permanently spot welded to the collars which were in turn spot welded to the outer cylinder of the collector. The inner ends of the struts were inserted into the inner collars but not welded to the collars. Thus the struts were allowed to slide relative to the collars under appropriate engine operating conditions. The amplitude of this sliding was very small.

The four collector struts covered the main bearing support struts as well as all tubing running from the No. 2 bearing housing to the outside of the engine.

The outer cylinder of the collector was reinforced circumferentially by several "hat" section stiffeners. On the downstream end of this cylinder was located the flange for attaching the collector to the No. 2 bearing support outer casing. The forward end of the exhaust collector cylinder spigoted over the second stage turbine housing liners. The entire collector and end-cone were fabricated from a 25 percent chrome – 20 percent nickel (AMS 5521) stainless steel.

8. Turbine Housing. The turbine housing was a machined steel cylinder that supported the O.D. of the first stage turbine nozzle, the second stage turbine nozzle, and the aft outer cylinder of the combustion chamber liner.

The aft end of the housing was flanged and fastened between the flanges of the combustion chamber outer casing and the No. 2 bearing support casing. From this point, it extended forward within the conical section of the combustion chamber outer casing. Its support for the first stage turbine nozzle was established by making contact with the ends of the nozzle vanes. Positioning and torque loading of the outer shroud of the first nozzle was through eight radial pins bolted to the housing.

The second stage nozzle spigoted on both ends of its outer shroud with mating lands on the housing. There were six radial retaining pins inserted through the combustion chamber outer casing and turbine housing into reamed holes in the outer shroud of the second stage nozzle.

Opposite the two turbine wheels were turbine housing liners. There were sixteen liners per stage, both double-end spigoted to the housing. Each segment of liner was notched and retained by a pin to control spacing during expansion. The most important function of the segmented liners was to shield the turbine housing from hot gases, thereby minimizing radial expansion of the housing. This caused turbine tip running clearances to be less than if there were no segmented liners.

The second stage segmented liners had a platform over which the exhaust collector outer cylinder was supported radially yet permitted free expansion longitudinally.

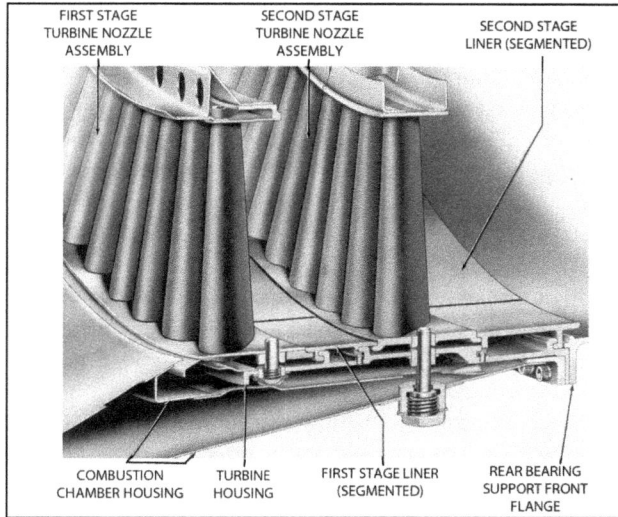

Figure 9 Turbine casing assembly cutaway, WAGT P47227A, 7/3/1952 (Redrawn). *Courtesy Hagley Museum and Library*

9. Exhaust Nozzle. The exhaust nozzle was composed of a cone and a pair of moveable "gates". The cone was a sheet metal part with a formed flange on the upstream end that attached to the No. 2 bearing casing. On the vertical centerline and about midway along the cone were welded two brackets for the pins about which the gates pivoted. At the downstream end of the cone was a flange to which was attached a segmented knife-edge seal. The seal segments were radially spring loaded and followed the inner surface of the contour of the gates as they changed position. The purpose of the seal was to minimize the leakage of exhaust gases between the gates and the cone. The seal segments were stainless steel with a brass rubbing portion brazed on the O.D.

The gates were segments of a true sphere and were symmetrical about the engine vertical centerline. On the horizontal centerline near the front edge on each gate was a bracket to which was connected the gate actuating linkage. Reinforcement of the gate sheet metal was accomplished by box, hat, and tubular section members placed around the edges of the gate and through the actuator bracket. With

the exception of the tubular stiffener and the pivot pins, all material was 20 percent chrome – 20 percent nickel – 20 percent cobalt AMS 5532. The pivot pins were Discaloy (PDS 49-4), a Westinghouse developed alloy.

10. Turbine Rotor Housing. The turbine rotor housing was a straight cylinder that connected the inner support of the first stage turbine nozzle with the inner cone of the diffuser. It also formed the inner passage wall for the combustor. The housing was reinforced internally by hat-section stiffeners against a collapsing pressure. The material used was 18-8 half-hard stainless steel.

11. Firewall Seal. A firewall seal completely encircled the engine at the downstream end of the diffuser. The seal was made of stainless steel sheet metal (AMS 5510). Through the bottom portion of the seal passed all the tubes, cables and linkages that went to the aft portion of the engine.

12. Exhaust Nozzle Actuating Yoke. The yoke was a two piece elliptical cross-section semi-circular frame. The two parts were hinge-jointed at the bottom with an offset for attaching to the exhaust nozzle actuator. About halfway up from the bottom on each yoke were bosses that connected to the yoke mounting brackets which were on the frame of the No. 2 bearing casing. This connection was pinned through a ball and socket joint. At the top on free ends of the yoke were sections of linkage that connected the yoke to the exhaust nozzle gate brackets. Both ends of these links had ball and socket joints. Full swing of the yoke permitted a 53 degree change in exhaust nozzle area from the minimum. Material in the yoke was heat-treated 12 percent chrome sheet (AMS 5504).

B.   Flow Path Elements

The flow path elements were considered to be all those parts that had any effect in changing the thermodynamic and aerodynamic characteristics of the gas flowing through the engine. Those parts were the compressor rotor, compressor inlet guide vanes, compressor diaphragms, compressor outlet guide vane, combustion chamber liner, and turbine. The diffuser and exhaust nozzle also fell into this category but were discussed above.

1. Compressor Rotor. The compressor rotor had eleven stages with a constant O.D. and a varying hub diameter. Of the eleven discs, the first seven were aluminum alloy (AMS 4135), the eighth stage was cold worked 18-8 (PMS 6599-2) steel, and the last three stages were 12 percent chrome (AMS 5615) steel. The use of steel on the aft four stages was mandatory from strength considerations at elevated temperatures. An 18-8 steel disc between the seventh aluminum and ninth 12 percent chrome steel discs provided a less abrupt transition in coefficients of thermal expansion between mating parts.

Each disc was machined separately and had a Gleason curvic clutch 13 inches in diameter cut on a cylindrical land on both sides of all discs except the front side of the first stage. These curvic clutches mated with one another as the discs were stacked for assembly on a short steel coupling. When bolted together with sixteen .5 inch diameter through-bolts torqued to a predetermined elastic stretch, the discs became an extremely rigid rotor. The ease with which the discs and coupling assembly, and the accuracy of alignment of discs achieved by use of the Gleason curvic coupling was excellent. On the front side of the first stage disc was bolted

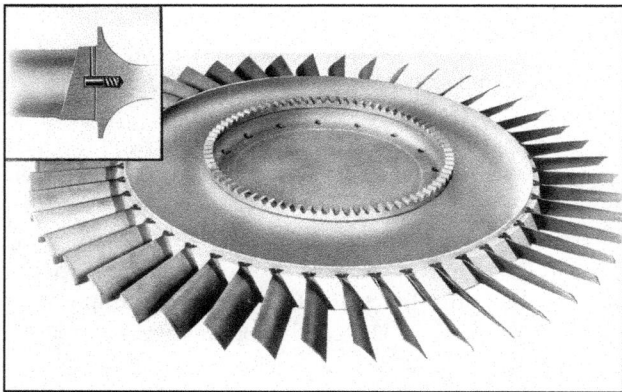

Figure 10 10th Stage compressor disc with inset showing blade locking system using a spring loaded shear-pin, WAGT P43830A 1/12/1951. *Courtesy Hagley Museum and Library*

a steel (AMS 5615) stub shaft and an aluminum seal ring extending from the through bolt circle to the hub O.D.

Blade grooves in the discs were broached to within close tolerances in order to control the amount of clearance between the grooves and the blade roots. The clearance could be from .001 to .010 inches in the aluminum discs and from .000 to .009 inches in the steel discs. A limited amount of looseness had a desirable damping effect. The blade groove shape in the discs was of the dove-tail type.

The compressor blades were manufactured from precision forgings of PDS 2881 steel which was a 12 percent chrome type alloy. The blade roots were milled and the airfoil sections were ground by machine and by hand finishing operations. There were a total of 512 blades in the rotor in varying amounts from 19 in the first to 75 in the last. The blades were locked in the grooves by spring-loaded aluminum shear pins that allowed very little axial displacement. To remove the blades, the pins were sheared by striking the blade root with a suitable tool. Blades could be removed or replaced without damaging the discs or blades.

An unavoidable result of using loose blades in discs was a tendency to gall in the areas where blades bore against the disc. To minimize this condition almost to a vanishing point, a molybdenum-disulfide base lubricant was applied to all the blade roots.

Dynamic balance of the rotor was simplified by assembling blades in a patterned sequence according to their weight and by controlling manufacturing tolerances of the disc contours and curvic clutches. Compensation for any unbalance was made by adding balance weight to the long through bolts when the entire engine rotor was balanced. This was accomplished by using nuts of different weights as necessary on the long through bolt on both ends.

2. Compressor Inlet Guide Vane. The inlet guide vane was composed of twenty-four hollow sheet metal vanes welded to ducted shrouds segmented into sections for channeling hot anti-icing gases through the vane assembly. Its support on the O.D. was described above. Axial restraint of the inner shroud was by six .25 inch bolts which passed through brackets on the inner duct and fastened to the No. 1 bearing housing. Torque on the inner shroud was absorbed through brackets by two shear pins also fastened to the No. 1 bearing housing. The division of the inner and outer ducted shrouds was such that hot gases flowed radially inward through certain vanes and then radially outward through other vanes. The flow and temperature of the hot gases was regulated by the anti-icing valve located on the combustion chamber outer casing. Maximum temperature was approximately 700°F with a pressure not in excess of 25 psi. Inside each hollow vane was a thin sheet metal baffle that guided all the hot gas along the concave surface where ice was most likely to form. Expanded anti-icing gas was conducted to a port which was a "customer's connection" and presumably discharged overboard from the airplane. All material used in the vane assembly was 18-8 stabilized stainless steel (AMS 5510).

3. Compressor Diaphragms. There were ten stages of compressor diaphragms each consisting of two 180° halves. They varied in vane length and in number of vanes to suit each rotating stage. There were a total of 768 vanes varying from 38 to 124 per stage. All vanes were rolled from bar stock to the correct airfoil shape, then cut to length and finally coined to give the desired twist. Each vane when sheared to length had a peg left on either end that fit into punched holes in the inner and outer shrouds. These pegs were then heli-arc welded to the shrouds. In addition to welding the pegs, all vanes in the first six stages were fillet brazed to the inner and outer shrouds, and in the remaining diaphragms, only the four end vanes were so brazed. Material in both the vanes and the shrouds was of the 12 percent chrome steel type (PDS 2896) and was heat treated after welding for high strength. On the inside of the inner shrouds were welded light gauge (.032 inch) sheet metal seal strips. These strips rode over lands on the compressor

discs with a minimum clearance to reduce leakage between stages. The outer shrouds fit into recesses in the compressor casing to form a flush wall as well as to retain the diaphragms axially. Radial retaining screws threaded into the outer shrouds.

4. Compressor Outlet Guide Vane. The outlet guide vane had two rows of straightening vanes in shrouds tapering to a slightly diffusing passage. It was similar in construction to the other compressor diaphragms except that it had a 360° ring. It was spigoted and bolted to the forward flange of the inner cone of the diffuser through a flange welded on the aft end of the inner shroud. The front edge of the outer shroud fit into a recess in the compressor casing which partially restrained the shroud in both radial and axial directions. Between the aft end of the outer shroud and the overhung portion of the leading edge of the outer diffuser cone, was an annular split approximately 3/16 inch wide through which all compressor bleed air passed into the bleed manifold. On the inner shroud forward end was a double leg labyrinth type seal opposite to which was another of the same kind of seal cantilevered on a support extending from the vane attachment flange. These two seals rode over and under the eleventh disc aft seal land. All material in the vane assembly was 18-8 stabilized stainless steel (AMS 5510).

5. Combustion Chamber Liner. The combustion chamber liner was a single annulus one piece all welded assembly. The two concentric walls were composed of a number of cones and cylinders that bound a gradually expanding volume along approximately the first 85 percent of its length. Beyond this, the walls tapered more sharply inward to the mate with the annulus of the first stage turbine nozzle. At the upstream end, the walls were connected to a "U" shaped header. The header was the main supporting member which, when brought into engagement with its mating part on the diffuser by a bayonet joint, caused the entire combustor liner to be a cylindrical skirt extending from the first stage turbine nozzle to the end of the second nozzle. The skirt lay between the conical section of the combustion chamber outer casing and the turbine housing. This skirt functioned as a baffle to guide a blanket of compressor discharged cooling air over the turbine housing and its mating parts. The air, after cooling the turbine housing, flowed into the turbine at the leading edge of the first stage nozzle outer shroud. The downstream end of the liner inner wall rested on a shelf that extended a short distance upstream from the inner support and shroud of the first stage turbine nozzle. The ends of both walls were restricted in radial movement but were free to expand axially during combustion.

Adequate endurance of the liner walls was achieved by blanketing each part of both walls with a layer of compressor discharge air. This practice was also applied to the header which in addition to being air-cooled had perforated baffles between fuel nozzles through which the same cooling air flowed to restrict carbon formation. Certain of the sections of both walls were slotted to add rows of holes arranged in a definite pattern for the introduction of air to accomplish combustion within the liner and to inject diluted air to give the desired temperature pattern immediately ahead of the turbine. Material in the header assembly was 18-8 steel (AMS 5510), and the remainder of the liner was 25, 20 stabilized stainless steel, (AMS 5521)

6. Turbine. This group included the turbine shaft and coupling, first and second stage turbine wheels, and first and second stage turbine nozzles.

The turbine shaft was a 13 inch diameter cylinder 43 3/8 inches long., which had a .10 inch thick wall. The material was a 12 percent chrome steel forging (AMS 5613). Both ends of the shaft connected to couplings. The coupling at the turbine had a flange and spigot on one end to connect to the shaft and a curvic clutch to mate with the first stage turbine disc. Within the coupling there was attached a sheet metal baffle to prevent heat from being freely transmitted to the turbine shaft from the hot turbine disc. That device together with another baffle external to the shaft (running radially inward from the first stage turbine nozzle support flange) effectively blocked the flow of heat to the shaft and minimized "shaft bow" after engine shut down. The coupling material was an alloy forging of Discaloy (PDS-48-7). A turbine bolt retaining ring located with the coupling was threaded to receive the upstream end of twenty turbine bolts. The turbine through bolts passed through both turbine discs. Nuts on the downstream end of the turbine through bolts bore against the aft face of the second stage disc and caused curvic clutches on the turbine discs to be drawn tightly together.

The first stage turbine disc had a curvic clutch on each face and on the same diameter as the compressor curvic clutches. The O.D. of the disc was 21.684 inches but the hub diameter was actually formed by the shoulders of the 62 blades that touched one another. The blade groove was a two prong "Christmas tree" form. Material for the disc was a Discaloy forging.

The second stage turbine disc had only a forward curvic clutch. It had 32 blades and a 21.830 inch hub diameter. The blade root was a three serration "Christmas tree" form. On the aft face was an 8 1/8 inch diameter flange to which was bolted the turbine stub shaft.

Figure 11 1st Stage turbine disc with inset showing blade locking mechanism, WAGT P43832A, 7/30/1953. *Courtesy Hagley Museum and Library*

The turbine stub shaft tapered in two sections of 45 and 14 degree angles respectively. The 14 degree tapered portion was 2 1/16 inches long and mated with an internally tapered sleeve 3 5/16 inches long. The internally tapered sleeve had a stepped cylindrical outside diameter. One step which received the rear bearing inner race, the other step was a surface over which ran the spiral groove bronze oil seal for the No. 2 bearing housing.

The internally tapered sleeve was held on to the stub shaft by a nut screwed onto the threaded portion at the aft end of the stub shaft. Fitting over the retaining nut and spigoting to the bearing inner race was a pinion gear for driving the rear oil pump. It was held on to the retaining nut by six small bolts. The material in the stub shaft and sleeve were both Refractaloy (PDS 47-12). The stub shaft was fastened to the disc flange by ten 3/8 inch Refractaloy studs and nuts.

The first and second stage turbine blades were precision castings with ground roots. The fit of the blade root in its groove could be from .000 to .133 inches loose. The O.D. of both stages was ground to a size which resulted in a "cold" tip clearance of between .100 to .133 inches. The first stage blade roots were pre-stressed by applying to them

a "shot blast" operation. This was followed by coating the roots with graphite for lubricating purposes. Locking of the first stage blades was by means of steel pins 1/8 inches in diameter inserted from under the seal land and peened into a notch in the base of the root of each blade. The second stage root did not have any special treatment and it was locked by peening metal from the disc into recesses ground into the blade platform. Material for all blades was "Stellite – 31" (AMS 5382).

The first stage turbine nozzle was composed of 56 hollow, precision cast vanes held by, but not welded to, inner and outer 360 degree shrouds. The inner shroud was welded to a machined support that was bolted onto the aft end of the turbine rotor housing. The ends of the vanes protruded 3/16 inches through the inner shroud and rested on a machine land on the support. A tack weld on the ends of each vane in this space retained the vanes. On the O.D. the vanes similarly extended beyond the shroud 3/16 inches and had small cast-on lips that kept the shroud in place radially. The ends of the vanes when hot, expanded and contacted a sheet metal ring welded to the front end of the turbine housing. The nozzle was then restricted in radial movement at both ends. The outer shroud was further retained by a spigot on its downstream end mating with a shoulder on the first stage turbine segments, and by eight radial retaining screws fastened to the turbine housing for torque loading. Both shrouds were AMS 5521 material.

The second stage nozzle was a 360 degree ring cantilevered on the outer shroud, and split at the inner shroud in ten equal parts of six vanes per section. The vanes were solid precision castings welded to both shrouds. Seal strips were spot welded to the inner shroud to minimize leakage around the nozzle between stages. The outer shroud was spigoted on both ends to the turbine housing and was free to expand radially with the limits required. Six radial retaining pins absorbed the torque loading. The vane material in both nozzles was "Refractaloy #80" (PDS 9296).

Six different combinations of compressor, combustor and turbine were tried before the proper temperature distribution throughout the engine was achieved. The final qualification test results were:

	Qualification Engine		vs.	Model Specification	
	Thrust	**SFC**	**RPM**	**Thrust**	**SFC**
Military	7,600	.975	7,260	7,500	1.04
Normal Rated	6,900	.948	7,260	6,800	0.99
Cruise #1	6,200	.935	7,260	6,120	0.96
Cruise #2	5,100	.927	7,050	5,100	0.94
Idle	420	3.17	300	450	4.00

C. Lubrication System

Lubricating oil, MIL-0-6081, grade 1010, was drawn from the oil reservoir (located on the engine center line between the accessory gearbox and the front bearing support) into the main and auxiliary elements of the gearbox driven lube pump. The main element drew oil from the front bottom portion of the reservoir while the auxiliary element drew oil through a swivel pickup located in the rear portion of the reservoir. The discharge of both elements passed together through the pump pressure relief valve which bypassed any excess oil back to the main element inlet. As the capacity of each pump element (at military rpm) was sufficient to supply the entire engine oil requirements, the relief valve would begin to bypass at approximately 50 percent of military rpm. The oil pressure would be constant in the range from 50 to 100 percent of military rpm. Before leaving the pump, the oil passed through a 10 micron filter (AN6235-4). A 50 psi relief valve was in parallel with this filter so that if clogging of the filter increased the filter pressure drop to more than 50 psi, oil would bypass the filter.

The oil was discharged from the lube pump through the pump flange through a cored-in passage of the accessory gearbox to a fuel-oil cooler mounted on the aft face of the gearbox. The oil made two passes through the cooler. At the end of the first pass, approximately one gallon of oil was bled off through a cored-in passage of the accessory gearbox, through a knife-edge filter to a pressure regulating valve. The pressure regulating valve reduced the oil pressure to 25 psi above the gearbox cavity pressure. From there the oil was distributed to fourteen oil jets, each having a .025 inch diameter orifice which directed the oil upon highly loaded gears and bearings.

At the cooler outlet, part of the oil was piped aft to the high pressure pump and the rear bearing, and part to the front bearing. The front bearing supply passed through a knife edge filter mounted on the front bearing housing. In the bearing housing, the oil was distributed to four oil jets. One .047 inch diameter jet lubricated the front of the thrust bearing. A second .047 inch diameter jet delivered oil to the vibration damping ring, and two .025 inch diameter jets sprayed the rear of the thrust bearing and the face of the rubbing seal respectively.

Normally, the rear bearing oil supply flowed from the cooler outlet, through the high pressure pump (located in the tail cone aft of the rear bearing), through the exhaust nozzle actuator to the rear bearing. To assure rear bearing oil flow in the event of either the high pressure pump or the exhaust nozzle actuator failing, the pump inlet and the actuator discharge lines were connected so that oil could flow to the rear bearing directly from the oil cooler. Within the rear bearing housing, the oil flow was distributed to five

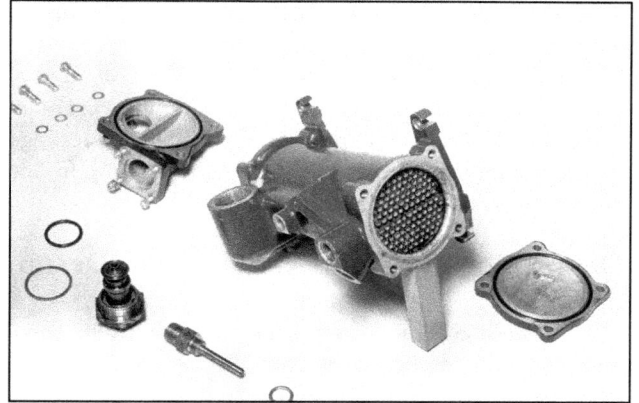

Figure 12   Oil heat exchanger disassembled, WAGT P43895, 1/19/1951. *Courtesy Hagley Museum and Library*

.047 inch diameter oil jets. Two of those sprayed oil on the front of the roller bearing, two sprayed oil on the rear of the bearing and one supplied oil to the vibration damping ring. The lube line to the rear bearing housing contained a knife edge filter and a 5 psi check valve. The check valve shut off the rear bearing oil supply during engine shut-down at approximately 900 rpm, thus insuring that the bearing housing was scavenged dry and eliminating any oil smoking after shutdown.

The rear bearing housing was scavenged by an element built into the high pressure pump. As the capacity of the scavenge pump was approximately four times as great as the lube supply, air was drawn through the rear bearing labyrinth seal. This caused a pressure gradient across the seal that prevented any oil from leaking out. The scavenged oil and air were pumped through a passage in the gearbox driven lube pump where the pump joined the combined scavenge discharge from the No. 1 bearing and the gearbox.

The front bearing housing was scavenged by an element of the gearbox driven lube pump. The gearbox was scavenged by three separate elements of the gearbox driven lube pump. The gearbox scavenge inlets were located at the bottom front, bottom rear, and upper center portion of the gearbox. As each scavenge element had sufficient capacity to scavenge the entire gearbox lube flow, the gearbox would be adequately scavenged during any condition of diving, climbing, or inverted flight. The discharge from these four elements, plus scavenge flow from the rear bearing, was returned to the oil reservoir through a curved baffle plate (built into the reservoir) which separated the air from the oil with a minimum of foaming. The air was vented from the reservoir into the gearbox and then overboard. In passing overboard, the air was passed through a centrifugal slinger which removed any suspended oil droplets. The gearbox overboard vent contained a relief valve which maintained the gearbox cavity and oil reservoir pressure constant at three psi. The relief valve had a small by-pass orifice through which air passed on engine shut-down. The function of the orifice was to allow the gear box internal pressure to reduce to ambient pressure after the 3 psi relief valve closed. This action prevented any static oil leakage which would be caused by retaining pressure in the lubrication system after engine shut-down.

Description of Engine Aerodynamic Design

This is a general description in large part, with some details added to the information in the Westinghouse documents from various NACA test reports.

Air entered the engine through the split inlet duct to the inlet guide vane of the compressor. In designing this duct it was necessary not only to keep the duct flow losses to a minimum but also to control the flow distribution at the downstream annular end of the duct. The latter was important from the standpoint of the effect on compressor rotating blade vibration. Duct losses were found to be in the order of a maximum of 1.6 percent measured on a basis of total pressure loss. It was also found by test that no dangerous resonant excitation existed in the early stages of the compressor due to the flow through the inlet duct.

Next in the flow path was the compressor. The inlet guide vanes, all diaphragms, and the outlet guide vanes were considered part of the compressor. The compressor was an eleven stage axial flow type with a modified symmetric flow pattern. The design pressure ratio was 4.61 with an air flow of 136.2 lbs/sec. The tip diameter varied from 32.150 inches at the front stage to 32.180 inches at the aft end. The hub-tip ratio at the inlet was .6 and at the outlet .84. Maximum rotor rpm was 7,260 which gave a tip speed of 1,025 ft/sec. Maximum design efficiency was 85 percent.

Tip diameter and speed were selected on a basis of compressibility losses as compared to existing Westinghouse compressors (J34). Air flow, pressure ratio, and efficiency were arrived at by making a number of cycle, weight, and aircraft application studies. The number of stages was determined by use of an enthalpy (energy) rise coefficient within the range of past practice. Blade chord lengths were established by stress considerations, taking into account camber, pitch-chord, and thickness-chord ratios. Blade spacing was selected by use of a formula involving a lift coefficient with a value intended to give minimum profile and secondary losses. The blade selection was from the NACA 65 series airfoils, modified at the trailing edge to suit the J40. Small compromises were made from the theoretical blade configurations in order to group blades (make more than one state of blades from a particular forging) and thereby save manufacturing costs. Grouping was done within the limits of one degree for outlet blade

angles (twist) and three degrees for camber angle from the required values.

The inlet guide vane, because of its expanding effect on the flow, was designed on the basis of turbine nozzle information. The aerodynamic requirements were based on the selection of a lift coefficient which, when used with an equation involving inlet and exit angles and pitch-length ratio, gave the pitch distance. The remaining pitch-length ratio gave the general blade section shape, camber line curvature, and trailing edge thickness, which were determined largely by the design and mechanics of an anti-icing system.

The diaphragms (guide-vanes) including the straightening vane, were defined by the same formulae and design criteria used for the rotating elements. The straightening vane, however, was divided into two rows of vanes for more effective turning of the air which exits the last compressor rotating stage at a high angle from the axial direction.

In the design of the diffuser, the two factors of importance were efficiency and exit flow distribution. Exit flow uniformity was vital in creating the proper temperature distribution at the combustor outlet. The efficiency of the diffuser was a function of area ratio and equivalent angle. The area ratio was the ratio of the outlet to the inlet areas and was entirely determined by the requirements of the combustor inlet and the compressor outlet annuli. For the XJ40-WE-6 this ratio was 2.24. Using this ratio and an assumed efficiency of 80 percent, the equivalent angle was found to be 25 degrees. The distribution of flow was intended to be equal from each annulus.

The function of the combustor was to provide a continuous energy release resulting in a flow of hot gases to the turbine with a precise temperature gradient from turbine hub to tip under all operating conditions. The size of the combustion chamber liner was determined from combustor altitude stability or pyrogenic limits, pressure drop, and outlet temperature distribution requirements. These factors were related on a dimensionless basis in terms of Reynolds and Mach numbers, temperature rise ratio, pressure drop, and mixing length which was the ratio of combustor width to length. Dimensionless design curves were made originally for the Westinghouse 19-XB-2B single annulus combustor and were used as a basis of design for the J40 model. The design points fixed by Model Specification requirements and engine geometry were (compressor ratio 4.61 to 1) 60,000 feet altitude operational limit at 150 knots I.A.S., 1425°F turbine inlet total temperature, 3.4% pressure loss, and minimum combustor inside diameter of 18 inches.

Other important considerations in the design of the combustor were coking, stability, efficiency, and ignition.

Coke formation, which depends somewhat on the type of fuel used, was controlled by forcing a blanket of compressor air to flow over the area where coke might deposit. Stability, efficiency, and ignition were evaluated by testing and improved by development to meet the initial requirements.

The high guaranteed engine performance put exacting requirements on the turbine design beyond anything previously attained by Westinghouse engines. An efficiency of 90% was selected to be accomplished by a two stage reaction turbine of constant hub and tip diameters.

The turbine was designed for an isentropic velocity ratio (ratio of peripheral velocity to theoretical gas velocity) based on an enthalpy drop of .59 at the mean diameter. That value was kept close to that found yielding peak efficiency on the J34 turbine. The axial leaving velocity Mach number was selected at .45 (69% critical flow) after careful consideration of turbine diameter (which involved weight), and losses in the exhaust annulus (which were affected by axial velocity). The distribution of work output per stage was fixed at 60 and 40 percent respectively for the first and second stages. The total work output under static conditions was 15,480 hp. The reaction (pressure drop) was made positive at all radial stations to avoid diffusion loss in the blade rows. Blade efficiency was kept to a practical minimum and was actually reduced during operation by air cooling the turbine housing so that the tip clearance became smaller as the turbine parts became hot.

Just as in the case of the compressor, the ideal blading design from aerodynamic and thermodynamic standpoints was not possible and had to be compromised by stress and casting standard limitations. Consequently the pitch-chord ratios and hub and tip blade angles became the starting point for design. Actual shapes and blade sections were then determined on the drafting board after due consideration for strength and casting limitation and after providing the proper kind of flow passages to give an efficient overall flow between the turbine inlet and exit.

Another important factor which had to be taken into account was the matching of the turbine with the compressor to avoid, as nearly as possible, any compressor stall. The effect was related to pressure ratio and flow number which in turn were associated with work distribution and reaction. These characteristics were checked at sea level over a wide range of values and found not to be critical on stall under steady state operating conditions.

The variable area exhaust nozzle with "eye-lid" or "clam-shell" type gates had to be designed with particular attention to the shape of the trailing edges of the gates and with emphasis on the seal between the gates and the exhaust nozzle cone. It was found that the trailing edges of the gates must flare out smoothly and that in the closed position the exit orifice formed by the gates must be as nearly circular as possible. The importance of a good seal was demonstrated by the fact that it was possible to cause a change in nozzle velocity coefficient on the order of 3 percent between a sealed and an unsealed condition.

In order to meet the performance requirements of the engine, the values of .975 for velocity coefficient and 95% nozzle efficiency were found to be practical based upon experimental tests on a similar nozzle.

Comparison of Original Design Point Factors with Final Engine Configuration		
Factor	Original Estimate 4/15/1947	Final Production Configuration
Engine R.P.M. (Military Speed)	7,260	7,260
Air Flow – lb/sec	137.0	140.9
Compressor Pressure Ratio	4.61	5.1
Compressor Efficiency	85%	81.7%
Combustor Efficiency	97%	97.2%
Combustor Pressure drop	3.0%	4.2%
Turbine Inlet Temp	1425°F	1425°F
Turbine Flow Number	.242	.232
Turbine Pressure Ratio	2.11	2.39
Turbine Efficiency	90%	85%
Exhaust Nozzle Velocity Coefficient	.975	.920
Exhaust Nozzle Expansion Efficiency	95	84.6
Weight	2,680 lbs	2,974 lbs

Mechanical Changes Reflected in Design as a Result of Testing	
Mechanical Difficulty	**Corrective Action**
Compressor – Blade tip and seal rubs	Increased clearances
Compressor – Galling of roots	Coated roots with graphite film for lubrication
Fuel Nozzles – High frequency (70 cps) hunt	Orifice (.020 inch) inserted in line to balance diaphragm on master nozzle
Combustion chamber liner – Small buckling and cracking in walls. Distribution, radial circumferential. Coking of upstream bulkhead	Shorter steps in the walls. Selection of proper air entry hole arrangement. Cooling air baffles redesigned
Turbine Shaft – Temporary bowing due to heat after shut down causing vibration upon starting after an elapsed critical period	Heat baffles added at downstream end of shaft, internal and external, restricting flow of heat from hot turbine to shaft
First stage turbine nozzle support – Inner shroud and liner support cracks	Added reinforcing ring and rearranged welding. Changed shape of vane support at trailing edge of vane
First stage turbine nozzle vanes – small trailing edge cracks	Trailing edges cut back to thicker section (.027 to .050 inch)
First stage turbine blade – Cracks in root	Blade tip thinned down. Root redesigned 3-prong to 2-prong. Shot peened and graphite coat operations added to root
Turbine 2nd stage nozzle – Cracks in trailing edge of vanes at O.D. Seal strips cracked in spot welds at ends of cuts	Changed welding technique. Reinforcement brackets added
Second stage turbine liners – Small fatigue cracks	Relieved sharp corners. Changed location of slot
Turbine housing – Insufficient support for loose first stage turbine nozzle vanes at O.D.	Spot welded sheet metal band with controlled clearances around O.D. of front edge of housing
Exhaust Collector – Fatigue cracks and buckling of outer cylinder	Sliding strut with large end fillets incorporated. Added strategic reinforcements and improved method of cone
Exhaust nozzle inner cone – Buckled inward at hinge points	Reinforced with back-up plate
Exhaust Nozzle Gates – Several fatigue cracks and some distortion	Redesigned and added more stiffeners and increased shell thickness from .032 to .050 inch
Controls – vacuum tube failures	Three tube designs were improved (ruggedized) and vibration insulation added
Controls – stall free acceleration issues using the primary control	Very careful adjustment and modification to get all components to work consistently
Fuel System – high frequency surge action	Insertion of a .020 inch orifice plug in the line to balance diaphragm side of fuel nozzles
Ignition System – Minor modifications	Added wire for preheating of tube circuits and another for an emergency control signal
Lubrication – rear bearing housing leaks	Babbitt coated type seal replaced with bronze type, reduction in compressor discharge air to the seal area.
Lubrication – leakage in main bearings	Reduction in oil past damper ring via increased press fit between the retainer and the housing and adding an orifice in the passage to the retainer ring
Lubrication – front bearing	Improved front seal carbon-ring face-type to provide more positive sealing than the rear seal to prevent high pressure air flowing into the bearing housing and gearbox.

NACA Testing Results

Two early production engines were tested by NACA in Cleveland and there were two major findings requiring engine changes to be applied after the engine had passed its qualification tests:

1. Turbine Inlet Temperature Inversion – While operating the engine at simulated altitude (something WAGT could not do due to lack of appropriate test facilities) in the altitude wind tunnel it was determined that a turbine inlet temperature inversion problem existed. This was a phenomenon which resulted in increasing the temperature at the hub portion of the turbine blades beyond desirable limits. The condition existed only at altitude.

This was not entirely unexpected therefore Westinghouse had on hand an alternative compressor outlet vane type mixer. Inclusion of this mixer in the engine contained the temperature inversion effect within safe limits and it was introduced into previously delivered engines during overhauls.

2. High Referred Speed Compressor Stall – It was determined that the compressor experienced a steady state stall at high referred R.P.M. The compressor, which was aerodynamically the same as the Qualification Test engine, was termed a C-1 compressor design.*

3. Two ways of eliminating the high referred speed stall were determined:

a. By reorienting the turbine nozzle vanes to open the turbine, the compressor pressure ratio was reduced to the point where compressor stall was eliminated. This was the solution applied to all previously delivered production -6 engines during overhaul and involved some loss in performance by making the turbine less efficient. The performance loss was regained through higher fuel consumption. All production -6 engines operated at higher actual fuel consumption targets than the specification requirement with consent of BuAer. Some compressor stalling continued to be experienced after the modifications were made but later research was applied to curing it in the -8 models.

b. A new compressor aerodynamic design called the C-20 was developed. This required no reduction in compressor pressure ratio to operate stall free at high referred speeds. Sea level performance was not determined in time for inclusion in the -6 program, but such tests were being performed with the intent of using the C-20 in the -8 model.[1]

* WAGT used model numbers beginning with the letter "C" for compressor designs in the J40 series of engines. Compressors with numbers up to C-23 were mentioned in some reports. The C-23 was a 12 stage design. Details on other designs were not found.

Chapter 3 - J40-WE-6 Mechanical Description
Citation

[1] WAGT Report, May 20, 1951, Report A-1237, "Summary Evaluation Report on the XJ40-WE-6 Jet Engine Model, Navy Contract – NOa(s) – 9212" This report never mentions that the first production engine was delivered more than two years late. **B307VR&CRG72.3.2NACP**

Chapter 4

J40-WE-6 Production and Service

◆━━━━━━━━━━━━━━━━━━━━━━━━━━━━━━━━━━━━━◆

The production of the J40-WE-6 was accomplished through contract NOa(s) 10385 with a total of two XJ40 test and 24 production engines being built at the WAGT Essington, PA facility and some production engines being tested in Kansas City. (Note: *Confusingly, XJ40-WE-6 #2 was re-serialled as WE030003 after the Qualification Test and delivered as part of the 24 engines.) The Essington site was not a true production environment, as will be seen from some of the engine problems that occurred as the -6 was rushed to the airframe companies to begin airframe testing. At the end of the entire J40 program, all -6 engines were at the Philadelphia or Kansas City sites awaiting disposal instructions, no longer of use to BuAer.

The serial number block assigned was WE030001 – WE039999. The first production schedule, set in December 1950, was as follows:

Year	Month	J40-WE-6*	J40-WE-6
1951	February	1	
	March	2	
	April	2	1
	May		5
	June		5
	July		2
	August		
	September		3
	October		2
	November		1
	December		4
1952	January		4

*Prequalification configuration. Balance of the engines to approved final qualification configuration.[1]

From this point on, there was a constant dialogue and changing of schedules as WAGT attempted to also begin to finish development and begin initial production of -8 engines in parallel with the -6 at the Essington site. BuAer was juggling airframe programs to get engines to whomever appeared in the best position to immediately use them. Since all the airframe manufacturers were awaiting engines

(except Douglas for the XF4D, which was proceeding with limited testing using a J-35.), the inconvenience of even small delays caused constant rescheduling of their priorities. The price was to be $114,980 per engine. WAGT had agreed to overhaul all prequalification configuration engines to final production standard at no cost to BuAer.

BuAer set the initial engine distribution in January as follows:[2]

Airframe Manufacturer	Production Sequence
McDonnell	1,2,4,7,10,13, 17
Douglas	3,5,8,9,11,14,15,16
Grumman	6,12,18

On January 3, 1951 Westinghouse sent BuAer a wire informing them they could not meet their schedule and proposed an alternative schedule (the wire has not survived in the files). BuAer responded that the schedule was totally unacceptable and would hold up four critical airframe projects that needed additional engines. A schedule of: July – 3, August - 5 and September – 1 could be accommodated. It is not clear whether these were engines expected earlier in the schedule of December 27 or engines being pulled forward. What is clear is that Westinghouse was struggling. Severe problems not uncovered in the qualification testing were appearing in the production engines.

The allocation to airframe manufacturers was changed again on February 8 as follows:

Airframe Manufacturer	Production Sequence
McDonnell	1,4,5,9,11,14,18
Douglas	3,6,8,10,13,15,16
Grumman	12,18
NACA Cleveland	2,7

In March, the above was changed to ship the number two engine to Douglas and the number three engine to NACA. This change was made as the engine to go to NACA required special instrumentation.

The BAR was authorized to accept J40-WE-6 production engines with the following restrictions:

1. All approved changes in design were incorporated,
2. Each engine was equipped with a verified booster fuel pump,
3. All turbine-out thermocouples were installed and operative, and
4. All other conditions for acceptance were met, with the following exceptions:

a. Power lever torques could reach maximum values not in excess of thirty-one pound-inches on the first six engines, and the first two engines could have standard tubes in the power regulator, provided Westinghouse agreed to back-fit with the finally approved power level torque "fix" and with high-reliability and ruggedized tubes at Westinghouse expense, and

b. The provisions of CID No. 14513 could be incorporated in the first ten engines to provide suitable clearance between compressor diaphragms and rotor.[3]

Operating instructions were issued to the airframe manufacturers in early February, describing the engine and covering normal operating procedures. Ultimately, many of the engines had to be operated on emergency power control on numerous occasions because faults developed in flight.

Engines began to be tested for delivery to the airframe manufacturers in March and immediately began to experience problems due to faulty parts, blade rubs (due to stator vane shifting), thermocouple and thermocouple harness issues that affected the electronic power control, and the electronic power control tubes, among many other things.

A severe setback was experienced to all the careful planning when engine WE010001 destroyed itself on the test bed during green testing due to the compressor experiencing blade rubs between the stator vane shrouds and the compressor blades. The damage was extensive and required a complete review of the compressor changes to date. Blade rubs were occurring in the second stage turbine as well. All new findings were reviewed to decide what was going wrong and what needed to be done.

WAGT's findings and recommendations were:

In general, BuAer should accept de-rated engines in the first batch of 10 until the changes recommended could be incorporated. The de-rated engines would be overhauled later to full rated power at no expense to BuAer. This would allow airframe testing to continue. De-rating was the result of increasing blade tip clearances in the turbine to prevent rubs.

Rubs were occurring due to:

1. Thermal distortion of the compressor housing due to the greater cooling area and rigidity of the bolted compressor flanges.
2. Shaft eccentricity due to uneven cooling at shut down.
3. Distortion of the turbine housing due to insufficient cooling of the outer shroud of the second stage diaphragm.

Corrective actions planned were:

1. The compressor housing components were to be changed to provide ribs of material parallel to the longitudinal axis of the engine and this would increase the cooling effect of the compressor housing and at the same time strengthen it. The added ribs would be approximately the same shape as the bolted flange, and together with the bolted flange, would constitute ribs spaced around the compressor housing at every 45 degrees.
2. Holes in the engine bearing housing and the vibration compressor damper ring would be drilled for retaining pins which were to prevent the vibration damper from rotating. It was expected that the change would further reduce shaft whip due to shaft bowing.
3. Additional cooling air was to be provided directly on the outer shroud at the second stage turbine diaphragm for increased cooling. The change was made in view of the fact that the current method of cooling was by means of air which traveled a longer distance through the turbine housing and was also mixed with some turbine gases.

The first corrective action required a casting change, but the second two were under test and when combined should allow the tip clearances to be reduced and full thrust again attained. Additionally, testing to allow the possible raising of the allowable top turbine inlet temperature above the present allowed 1,425°F would be undertaken.

With the airframe manufacturers desperate for engines, they told BuAer they would accept de-rated engines for early testing and install fully rated engines for any formal acceptance tests.[4]

It is not clear exactly how much thrust was provided by all of the de-rated engines, as it varied from engine to engine for some reason. McDonnell continued to use WE010003 de-rated at 7% of speed and 18% of thrust according to a WAGT memo. Later deliveries in June supplied one engine at full thrust and one de-rated 2% on

RPM and 5% on thrust, and in July one full thrust engine and one de-rated 3% on RPM and 8% on thrust. The serial numbers of these engines is not known.[5]

WAGT report A-1157 was produced in early June to explain in detail how the compressor blade clearances were measured and the clearance set based on thermal clearances, shaft bowing and vibration. The de-rated engines generally had their governor circuits set to run at slower than design maximum RPM and the turbine blades in the second stage had increased clearances as a precaution. Both changes lowered available thrust at full throttle.

The control characteristics of the power control system demonstrated behaviors that made it a difficult engine to "fly". When moving the power control level from one position to another and then back to the first position, the engine rarely returned to the exact same thrust level, varying enough for the pilot to have to move the control again to return to the power desired. In unchanging flight, the thrust of the engine "hunted" up and down as the governor tried to balance the RPM and turbine temperatures, the two functions fighting each other. The two behaviors made precision flying very tiring for the pilot and formation flying all but impossible. In addition, the tubes of the electronic portion of the system aged and changed their response over time and also had a high rate of failure, forcing the pilot to fly using the more limited hydraulic emergency power control. The -8 planned to use a "simplified" hydraulic power control and eliminate the electronic components, so development of the electronic power control was stopped except for efforts to reduce the tendency of tubes to either burn out or break from vibration on the -6.

The rate of development changes still coming out of WAGT to address both known deficiencies at the time of the qualification tests and upgrades to the early engines in the field meant that for the first year no two engines were exactly the same. The airframe companies were helping to develop the supposedly "qualified" engine as well as trying to test their own hardware, a frustrating situation. McDonnell sent information to WAGT on every problem they were encountering on the test stands and WAGT finally sent a long response, presented in its entirety below.

(Interestingly, some of the -6 engines shipped had "dummy" afterburners installed. This allowed the engine to be installed in an airframe designed to take the -8 model, not yet available. The dummy afterburner installation was not part of the qualification testing and no test reports on its functioning and reliability were referenced or found.)

Planned Corrective Actions to Address Early Usage Issues:

"Summary of corrective action taken by Westinghouse on J40-WE-6 engine items reported in engine report No. 2 as encountered during preliminary test stand operation by McDonnell Aircraft Corporation.

A.   Engine
a.   Construction Items
i.   TLN Ignition Plugs
1. Nature: Ignition plugs difficult to maintain.
2. Cause: Plug seats too low in socket requiring a special tool for servicing.
3. Correction: A change will be accomplished to lengthen the barrel of the plug which will bring the nut outside of the combustion chamber outer casing. It is planned that this new ignition plug will be incorporated in the first J40-WE-8 production engine.
b.   Exhaust Nozzle Actuator Linkage
i.   Nature: The exhaust nozzle actuator push-pull rod end interferes with the yoke in the closed nozzle position.
ii.   Cause: The clevis bracket on the yoke does not provide adequate clearance for the end of the exhaust nozzle actuator push-pull rod.
iii.   Correction: The end of the exhaust nozzle actuator push-pull rod was ground to eliminate interference with the clevis bracket on the yoke of the dummy afterburner of engine serial number 030002. This interference was encountered on this dummy afterburner only. All later dummy afterburners are fitted with a wider clevis bracket on the yoke of the dummy afterburner eliminating the need of grinding the rod end. The same interference was encountered when the dummy afterburner of the engine serial 030002 was installed on engine serial 030005 but was corrected by McDonnell by grinding the yoke clevis bracket and removed any further interference should this yoke be used on a later engine.
c.   Exhaust Nozzle Actuator
i.   Nature: Exhaust nozzle actuator push-pull rod interference with exhaust collector blanket in closed nozzle position.
ii.   Cause: Insufficient clearance provided for exhaust collector blanket.
iii.   Correction: The J40-WE-8 and -10 engines provide clearance for an exhaust collector blanket, but the J40-WE-6 engine was not specifically designed for blanket use. To provide clearance for a blanket on the J40-WE-6 engine would require a basic engine change. This change would involve a redesign of the yoke and connecting linkage or repositioning of the exhaust nozzle actuator itself. It is believed that the more desirable course of action is to custom fit the blanket to the present J40-WE-6 design.

d.   Emergency Fuel Control Adjustment Cap
i.   Nature: Small fuel leak in emergency fuel control adjustment cap.
ii.   Cause: Improper finish and alignment of parts.
iii.   Correction: McDonnell replacement is adequate. A new acceleration control valve housing has been designed which incorporates a finish that will eliminate leaking in the area of the fuel control adjustment cap. All engines in the field will be back fitted with this new housing at overhaul.
e.   Oil Reservoir
i.   Nature: A leak was found in the oil reservoir.
ii.   Correction: See statement of corrective action forwarded to McDonnell under paragraph 1g of Westinghouse Letter W2539G dated August 29, 1951. (Letter not located – Auth.)
f.   Oil Reservoir
i.   See paragraph 1e above.
g.   Afterburner Diffuser
i.   Nature: Cracks in afterburner diffuser at connection to afterburner.
ii.   Cause: Undetermined
iii.   Correction: Hairline cracks in the dummy afterburner diffuser aft flange seam weld area were stop drilled. Tests are being conducted to improve seam welding techniques in an effort to eliminate cracks in the seam weld areas.
h.   Oil Reservoir
i.   See paragraph 1e above.
B.   Control System Malfunctions
a.   Acceleration Control
i.   Nature: Compressor stalls occurred during rapid acceleration.
ii.   Cause: Acceleration rate too rapid causing compressor stall
iii.   Correction: This condition was present at all times before Westinghouse installed an adjustable acceleration control which allows adjustment to acceleration rates below the compressor stall region. All engines will be equipped with an acceleration control, either by back-fitting engines in the field or by installation of new engines. (This control, if utilized, was anticipated to add 4-5 seconds to the acceleration to full throttle below 10,000 ft altitude. Actual flight tests later showed it improved accelerations times by 6-8 seconds.)[6]
b.   Exhaust Nozzle Actuator
i.   Nature: Exhaust nozzle operation sluggish.
ii.   Cause: Foreign matter in the hydraulic system entered the actuator and it gradually became inoperative.
iii.   See statement of corrective action forwarded to McDonnell under paragraph 2a of Westinghouse letter W2539G dated August 29, 1951. (Not located. However, a filter was installed in the hydraulic circuit to the actuator after this time to prevent this problem. Auth.)
c.   Exhaust Nozzle Actuator
i.   Nature: Inadequate range of adjustment provided on exhaust nozzle actuator.
ii.   See statement of corrective action forwarded to McDonnell under paragraph 2j of Westinghouse letter W2539G dated August 29, 1951. (Not located – Auth.)

d.   Engine would not accelerate from minimum fuel flow. McDonnell responsibility, no Westinghouse action necessary.
e.   Thermocouple Leads
i.   Nature: Engine shutdown from 65% rpm steady operation and would not start again.
ii.   Cause: Thermocouple lead shorted to ground in engine junction box.
iii.   Correction: Westinghouse has notified supplier to tighten manufacturing control in an attempt to prevent a recurrence of this item. This condition was probably caused by improper install of the chafing sleeve around the cables.
f.   Power Scheduler
i.   Nature: Pins in the electrical connector to the power scheduler were found pushed in.
ii.   Cause: The pins in these electrical connectors are set in rubber and can be pushed in if enough force is applied.
iii.   Correction: No change in electrical connector design can be effected at present because a better connector has not yet been proven. Other designs of connectors are being investigated in an effort to find an improved design.
g.   Exhaust Nozzle closed at 50% rpm on primary control system and would not open. McDonnell responsibility, no Westinghouse action necessary.
h.   Thermocouple Leads – See paragraph 2e above.
i.   Thermocouple leads shorted in extension. McDonnell responsibility, no Westinghouse action necessary.
j.   Exhaust Nozzle Actuator
i.   Nature: Exhaust Nozzle Actuator failed to operate on the emergency control system.
ii.   Correction: See statement of corrective action forwarded to McDonnell in paragraph 2a of Westinghouse letter W2539G dated August 29, 1951. (Not located – Auth.)
C.   In reference to two items mentioned in the test of Mr. Baldwin's letter:
a.   There is at present no program for the elimination of manual adjustment for stall prevention (referencing the compressor stall control above). The J40-WE-6 engine, of which only a limited number (23) are being built, is an interim engine to be used in experimental aircraft until the J40-WE-8 engine becomes available. This later engine will not use the electronic control system, but will use a simple hydraulic control system instead. In view of the small number of J40-WE-6 engines to use the electronic control, Westinghouse did not plan to use an automatic stall limiter in place of the manual adjustment.
b.   The information on H-2 hydrolube was recognized as an item requiring investigation. Tests would be conducted in the laboratory and on the engine to determine the full effect of hydrolube on the electrical harnesses of the engine. No decision of action to be taken would be made until the results of these tests had been obtained and analyzed. As these hydrolube tests progress, McDonnell was to be kept advised of results."[7]

It was not only shipped engines that were showing problems. The BuAer supplied AiResearch starters gave trouble on Serials WE030008 and WE030011, exploding during startup and damaging the engines. WAGT was able to report that it had two starters of that particular type left for testing. The failed starters were returned to AiResearch for analysis and a detailed service instruction list was sent out from WAGT to the service reps in the field on proper operation. The exact cause of the failures was not known but was under investigation.[8] Two more starters failed during testing, one at Essington and another in Kansas City and on both occasions the engines were still turning over at low RPM when the starters were engaged. A broken pawl on one starter strongly suggested that re-engagement of the starters while the engine was turning was the basic problem. It was suggested to BuAer that starters that could re-engage the engine while the engine was still turning was a desirable feature and should be considered.[9]

Two of the first production engines WE030003 and WE030010) were sent to NACA in Cleveland and tested in their high altitude wind tunnel. These test results and those from McDonnell from actual flight tests were analyzed. They showed that the engine had compressor stalling problems if the total compressor ratio exceeded 5.4 in certain conditions. As a result, WAGT sent BuAer a memo placing certain operation restrictions on the engine:

1. Military rated turbine outlet temperature permitted at a compressor inlet total temperature of plus twenty degrees Fahrenheit and above.

2. Military rated turbine outlet temperature minus fifty degrees Fahrenheit permitted at a compressor inlet total temperature no lower than plus ten degrees Fahrenheit.

3. Normal rated turbine outlet temperature permitted at a compressor inlet total temperature not lower than minus five degrees Fahrenheit.

4. The maximum altitude permitted with these limits is forty thousand feet.

5. An alternate suggested limitation is to observe compressor total pressure ratio and limit this ratio to a value of 5.4.[10]

The unapproved Service Instructions manual, dated December 15, 1951, was released to BuAer prior to its formal approval, for information purposes only. This was required by the engine users to gain some knowledge of the engine in anticipation of the broader group of personnel now being drawn into the flight test activities.

The number one bearing was reported to leak oil at shutdown. (*In the XF3H-1, which had a significant tail down attitude when parked.*) It was suspected that the oil scavenge pump was shutting down slightly before the oil feed pump,

leaving as much as .3 pints of oil in the bearing This would overlap the bottom of the bearing seal, which was not a positive seal and had a slight leakage path through it. The oil could leak into the engine after shutdown. Tests were run to verify this analysis but those tests on three engines could not duplicate the problem. As a precaution, a sump was planned to be added to the bearing housing to contain slightly over .3 pints of oil to prevent the leak.[11]

During development, WAGT had upgraded the compressor blades to 12% chromium steel from the original 18% chromium 8% nickel steel to obtain increased

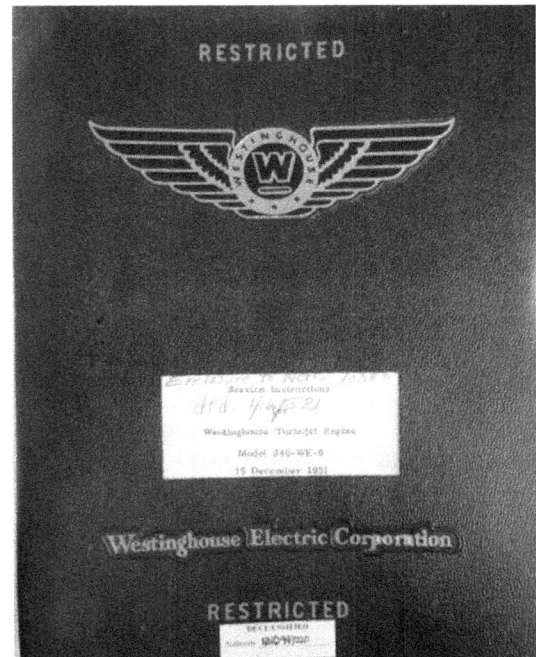

Figure 1 Cover of the J40-WE-6 Service Instructions Manual, *National Archives*

improved yield strength and damping qualities. During overhaul of WE030005, it was discovered that one sixth stage compressor blade was of the old material. In spite of making a case that there was little risk of a failure in the existing engines should others contain the correct blades (with the possible exception of WE030010 at NACA), WAGT had to recommend that every engine be limited to ground running only until every blade in every engine was checked. The check involved removing the top compressor housing and testing each blade with a magnet. The 12% chromium steel was magnetic and 18/8 steel was not. After this report, immediate action was taken to verify proper blade installation. The report does not discuss how the older blade got into the engine undetected.[12]

A full teardown report on WE030005 was issued with a recommendation for the initial overhaul periods for the -6 in general. The engine had run 54.49 hours of which 5.97 hours were at military power.

Some cracking of one inlet guide vane and some second stage nozzle vanes was noted. The combustion chamber liner had some buckling on the second step of the inner wall and was repaired. Eleven compressor blades were replaced due to nicks from FOD. A new style diffuser was installed. All other parts were in good condition.

Based on the above, WAGT recommended 100 hour overhaul periods on the next 3 engines of which not more than 15 hours would be at military power. Engines with the old diffuser installed (WE030006, WE030007, WE030008) should be bore scope inspected for signs of fatigue failure at 50 and 75 hours.

Any engine reaching 150 hours should have the thrust bearing replaced, regardless of last overhaul period duration.[13]

The teardown of WE030005 had been instigated to investigate the cause of rough cutovers from the primary to the emergency power control. A separate report on this condition was issued by WAGT (A-1286). The incident triggering the investigation stated:

"On 10/8/51, during flight testing on engine 030005 in the McDonnell XF3H at Pratt, Kansas, a switchover from the Primary to the Emergency control was made at 10,000 ft. This resulted in very rough engine operation accompanied by a loud report, and the pilot in the chase plane reported that an 80 foot flame came from the exhaust nozzle. This had also occurred on a previous flight.

"It was decided to continue the flight tests, but no more primary to emergency switchovers were made. At the end of 50 hours operation, the engine was returned to Westinghouse for overhaul and investigation of this switchover phenomenon."

The investigation showed the Emergency Fuel regulator had a small amount of Permatex gasket cement that had entered into the pawl stem low pressure end, binding it fast in its liner. This meant that the pawl, which served the purpose of transmitting the motion of the altitude bellows to a servo valve which reduced fuel available as the atmospheric pressure decreased at altitude, could not function. The result was a large amount of fuel available at cutovers in excess of requirements. The reason for the Permatex cement was assumed to have been to stop a fuel leak, but why it was used in violation of established procedures and why it was not caught prior to installation on the engine was not explained. The engineering drawings for the part were modified via CID 14937 to eliminate bolt holes from being drilled into a cavity that might have caused the leak. All engines in the field and at Essington were checked for the condition and none were found affected.

The rough running on the primary control was determined to have been caused by the Fuel Valve Motor becoming saturated with fuel, giving it a very low armature resistance and causing it to behave erratically. Changes were made to move a vent to prevent fuel from entering the motor. Clearances were loosed by .030" on the throttle valve line to reduce possible binding. The Fuel Actuator was replaced and CID 14720 was issued to change the drawings relating to the drill depth of the bolt holes.[14]

Parts Shortages for Production

Parts shortages were brought to the attention of BuAer in January and at a conference, WAGT requested BuAer's assistance in obtaining materials to produce the parts. BuAer initiated the "BRICKBAT" program, forcing all vendors supplying raw materials or finished components to verify that the purchase orders placed with them would be delivered on time or supply specific reason why not. This program included all models of the J40. The Korean War found suppliers with greatly increased orders for limited and rare materials and coming under pressure to react, with each government function arguing their orders took precedence over the others. Possibly the most embarrassing finding was that two critical alloys comprising 15-17% of the metal in the engines, "Discalloy" and "Refractaloy", had been specified for many of the parts external to the engine requiring resistance to heat greater than Iconal alloys could provide. By June the BRICKBAT program uncovered many vendors awaiting supplies of either or both of these raw materials to enable them to produce their components. It was becoming a critical situation. It turned out that both alloys were a product of another division of Westinghouse in East Pittsburg, a laboratory being the production facility limited to production of only one of the alloys at a time. They were the only source for the metals. A second source for Refractaloy (Universal Cyclops Corporation) was initiated once this was uncovered, but the record shows that WAGT never brought this to the attention of BuAer, BuAer finding out this situation through their investigations into the shortages.[15]

Manufacturing Errors / Inventory Control Problems

Poor manufacturing and inventory controls were the cause of the -6 engines being produced having two different versions of approved model power regulators. Testing had shown that a modified version of the power control used on the J34-WE-32 and -38 engines was required to prevent compressor stalls between 50 and 75% RPM when a throttle slam was initiated by the pilot. The changes resulted in the "-16" power regulator version being designed. However, for some reason not explained, not all "-16" power regulators had the necessary modified circuitry built into them. WAGT had all regulators inspected both on existing engines and in the inventory to determine which lacked the necessary modifications. All those found with incorrect circuitry had to be removed from the engines and returned to WAGT for modification at no cost to BuAer. It was planned to replace all "-16" regulators with the "-15" model upon overhaul. The reason for that change was not explained, but it appears that the "-16" regulator was actually a "-12" regulator with circuitry modified after manufacture, whereas the "-15" would be manufactured with the correct circuitry in the first instance.[16] It was left to BuAer to determine which regulators on which engines would be changed out first. It would be as late as March before all the erroneous regulators were returned, modified and reshipped to the airframe manufacturers to be reinstalled on the engines affected. All -6's from WE030017 on had the -15 regulator. The prior serial number engines had the -16 regulators replaced with the -15 model on overhaul.

BuAer approved the change process for the regulators, but later questioned whether a change to the fuel valve control sensitivity would not have produced the same result by decreasing the sensitivity of the circuit, thus reducing the requirement for a separate acceleration control. WAGT responded in detail with an explanation showing their reasons for leaving the fuel valve control as it was and using the acceleration control solution. They also pointed out that there was little likelihood of getting statistical data on the fuel valve performance from actual air testing, and, as the control was only going to be used on the -6 and not on the follow-on models, it made little sense to invest the resources to do such a study.[17]

WAGT had recommended that voltage regulators installed on the -6 be removed as it was felt they were not needed. BuAer followed up and found that WAGT had never forwarded the CID's necessary to effect this change to BuAer for approval and no action had actually been taken. A memo agreeing with WAGT's position confirmed that flight testing had shown the regulators were not needed on the -6. No actual CID was referenced or located. It is not clear if the regulators were ever actually removed.[18]

NACA Detailed Test Results

The rigorous testing of the -6 engines in the Lewis altitude test cells of the NACA Cleveland facility turned up the same surge problem in the compressor that McDonnell was observing in actual test flights, namely a severe surge condition at high equivalent engine speeds. Westinghouse did not have the ability to provide the cold air required for

high equivalent engine speeds at their facilities, hence the problem had gone undetected until flight test began. The findings were that the simple radial equilibrium requirements were not satisfied in the inlet and exit stages. A basic redesign to satisfy these requirements called for changes in the rotor-blade section angles. Westinghouse had already come to the same conclusion and had a partially completed engine with re-slotted rotor discs available for later testing to determine how to quickly eliminate the surge in the engine with both the easiest and quickest changes necessary. Blade angles, stator blade angles and exhaust nozzle outlet area changes were investigated.

The original design of the compressor had a rotor tip diameter of 32.14 inches and an inlet hub-tip ratio of 0.6. It was initially designed to handle a specific air weight flow of 24.2 lbs per second per square foot at a pressure ratio of 4.6:1 and a shaft speed of 7,260 rpm. Mismatching between the compressor and turbine was encountered at the start of the development program and the compressor stator and rotor blades were reset to shift the design point for the original production engine to a pressure ratio of 5:1 and the airflow weight to 24.8 lbs per second per square foot (140 lbs/sec total). The first rotor tip section was stalling and the hub section was approaching a turbining condition at design speed. The surge restriction placed a thrust reduction of almost 20% on the engine when operating within the surge zone. When the initial compressor design point was shifted to a higher pressure ratio and weight flow, the margin that existed between the high-speed stable operating line and the surge limit was apparently eliminated.

In the redesign, the blade-angle settings in the first rotor were changed so that the hub section was opened approximately 3 degrees and the tip section was closed approximately 4 degrees. This loaded the hub section and unloaded the mean and tip sections. The second stage rotor was twisted closed at the tip approximately 4 degrees so that this section was unloaded while the hub-section angle remained unaltered. These changes satisfied the design specification. (These extensive reports also covered many other changes tried on the engine to improve surge resistance, fuel economy and thrust. This information was used in modifying the -8 design compressor during development.)[19, 20]

As the NACA investigation data came in, Westinghouse decided to take the fastest path to getting changes into the engines to remove the surge problem, but at the expense of a loss of performance and increased fuel consumption. This trade-off had to be made because there was no time to redesign the compressor on what was now understood to be a low production volume interim engine. Specifically, WAGT recommended:

1. Open the first stage turbine nozzle gauging by approximately 16 percent.

2. Install a compressor outlet air mixer in place of the second stage compressor outlet guide vanes to better eliminate temperature inversion in the combustion chamber.

3. The compressor would be limited to just under a 5.6 pressure ratio, the NACA determined limit of the compressor. If an appreciable amount of surge was encountered at the extreme operating conditions, a further reduction of engine pressure ratio could be accomplished by a change of the first stage turbine rotor, but such a change was not anticipated to be necessary.

These changes in the -6 would result in a thrust loss of approximately two percent and an increase in specific fuel consumption of slightly over two percent at military rating. The exact amount would be determined after a number of engines incorporating the changes had been tested. CID's incorporating the changes would be issued immediately. Investigation with NACA was ongoing relating to the -8 engine to remove the performance restriction on that engine, which shared many of the same designed components as the -6.[21]

Figure 2 J40-WE-6 First production engine. WAGT P43566, 11/20/50. *Courtesy Hagley Museum and Library*

Continuing Usage Issues

Douglas was reporting corrosion of the rear bearing support assembly and WAGT responded saying the alloy of the support assembly was chosen for its superior strength in high temperature conditions. They had observed the corrosion on their test engines and were working to find a method to control it. Various painting schemes had been tried and had failed. Metalizing the surface with aluminum metal sprayed onto a sand blasted surface was currently being investigated. Research would continue in light of both the -8 development program and the other higher thrust models under development.[22]

Engine WE030004 was returned to WAGT with hot spot damage in the engine. The hot spots had burned out some first stage turbine nozzles. The damage was discovered during field maintenance to replace the separate segment second stage turbine nozzles with the newer interlinked model in an attempt to correct a shrinking clearance between those nozzles and the front of the second stage turbine on this engine. The outer annulus of the combustion chamber had circular burned out sections and minor cracks were found directly in line with the burned first stage nozzles. The trailing edge of some nozzles had .5 inch cracks. The clearance problem appeared to have been caused by the hot spot conditions. The engine had 58 hours of operations at McDonnell since its last overhaul, 11 of which were flight hours. WAGT promised a full report when the engine was torn down.[23] The promised report never materialized, as WAGT finally responded with a memo in October saying that since they were overhauling the engine and replacing the compressor outlet mixer with the new design, it would not be possible to identify a single specific cause for the hot spots. Teardown had revealed the diffuser flow passages to be eccentric in that the entry area varied greatly around the circumference of the diffuser. This condition was known to cause adverse temperature distribution problems in the past. The largest contributor was the outlet guide vanes. Small variations in manufacture of this part were known to have a large effect on pressure distribution and consequently on temperature distribution. When the engine had been built and tested, it had shown it was at the limit of acceptable temperature hot spots. Since then, because of the subsequent damage found in WE030004, the allowable peak temperature had been reduced. The memo went on to explain that simply disassembling and then reassembling an engine with the same parts would produce a new variation of temperature distribution and this was not predictable. No engine with the new compressor outlet mixer had shown excessive hot spots.[24]

Blade rubs were the subject again when it was found that the stator vane diaphragm retaining screws were becoming loose. Observation of the screw, washer, compressor housing and diaphragm outer shroud showed the magnesium of the housing was crushed or abraded in the vicinity of the diaphragm retaining screw hole, and the washer under the retaining screw was deformed into a roughly conical shape. These things reduced the tension established by the proper torqueing of the screw during assembly. The diaphragm was then allowed some slight freedom of motion. To fix this, a 1/16" steel washer was inserted on the magnesium to prevent crushing and abrading. The original washer was increased in thickness to 3/32" and the inside diameter decreased to be a tight fit around the retaining screw and the outside diameter was increased. These changes would prevent "coning" on the washer and still allow the necessary play to accommodate thermal expansion and contraction of the engine. A program was recommended to reassemble WE030008 and test the torque after 5 hours of running and then if the torque was within limits, clear the engine for an overhaul limit of 100 hours. On new engines, (WE030017 being the first), the changes would be incorporated during manufacture and the torque rechecked every 30 hours of operation in the field. On existing engines, a check every 30 hours would be done and the newly designed parts incorporated at the next overhaul. The WAGT memo closed by pointing out that diaphragm distortion was a problem of engine assembly and resulted from residual stresses remaining in the diaphragm after construction. A program was in place to reduce the stresses. Any further progress in reducing these residual stresses would have a beneficial effect on the diaphragm axial clearances.[25] BuAer accepted

the changes and recommendations. A later memo discussed the manufacturing issues related to unreleased stresses in the diaphragm and the process changes to bring these under control for any diaphragm cleared to be used in manufacture.[26] This process showed that WAGT was still twisting stator vanes to the proper angle, sometimes multiple times during the annealing process, the stress created needing to be released. Engine logs would indicate which process was used on specific -6 and -8 engines so that replacement parts would match the same process. This was verbally approved.

During the investigation of engine flameouts (blowouts in WAGT speak), Grumman suggested that the engine speed be limited on the primary control system on the lower end during flight to 75% in lieu of Westinghouse's suggested fix of internal compressor changes. The Grumman solution would eliminate accelerations between 50 and 75% engine RPM, the range where most flameouts were occurring. The reduction in RPM control reduced thrust control in flight from a normal 6 to 100% to 34 to 100%. The change was to add a speed potentiometer to the primary control system to override the throttle setting in flight any time the throttle was reduced below 75% engine rpm. Tests showed it worked smoothly on both a test J34 and an in-house J40-WE-6, with no flameouts occurring during the testing. The WAGT report recommended it as an alternative to the WAGT proposed fix and that Grumman could use it, providing they supplied all the parts except for the throttle position switch. Apparently this approach was never used by anyone other than possibly Grumman.[27]

The pressure for more engines was unrelenting and an effort to improve delivery schedules was made with the Navy requesting that WAGT assemble 12 engines at Essington and have Kansas City do the production acceptance testing. The outcome of this request is not known.[28]

March found WAGT recommending to BuAer that the current flight restrictions on the -6 be removed for any engine overhauled that incorporated the two fixes identified during NACA testing, these eliminating the temperature inversion problem in the engine. First was CID 19119 (engine Notice 29) that changed the first stage turbine nozzle by opening it 16% greater than the prior nozzle. The second was COD 19121 (engine Notice 27) that replaced the outlet guide vane with a mixer unit of one row of guide vanes followed by a row of mixing vanes. These changes allowed the engine to operate up to guaranteed speeds and altitude (60,000 ft.). Both changes had to be installed by WAGT at Essington and were not field installable.[29] When WE030014 was production tested in Kansas City in April, the CID's were incorporated but as the BAR noted and although the engine met military power specifications, it did not meet those at lower powers and was out of specification on SFC's at all power ratings. He asked BuAer to accept the engine in view of the fact that McDonnell had immediate need of the engine.[30] The restrictions were lifted via telegram on May 6, noting that one engine had non-conforming parts but would be upgraded to the approved parts at its first overhaul.[31]

Field maintenance of the engine had to begin including an oil spray into the engine after shut down to coat the interior parts to prevent corrosion, similar to the J34. Blade rubs were abrading the seals and lining of the compressor leaving the exposed parts open to corrosion.[32]

Number One Bearing Issues

Afterburner development on the -8/-1/-22 had uncovered the fact that the original No. 1 bearing design could not stand up to the thrust loads imposed during A/B light-offs, the over pressure causing the bearing to wear far faster than anticipated in the initial design. An improved bearing was designed and WAGT planned to replace the -6 bearings at the planned 150 hour overhaul (and for the first nine -8 engines as well). In-flight testing of the actual thrust loadings of the bearings was needed and the new bearing design, anticipated to have an operational life 30 times that of the original, was designed using data extrapolated from sea level, static engine data.[33] (In fact the original -6 No. 1 bearings were perfectly suitable since it did not have an afterburner, but BuAer requested they be replaced at overhaul.)

Changing the Ignition System

WAGT had qualified the -6 using a high tension ignition system (TBN), but this was shown to be unreliable for engine restarts above 35,000 feet altitude. A replacement low voltage high energy (TLN10) system replaced the TBN system on all -6 engines after WE030016 (Engine Change 65), but it required a modification to the combustion liner to prevent interference with the spark plugs. Some field engines had been locally modified with the TLN system without the combustion chamber modification and experienced very rapid spark plug wear and failure (6 plugs in 60 hours of operation at McDonnell). In a long memo, WAGT explained the need for an adapter to be installed on the field engines if the TLN system was to be used without the combustion liner modification. BuAer complained that no CID had ever been issued for this adapter and officially it did not exist, citing it as yet another example of WAGT not following procedures for engineering changes to the engines. WAGT's long-winded explanation of the need for the combustion chamber modification to eliminate interference with the spark plugs using the TLN system in

engines prior to WE030017 also admitted that as long as the spark plugs were cleaned and dried between flights, the unofficial adapter appeared to work equally well in eliminating short spark plug life. All engines up to WE030016 had the TLN system eventually installed as part of overhaul along with the proper combustion chamber modification.[34]

WAGT wrote a response clarifying the sequence of events in an effort to refute BuAer's statement that they were going around official procedures by introducing the adapter use. They pointed out that a request to use an adapter had been submitted in April to BuAer, but that none had actually been installed in field engines without approval. BuAer responded with appreciation for the clarifying summary, approved the off-set adapter design and requested a CID be submitted for formal approval and installation.[35] Eventually, four engines (WE030007, WE030009, WE030011, WE030012) were identified that might require the interim adapter and permission was given to install it if the TLN10 ignition was to be field installed on those engines.[36]

More Production Quality Issues

The string of quality issues with production components continued, with WAGT having to inform BuAer that some engines delivered had compressor discs in the 6th and 7th positions that had unverifiable physical hardness properties due to the findings that the lab tests on the materials had no relationship to the actual materials of the discs in question. Also, one lab test on one disc in the 7th stage position had been lost. The engines had to be grounded until the discs could be retested and replaced if necessary. All of the five engines in question were either at NACA or at the airframe companies. All other compressor discs in the engines were verifiable and not involved. WAGT recommended the engines be restricted from further operation until the discs were properly verified as to hardness properties and possibly replaced if they failed. Two engines had already been tested and no problems were found, so they could be safely used. All discs in inventory in Kansas City and Essington would be re-verified.[37] The completed investigation in late June showed several discs were unaccounted for and were finally traced to the NACA test engine WE030003 in stages 1, 6 and 7. They had run 400 hours in those positions without problem and WAGT recommended to BuAer that testing of those discs was not necessary given their satisfactory performance and the upcoming overhaul of this engine at the end of NACA testing.[38]

Discrepancies between test equipment and procedures between the Kansas City and Essington locations resulted in a memo to BuAer asking that the procedures and results used be approved, even if they were different. This memo has a handwritten note attached that reads: *"J40-6 spec. does specify the AN/APR 4 receiver for radio interference testing. Therefore, Westinghouse procedure outlined herein considered satisfactory. Copy retained. PP-413. Proper liaison between Essington and Kansas City would have prevented this situation. If WEC has this poor liaison and cooperation between their own plants, what will it be between WEC and licenses?"* [39] *(The latter refers to Ford, under contract at that point to be a second production source for the -10.)* However, BuAer did authorize the use of other equipment and supplied it to WAGT at both locations, an AN/UMR28 receiver in lieu of the specified TS/277AR receiver.

Engine WE030009 required compressor rework (the reason not specified) and it was field repaired. (It was revealed at the congressional hearings much later that engine FODing was a problem at McDonnell during this period.) The incident revealed that field rework guidelines and limits had not been established, so the specifications for the J34 were substituted and BuAer agreed to the interim use of those specifications. A service bulletin specifying the J34 Bulletin 165 and Handbook of Service Instructions AN02-110-BA-2 and AN02B-110BC-2 be used in the interim was issued.[40] Any replacement of the first four stage compressor blades required an engine to be returned to WAGT, as the entire compressor assembly had to be broken down to allow enough space to get the old blades out and new ones inserted. The other stages could have a limited number of blades replaced, but complete rebalancing of the compressor/turbine assembly shaft had to be done at a WAGT facility.

Further NACA testing had revealed that exhaust nozzle areas had a much greater effect on the performance of turbojets than had been previously recognized. WAGT's report A-688 was the only previous report dealing with the exhaust nozzle area calculations that the Navy had seen. Based on their findings, NACA recommended BuAer consider switching the -6 to a petal type variable area exhaust nozzle from the eyelid type. Actual development of the -6 (except for engineering changes required to allow the engines to operate safely) had been halted by this time, so nothing came of this recommendation. It was noted for the later models, however. This was a good example of new basic research revealing that Westinghouse's fundamental base of knowledge was not keeping up with the industry.[41]

May found the Navy requesting confirmation of WAGT compliance with their directive that the automatic ignition features in all J40's be removed. In all -6's, this was complied with when the "-15" power regulators replaced the "-16" model on any engine that originally incorporated the "-16" version. All other models of the J40 did not have an automatic ignition feature. Automatic ignition restarted

the engine without pilot action after a flameout. Removal included adding a switch to the cockpit requiring the pilot initiate ignition manually in the event.[42]

June found WE030014 suffering an apparent compressor stall during flight and a failure of both the accessory gearbox and starter. Two jolts and two balls of fire were observed by the chase plane, followed by rough running and a steady rise in the No. 2 bearing temperature. The plane landed safely, but the engine quit as soon as the throttle was retracted to the idle position. Inspection showed the air turbine starter had broken off at the neck down portion of its mounting flange and was resting on top of the engine oil reservoir. The starter flange was still attached to the gearbox. The oil in the gearbox was extremely contaminated and contained ball bearings, two bolt heads, a large number of metal chips, parts of a bearing race, and a number of steel rods approximately .5 inch long and .032 inch diameter. The main oil pump filter was found to be completely loaded with metal particles. The oil sump was full except for a quarter quart used during the flight. It was believed the temperature rise at the number two bearing was caused by the filter becoming clogged with metal particles.[43] WAGT followed up the initial report stating that although the engine incorporated the 16% opened first stage turbine and had been expected to be stall free, NACA had not exactly duplicated all flight conditions such as ram, altitude, etc. and might still have compressor stalls under similar flight conditions to the one of the flight when WE030014 suffered its failure. Flight limitations might have to be established (again) on -6 engines incorporating the 16% opened nozzle. Contractors would need to approach the flight conditions in question with care until further testing was complete.[44] Further investigation by WAGT demonstrated that by putting the power regulator from WE030014 on one of the NACA engines they caused exactly the same surges at the same aerodynamic point where the aircraft had encountered similar problems. A short circuit between the power control and the engine was suspected, as changes to that cable to prevent short circuits from occurring appeared to eliminate the surging. No further surges could be induced on the engine after the changes were made. WAGT advised BuAer that they could remove their warning to the engine operators regarding caution approaching the aerodynamic conditions in which WE030014 had encountered a problem. BuAer, in a written note, seemed unconvinced by the argument that the surge was electrically caused and noted that the surges were always occurring at the same aerodynamic point of operation.[45] The engine was finally repaired at Essington, re-tested in Kansas City and the BAR requested shipping instruction from BuAer on August 22.[46]

Debate on the cause of the WE030014 failure continued after WAGT issued Report A-1419 naming the primary cause of the engine failure as the failure of the starter motor.[47] BuAer took exception, pointing out that the physical evidence indicated that the starter might have experienced high "g" accelerations beyond its guaranteed limit of 25g's, as similar starters in other aircraft had been breaking due to that problem. Also, the physical evidence showed the starter was not being driven by the gears in the gearbox at the time of failure. BuAer felt strongly the primary cause of the failure of the starter and gearbox was the strong engine surges experienced and wanted WAGT to explore the cause of that and respond.[48] WAGT wrote a memo exploring the possibility that vapor lock might have occurred due to the altitude, slow running of the engine (the fuel pump operating below full speed) and the large amount of fuel burned on the mission in the fuel tank, lowering the head pressure. This memo outlined an investigation approach to try to isolate this if a problem under these circumstances existed.[49] Aside from determining just what had occurred, the findings would decide who would have to pay for the engine's repair in the final analysis. WAGT pursued this further in a memo covering all the laboratory tests of the fuel pump, including test cell testing of the flameout point of the engine at various throttle settings as the pump input pressure was reduced. In all such tests, the pump met the specification. WAGT suggested the only way to settle the fuel pressure issue (i.e. vapor lock) was to test the engine in the airframe and they desired McDonnell to carry out such tests, outlining the test protocols for McDonnell.[50] In response to WAGT report A-1419 on the damage to WE030014, BuAer reported that they had knowledge of identical starters on other types of aircraft due to the starter experiencing excessive g-loads above its design specification.[51] WAGT in a further memo discussed all the possibilities mentioned in report A-1419 and eliminated the starter as a prime cause of engine failure, concluding that the problem must have occurred because of engine stalling and efforts needed to be put forth to eliminate stalls.[52]

Alternate fuels had been studied and a report on compatibility showed similar performance of aviation fuel, JP3 and JP4(test batches), with the following exceptions: lead buildup in the combustion area should be expected using aviation fuel but should not be an issue if the regular inspection schedule was followed; the combustor life would be shortened if aviation fuel was used; JP4 was proving incompatible with the rubber in Aeroquip-302A fuel hoses in some tests, leading to particles in the fuel system, but further testing was required to resolve contradictory results between tests. WAGT advised BuAer not to allow JP4 to be used in any -6 engine intended for flight use until the situation was clarified.[53] The use of 100/130 aviation fuel in lieu of 91/96 was approved if 91/96 was not available. JP4 was not authorized to be used in the engines. Also, BuAer

noted the similarity between JP3 and JP4 and requested data on whether hoses might be damaged by JP3 as well. Rubber particles were found in the various systems from time to time and their source was never completely identified.[54] WAGT responded in August that JP4 was damaging the Aeroquip-302A hoses on their test engine with definite deterioration having been observed. They were switching to the Aeroquip-302-02 hoses, which were far more resistant to fuels. The hoses would be changed out at overhauls on existing -6 engines or in the field using kits for the purpose as hoses became available. The -8 engine had never been tested with avgas grade fuels and it was expected that clogging of the afterburner ring orifices could be expected. Use of such fuels was not advised but if such fuel was used, field observations on clogging were requested to be forwarded. (Although the memo subject is specific to the -6, the memo also includes information on the -8 then in development and about to attempt its qualification tests. At this point in time, many of the documents include content with more and more overlap across contracts and models, making the tracing of activities for a given model at any point in time difficult. At the time the contracts were underway, this must have been a challenge to all involved as well.)[55] BuAer followed up in September asking for details on Aeroquip-301-02 testing with JP4 (in general and if used in the -8 acceptance test) and when the CID could be expected along with detailed engine starting instructions if using alternative fuels.[56]

Anti-icing testing was being planned by BuAer and discussions were underway to identify -6 engines that might be used. The two engines at NACA or the house test engine WE030002 were identified as candidates, with the NACA engines the best bet, requiring overhaul before use and addition of special test sensors specific for the icing test.[57] In September, house engine WE030003 was selected, as there were problems using the NACA engines, one having excessive running time on it and requiring a major overhaul and the other lacking a power control unit, it having been removed to keep the other engine operating.[58]

Flight testing brought increasing concerns about the bursting strength of the compressor and the need for some sort of warning system in the cockpit to allow the pilot to avoid dangerous over-pressures on the engine during flight. In the -6, an addition to the electronic control system that measured a differential between the compressor outlet total pressure and compressor inlet total pressure was under consideration. The simplified control system (for later models) would sense the compressor outlet total absolute pressure to trigger a limiting device. BuAer was requesting some type of cockpit warnings be provided for the various models, including the specification that the pilot should be able to safely operate the engine within the specification powers at all times with adequate warnings of an approaching over pressure condition.[59]

Improvements in field support for the -6 and later models were needed, particularly the ability to do field repairs to the number one compressor disc and blades which were most likely to be damaged by bird strikes, ice, or foreign object damage of any kind. As noted earlier, the compressor design did not allow blade changes on the first four compressor discs without removing the entire compressor and turbine shaft assembly from the engine and disassembling the compressor, a task which could only be accomplished at a WAGT facility in Essington or Kansas City. BuAer requested WAGT explore a way to remove and rebalance the number one compressor disc, replace blades and/or the disc, and rebalance the disc for reinstallation in the field. CID 9021 was generated for Tool 243778 S1 and 243623 S1 and CID 9022 for Tool 243706 S1 and a request made to build sample tools. As with many of these smaller type changes, tracking implementation after the tools were produced has not proven to be possible with the remaining documents.[60] This subject was pursued with WAGT by asking what it would take to modify later J40 designs to be a three bearing engine to allow field operating units additional major repair capabilities. WAGT argued strongly to retain the two bearing design for a number of reasons, pointing out that actual disc replacement in an engine due to damage or other cause was very rare, the circumstances requiring this action on test engines having been resolved through re-design of components, with no recurrence of the problems afterward.[61]

The wandering of temperature along with thrust pulsations on the engines found Douglas requesting permission to set the exhaust nozzle to a 60°F day instead of the standard 100°F day, allowing them to operate the engine to higher thrusts at low level and gain significant additional air frame flight test data while the pulsations and temperature wander were investigated. Douglas was asked to justify the request and demonstrate that the additional data outweighed the engine risks involved. The request was denied, and the request's timing, coming when overpressure concerns for the compressor were being voiced, is interesting and demonstrates the conflicting pressures WAGT was encountering.[62]

Engine starter requirements were under investigation and BuAer had changed some starter operating specifications and requested information on the various engine starter requirements for all the J40 models. The Navy requested a lower initial ignition point and WAGT said on their operating engines it was 500 RPM and their performance curves assumed that as the ignition point. The lower ignition point would lower the load on the starter sooner. The other requirement was defined as the starter cutout speed above which starter assistance was not required to make the engine accelerate to idle speed within

five (5) seconds at static sea level standard conditions. Earlier curves had shown this to be 1,800 rpm, but WAGT stated that when data from engine acceleration tests became available, this would have to be revised and estimated 2,800 rpm would be a satisfactory cutout speed. The entire table showing the polar moment of inertia for all the then current models is included here:[63]

J40 Rotor Polar Moment of Inertia	
Model	LB-FT²
J40-WE-6	717
J40-WE-8	717
J40-WE-22	758
J40-WE-1	717
J40-WE-10	942
J40-WE-12	942
J40-WE-5	942

Actual performance of the -6 came into question when McDonnell requested the results of the NACA testing data for "independent comparison" to the WAGT values. WAGT agreed to meet at any time with the McDonnell representatives and requested BuAer withhold release of the NACA data to McDonnell until they had met with their representatives. The memo pointed out that the data in question was taken during turbine inlet temperature distribution investigations, the emphasis being placed on determining the turbine inlet temperature distribution variation with altitude rather than on determining the pumping (i.e. – thrust) characteristics of the basic engine. As such, they were inadequate for airframe contractors to use for comparison purposes. A meeting with McDonnell's representatives would allow them to *benefit from the data and have the advantage of this contractor's background and experience in interpreting the wind tunnel data."*[64]

In July, BuAer agreed with a WAGT request to waive a qualification test on the gearboxes produced from Lycoming Spencer since they would be functionally and materially identical to the gearboxes produced by the prior vendor, Nutall. The waiver only applied to the -6 and -8 gearboxes, since the -22 gearbox would be different in some respects to support the constant speed generator requirement.[65]

Douglas had been reporting diaphragm failures on the micro-switch on the dual fuel pump warning light. Investigation showed three possible causes: 1) Rough pushrods chafing the center part of the diaphragm; 2) Sharp edges on the anchor washer or pump housing, which cut through the diaphragm; and 3) Radial slices in the diaphragm step (cause not determined and not duplicated in test stand operation). WAGT changed the diaphragm design from step convolutions to the loop type and added a sliding "O" ring seal on the pushrod to stop overboard fuel leakage in the event of a diaphragm failure. Until the switches with the new diaphragm could be installed, the engines were to be checked with the boost pump running to check for leaks at the switch drain hole. If fuel was detected, the switch would be changed or removed and a plug installed in lieu of the switch until the correct part was available.[66]

Feedback from the development of the -8, which used an overlapping segment type of exhaust nozzle seal, resulted in such a seal being developed for the -6 and supplied as part of CID 19257 on all new -6 engines, with the previously produced engines being field modified as needed or at the next overhaul. Douglas was sent a new seal for a -6 in July.[67] Douglas had previously reported a seal brazing failure which they had investigated themselves and reported the findings to WAGT. WAGT sent a memo to BuAer stating that they considered the failure a unique occurrence and since they were replacing the exhaust nozzle seals on all -6's in the field with the new type, would not investigate the Douglas type failure of the base plate further.[68]

Douglas reported another quality problem on WE030009 back in May when a misalignment of the bolt circle on the air bleed boss caused a small portion to crack out, exposing the metal to corrosion. They made a recommendation that the location of the elbow attachment bolt hole circle be centered on the mounting pad. WAGT took action in their Quality Control and Production Departments to prevent any future occurrence of the problem. The engine remained in service.[69]

Multiple instances of control cable failures were reported. This was traced to the control cable conduit failing because of fatigue caused by vibration against the thermocouple balance box. The solution was issued under CID 19350 to add a bracket to anchor the conduit and prevent the vibration. This was Engine Notice N50 once released.[70]

Engine WE030020 shipment authorization to be shipped to Douglas at Edwards Air Force Base was issued on September 26, 1952.[71]

With the Douglas XA3D-1 in flight test, it was reported that the two engines were generating pulses of a steadily increasing intensity at approximately 30 cycles/minute followed by two explosive reports with flame from the engines, the engines then repeating the entire sequence over and over. This had not been experienced with single engine installations and was outside of Westinghouse experience. WAGT stated the explosive reports were almost certainly induced by compressor stall, but (the engines) had not experienced any detrimental effects from it. However, the pulsating noise was considered likely to be from compressor stall and there was a strong degree of hazard associated

with it, experience showing that steady state stall conditions would produce engine failure.

They recommended that the pulsating noise be eliminated before further operation of J40-WE-6 engines in the A3D airframe continued. The two engines that had experienced pulsation should be overhauled before further operation. They listed three general actions that could be taken to eliminate the pulsations:

1. Alter the A3D inlet duct to overcome the irregularities of flow and pressure within the duct.

2. Alter the compression ratio by installing the opened first stage turbine nozzle per Engine Notice 29.

3. Reduce turbine outlet temperature of the engines associated with the A3D by running the engines on the ground with advancing throttle until pulsations were noted, reduce the throttle until it stops while reading the exhaust gas temperature, then use the set screw on the engine power regulator to adjust the engine top turbine outlet temperature to about 15°F below the value which was read when the pulsating noise began.[72]

Overhaul limits were revisited in October, with WAGT clarifying to the field organizations that the current limits were 100 hours all conditions with 15 hours military power. Recent overhaul of WE030005 had confirmed that 100 hours was justified. BuAer's request to extend that limit to 100 hours all conditions and 15 hours military was not yet supported, but WAGT had three engines in for overhaul and two high time engines about to enter overhaul and they would revisit the limits based on the conditions of these three engines.[73]

In November, WAGT was asked to confirm what action they had taken to eliminate fuel hose interaction and interference found on early engines between the hose and other components. No actual damage was referenced, but WAGT responded that Engine Change N-53 had called for the installation of a new hose, CID 19141. All engines from WE030017 on had the new hose and all other prior engines were being upgraded at overhaul.[74]

A very serious problem of fires in the aft engine compartment of the XF3H-1 on four occasions was reported and Westinghouse summarized the findings and planned actions:

1. *Tests showed that leaking fuel accumulated in the engine shell after a false start. The fuel could leak into the airframe compartment from the anti-icing valve bolted flange, the turbine inlet and outlet temperature probe boss, the turbine outlet pressure rake bosses, and the vertical flange connections at both the inlet and outlet ends of the rear bearing support assembly. In addition, the dual fuel pump had been known in several cases to leak fuel when*

the diaphragm failed. The one known source of oil leakage was the packing seal of the exhaust nozzle actuator.

2. The fuel leak positions would be supplied with improved high temperature gaskets when they passed appropriate tests.

3. The rear bearing support vertical flange would have a drain(scupper) as providing gasketing was not feasible due to the lack of clearance, the drain being ported overboard the airframe.

4. The rear bearing support flange would have a scupper and tests would be run with gasketing, but it could not be predicted if it would be successful at that location.

5. The leaking diaphragm of the warning light switch on the dual fuel pump already had a fix being put in place in the form of an "O" ring to stop fuel from flowing out if the diaphragm ruptured.

6. The oil leak would be handled with a drain and the airframe companies could port the drained oil overboard the airframe. Finding a good packing seal to prevent all oil leakage was proving a problem at that point, as there had been little success in experiments to date, but research was continuing.[75]

BuAer accepted WAGT's action plan and approved the plan for the -6 and -8, but reserved approval on the -22 until a complete engine survey was accomplished.[76]

Engine WE030006 had suffered a gearbox failure in May while powering the XF4D aircraft 124586. The engine was returned to Westinghouse for investigation of metal particles found in the gearbox and main oil filter element. The trouble was traced to a failed bearing inside the gearbox, it being found that because of a machining error, the bearing was forced to carry thrust loads which it was not designed to carry. It failed, leaving particles and broken components in the gearbox. Ultimately, WAGT got around to issuing report A-1420, dated October 28, 1952, describing the problem and their findings. No action was taken as a result of this failure, it being considered a singular occurrence that had not been repeated in any of the other gearboxes. How this machining error escaped detection during the component inspection quality check was not addressed.[77]

The acceptable primary and alternate fuel situation sorted itself out in November and BuAer issued a memo to all field users of the J40 stating the fuel types and their order of preference of use for J40 engines: (1) MIL-F-5624A, Grade JP-4; (2) MIL-F-5623A, Grade JP-3; (3) MIL-F-5572, Grade 80; (4) MIL-F-5572, Grade 91-96; (5) MIL-F-5572, Grade 100-130; (6) MIL-F-5572, Grade 115-145. On -6 engines, Engine Notice N53 had to be incorporated on any engine prior to use of JP-4. This notice called for the replacement of the original fuel lines with Aeroquip 303-02 hoses, the 303-02 being selected over the previously tested 302-02 because of its greater operating temperature range of 275ºF to -67ºF vs. the latter's 275ºF to -40ºF. Any engine using other than the primary preferred fuel had to follow the starting instructions for those fuels that would be issued shortly.[78]

In addition, WAGT later advised raising the external temperature to which the hoses could be exposed to 300ºF as a result of testing showing this had no effect on the hoses. This was approved by BuAer.[79]

On December 29, 1952, BuAer sent WAGT a contract termination memo on the -6. This cancelled the last 3 engines under contract and WAGT was instructed to stop all work on Item 1, Lot 1 of the contract. This had no effect on contractual guarantees on the -6 engines already in service and the overhaul services being supplied for normal engine usage reaching the overhaul limits on a given engine.

In December, after much testing, WAGT asked that the maximum oil discharge temperature allowed for the various J40 models be raised. For the -6, it was to be raised to 300ºF using MIL-O-6081-1010 lubricant. (This memo also covered the -1, -8, -22, -24 and -26 J40 models and all models of the J46. It is another example of changes applying to multiple contracts and models being covered in a single memo.)[80] This approval was not given, triggering a detailed examination of all hoses, hose condition in testing and overhaul and expected field use of the engines. The details of this will be covered for the other engine models in the appropriate place.

Incorporation of an anti-icing valve on the -6 was estimated by WAGT to begin in April 1953. It is doubtful if any were actually installed on the -6 engines, but documentation is lacking.[81]

Field engines continued to experience problems requiring investigation back in Kansas City. WE030019, WE030020 and WE030022 had to be returned with a notation that WE030020 had suffered unspecified damage to the second stage turbine. Field repair was considered for that engine, but in the end the complete engine was returned to Kansas City. The investigation attributed the cracking at the blade lands to either or both of the following factors:

1. A marginal root design
2. Dimensionally inaccurately machined disc grooves

The finding caused all -6 engines to be grounded. Engine WE030021, which had not yet been delivered to the Navy, had a -8 design second stage turbine disc installed as the solution to the problem. WE030008, undergoing overhaul, similarly had a -8 disc installed as part of the overhaul. Other engines were immediately shipped to Kansas City in anticipation of needing a part change of some sort, even if different from that recommended. In addition, the investigation into a field method of rotor balancing was accelerated, but was not yet ready for approval. It was hoped this would be successful so that field impact of the disc issue could be minimized. The CID to authorize -8 disc installations in -6 engines was being worked on but had not yet been submitted.[82]

The mechanical limits of the J40 were covered in detail in a WAGT memo, the memo giving the design philosophy evolution of the engines from prior to 1950 covering the need and intent of pressure limiting devices under study. Interestingly, for the -6, the memo stated:

"The J40-WE-6 engine was designed for the maximum loads encountered under the ram conditions specified in the engine model specification while operating at military conditions. For this reason, no pressure limiting is required on this configuration. The engine operating limits curve for this engine (curve No. 369041) has been altered to change the compressor mechanical limit line to the engine mechanical limit line. These new limits should be observed during flight operation." [83]

The subject of the engine's mechanical limits was a much greater concern for the higher power later models, which could push the aircraft to much faster speeds at all altitudes. This subject is explored further in the -8 chapters.

BuAer questioned the strength of the Exhaust Nozzle Actuator Bracket on the -6, noting that it flexed when moved by the actuator. WAGT responded that they knew it flexed, but that it was more than strong enough as proved by the many hours of testing and actual flight experience without any breakage. On the -8 and -22, the bracket was being reinforced because the actuator on those models exerted more force than the one on the -6, not because of a design deficiency. [84]

Shortage of engines caused by the lack of -8 engines shipping in quantity was partially being compensated for by allowing -6 engines to be run 22.5 hours at military power as one of the overhaul triggers in lieu of the earlier lower limit of 15 hours. This created a problem when the BAR in Kansas City lowered the total number of military operation hours on WE030021 after overhaul because of the "excessive" time the engine had to be run prior to final buildup. The Navy assumed the teardown after green run inspected all parts to be "as new" prior to reassembly (final build), so could not understand why the BAR took the action he did. Field use of the engines showed the 22.5 hour limit was supported by the condition of the engines at overhaul. The BAR was requested to inform BuAer as to what steps were required to increase the stature of the teardown and inspection subsequent to initial acceptance test to the point where operation prior to final acceptance test could be overlooked as regards to its effect on subsequent engine operation. [85]

Engine preparation problems during assembly and shipping became evident, as apparently WE030024 was shipped out with a cap still installed inside the Emergency Fuel Regulator. The engine was shipped back to WAGT and WAGT was forced to respond to this oversight with a memo stating they had demanded that the shipping department come up with plans to make it impossible for such a plug to be assembled into an engine in the future. [86]

In April, BuAer decided not to overhaul WE030005 and WE030012 at the time of repair to replace their number two turbine discs, even though they would only have about 32 and 21 operating hours left before the next overhaul. From these hours would have to be subtracted whatever test hours were expended after the repairs to certify them flight ready. [87]

Further, WAGT issued a memo stating that two failures of house -8 engines showed that the compressor discs in all the engines were operating at temperatures appreciably above their design values. Review indicated the discs might have a very short life at engine inlet temperatures that could be encountered in normal field operation. They recommended grounding all engines until the source of the overheating could be resolved. [88] Even after all the other problems, this memo must have shocked the Navy, another grounding from such a serious engineering problem this late being unexpected. It affected all production models (-1, -6, and -8). Research would show that leakage paths from the last stages of the compressor were allowing hot air through small gaps in the disc curvic clutches and around the rods that held the spool together. This affected the sixth and seventh stage discs by heating them beyond their predicted and designed values. Seals had to be designed to prevent this, a subject covered in the -8 chapters. Engines had the seals installed (requiring the compressor be disassembled and modified, then assembled, balanced and the engine retested). The exact length of the grounding is difficult to determine, but may have been more than three months. It is not clear that any -6 engines actually had the seals installed apart from normal overhaul before the no-fly grounding was lifted. This over-temp condition must also have been contributing to the blade rub problem but the temperature differences were not consistent from engine to engine depending on the normal machining variations between engines.

On June 22, in response to a request to identify 18 engines to remain in service, WAGT produced a chart showing those recommended for continuing service and the status of all engines produced: [89]

	Westinghouse Recommendations for Fourteen J40-WE-6 Engines For Modification of Compressor Rotor Fix					6/8/1953

Note: These engines were chosen on the following basis – (listed in order of importance):
1. Availability
2. Engine Notices Incorporated
3. Low Operating Time

Engine	Status	TSN **	Military TSN**	TSLBU and/or TSOH	Military TSLBU and/or TSOH	Engine Notices Incorporated
020002	O.H. at S.P.	31.0	5.7	18.7	2.4	All Applicable Engine Notices are Incorporated
030004	O.H at K.C.	67.2	16.5	8.4	1.0	1A,2, 4A, 5, 8, 14, 17, 18, 20, 15A, 12, 6, 10, 11, 13, 16, 21, 22, 23, 25, 27, 29, 31, 32, 33, 35, 37, 38, 40, 41, 42, 43, 44, 45, 46, 47, 48, 52, 54, 50, 53, 51, 63, 67, 65, 55, 56, 66
030006	Torn down, Inspected, parts in Rework, O.H.	74.25	17.22	67.75	15.72	N-1, 1A, 2, 4, 4A, 5, 6, 10, 11, 12, 13, 14, 15, 16, 17, 20, 21, 22, 23, 25, 27, 29, 31, 32, 33, 35, 36, 41, 42, 43, 44, 45, 46, 48, 59
030007	Torn Down, Inspected, Parts Reworked	46.31	8.1	40.70	7.10	1A, 2, 4A, 5, 7A, 23, 24, 26, 32, 36, 42, 22, 41, 25, 33, 44
030008	AOG* Recently OH	108.7	16.0	19.7	4.9	All Applicable Engine Notices are Incorporated
030009	Torn Down, Inspected, Parts in Rework	35.8	5.25	29.78	4.07	2. 4A, 5, 15, 22, 24, 25, 26, 32, 33, 36, 41, 7A
030011	Torn Down, Inspected, Parts in Rework	61.6	13.20	55.0	12.5	4A, 2, 23, 6, 22, 7A – others presently being accomplished
030013	Torn Down, Inspected, Parts in Rework	37.6	8.6	29.3	4.8	1A, 4A, 7A, 15, 16, 21, 22, 23, 24, 25, 26, 31, 32, 34, 36, 41, 42, 44
030014	Torn Down, Inspected, Parts in Rework	86.7	23.56	35.78	12.72	1A, 4A, 6, 22, 23, 25, 27, 29, 32, 33, 34, 31, 36, 41, 42, 35, 16, 46
030015	Awaiting Parts for Build up. O.H.	83.01	29.80	55.61	20.2	N, 1A, 4A, 6, 7A, 22, 23, 26, 27, 29, 33,
030016	Torn Down, Inspected, Parts in Rework. O.H.	36.0	4.6	32.7	3.1	N, 1A, 4A, 5, 7A, 15, 16, 22, 23, 24, 25, 26, 32, 36, 37, 41, 42, 44, 45
030017	O.H at K.C.	112.3	24.4	9.6	.5	All Applicable Engine Notices are Incorporated

Westinghouse Recommendations for Fourteen J40-WE-6 Engines For Modification of Compressor Rotor Fix						6/8/1953

Note: These engines were chosen on the following basis – (listed in order of importance):
1. Availability
2. Engine Notices Incorporated
3. Low Operating Time

Engine	Status	TSN **	Military TSN**	TSLBU and/or TSOH	Military TSLBU and/or TSOH	Engine Notices Incorporated
030022	Rework in Field	35.9	21.4	24.9	16.0	All Applicable Engine Notices are Incorporated
030024	Rework in Field	43.5	13.1	34.7	9.2	All Applicable Engine Notices are Incorporated

* - To be Returned for Compressor Fix   ** - Includes Factory Green Run Time

Westinghouse Recommendations for Additional J40-WE-6 Engines For Modification of Compressor Rotor Fix beyond Fourteen if Desired						6/8/1953

Note: These engines were chosen on the following basis – (listed in order of importance):
1. Availability
2. Engine Notices Incorporated
3. Low Operating Time

Engine	Status	TSN **	Military TSN**	TSLBU and/or TSOH	Military TSLBU and/or TSOH	Engine Notices Incorporated
030018	AOG*	42.9	23.1	33.2	8.7	All Applicable Engine Notices are Incorporated
030020	AOG*	33.8	11.3	25.8	6.0	All Applicable Engine Notices are Incorporated
030021	AOG*	40.9	16.3	22.1	10.6	All Applicable Engine Notices are Incorporated
030023	AOG*	35.9	21.4	24.9	16.0	All Applicable Engine Notices are Incorporated

* - To be Returned for Compressor Fix   ** - Includes Factory Green Run Time

					Military	
				TSLBU	TSLBU	
			Military	and/or	and/or	
Engine	Status	TSN **	TSN**	TSOH	TSOH	Engine Notices Incorporated

Supplement
Westinghouse Recommendations for Additional J40WE-6 Engines
For Modification of Compressor Rotor Fix   6/8/1953

(Note: Following is the Status of Remaining J40-WE-6 Engines Not Included
In Recommendation Report for Compressor Fix dtd. 6-9-53)

Engine	Status	TSN **	Military TSN**	TSLBU and/or TSOH	Military TSLBU and/or TSOH	Engine Notices Incorporated
030005	At K.C. for Repair-Torn Down	125.2	16.5	67.5	9.7	N-1A, 2, 4A, 5, 6, 7A, 8, 10, 11, 12, 13, 14, 15, 16,17, 20, 21, 22, 23, 25, 26, 41, 46
030012	At K.C. Awaiting O.H.	95.9	17.3	73.3	10.7	1A, 4A, 6, 7A, 22, 25, 26
030019	Awaiting BuAer Disposition. High Military Time. Rejected for Overhaul	72.8	21.3	67.1	20.2	All Applicable Engine Notices are incorporated.
030001	Retired					
030003	Retired					
030010	Retired					
* - To be Returned for Compressor Fix						
** - Includes Factory Green Run Time						

Following this exchange, the Navy records show that Westinghouse was instructed to retire seven -6 engines, but the serials are not recorded. However, this appears to likely match the seven listed in the last two tables above that were still in service at that time. Westinghouse requested that WE030005, WE030012 andWE030020 be diverted for use as house engines for work on later models.[90] BuAer accepted the recommendations, with the exception of WE030018 being substituted for WE030022, serials WE030005 and WE0300019 were allocated to Westinghouse for house use. Parts from WE030020, WE030021 and WE030023 were to be used for overhaul and repair of the fourteen engines retained in operation.[91] WAGT was instructed to retain all re-usable parts from the retired engines to be used for repair and overhaul operations.

The tables above do not mention that WE030023 had suffered a gearbox failure in March. The gearbox was returned for examination and it was found to have been built to non-standard specifications. Insufficient end play was found on a gear shaft and this caused a large bearing failure. The WAGT analysis report (not numbered) came out as a memo and was not issued until August. WAGT changed the gearbox specifications to require the end play be measured and recorded on each shaft, with the record to accompany the gearbox. This change also affected J46 gearboxes, which were not the same design but could be subject to the same production problem.[92]

The engines continued to be used during airframe testing until shipments of the -8's and very early -22A's began to reach the airframe manufacturers. WAGT recommended the overhaul period of the -6 (along with the -1, and -8) be raised to 150 hours based on the condition of engines observed during 100 hour overhauls. It is not known if the 150 hour period was accepted for the -6, although the BAR in Kansas City recommended acceptance. BuAer documentation was not located in response to the recommendation.[93]

At the cancellation of the J40 contracts, all J40-WE-6 engines were offered to the Air Force for their use as spare parts for the -1, but none were required. The remaining engines were scrapped and none preserved for display or educational training.

Chapter 4 - J40-WE-6 Production and Service Citations

[1] WAGT attachment, December 27, 1950 revision, "Westinghouse Proposed Delivery Schedule" **B940VE1RG72.3.2NACP**

[2] BuAer Twix, 1/16/1951, engine allocation instructions to BAR. **B940VE1RG72.3.2NACP**

[3] Naval Speed Letter from BuAer, March 23, 1951, "Contract NOa(s) 10385; acceptance of J40-WE-6 engines" **B940VE1RG72.3.2NACP**

[4] BAR Memo, July 18, 1951, "Contract NOa(s) 10385; blade rub difficulties on J40-WE-6 engines." **B940VE1RG72.3.2NACP**

[5] WAGT memo, May 25, 1951, "NOa(s) 10385; Engine Delivery Schedule Our Reference WG-63300" **B940VE1RG72.3.2NACP**

[6] WAGT memo, August 30, 1951, "Contract NOa(s) 10385 WEC Turbo-Jet Engine Model J40WE-6. Acceleration Control, Survey of. Engine Change No. 68." **B940VE1RG72.3.2NACP**

[7] WAGT Report, December 4, 1951, "Information on the J40-WE-6 Engine." **B940VE1RG72.3.2NACP**

[8] WAGT memo, December 6, 1951, "J40 Air Turbine starters: AiResearch P/N 55000, Failure of on J40-WE-6 Engines Serial Nos. WE030008 and WE030011." **B937VE8RG72.3.2NACP**

[9] WAGT Memo, March 1, 1952, "J40 Air Turbine Starters; AiResearch P/N 55000. Failure of on XJ40-WE-6 Engine Serial No. 1" **B937VE8RG72.3.2NACP**

[10] WAGT memo, December 7, 1951, "J40-WE-6 Engine Operating Limitations" **B940VE1RG72.3.2NACP**

[11] BuAer Memo, December 19, 1951, "Contract NOa(s) 10385 – J40-WE-6 Engine Oil Leakage at #1 Bearing"; references WAGT report dated December 12, 1951." **B940VE1RG72.3.2NACP**

[12] WAGT Memo, December 28, 1951, "Contract NOa(s) 10385 – J40-WE-6 Engines Containing Compressor Blades of Improper Material." **B937VE8RG72.3.2NACP**

[13] WAGT Memo, January 2, 1952, "Recommended overhaul period for J40-WE-6 engines." **B937VE8RG72.3.2NACP**

[14] WAGT Report A-1268, January 28, 1952, "Contract NOa(s) 10385 WEC Turbo-Jet Engine J40-WE-6. Investigation of Rough Switchover, J40-WE-6 Engine 030005 at McDonnell Aircraft Corporation" **B937VE8RG72.3.2NACP**

[15] Office of Inspector of Naval Material (Philadelphia District), Memo of June 13, 1952, "Brick-Bat – NOa(s) 10385 and NOa(s) 10067 – Westinghouse Electric Corp., Essington, Pa., Serious delay in delivery, due Discaloy and Refractaloy shortages." **B942VE9RG72.3.2NACP**

[16] WAGT Memo, January 10, 1952, "Power Regulators, P/N 61F758-16, Modification of: (J40-WE-6 Engine on Contract NOa(s) 10385)" **B940VE3RG72.3.2NACP**

[17] WAGT Memo, February 4, 1952, "Contract NOa(s) 10385, J40-WE-6 Power Control System: Addition of Acceleration Control" **B940VE2RG72.3.2NACP**

[18] Naval Speed Letter, February 1, 1952, "Power Regulator Tube Filament Voltage Regulator for J40-WE-6 engines, request for information on." **B940VE2RG72.3.2NACP**

[19] NACA Research Memorandum RM SE52G03, July 25, 1952, "Effect of Rotor- and Stator-Blade Modification on Surge Performance of An 11-State Axial-Flow Compressor I – Original Production Compressor of XJ40-WE-6 Engine", Conrad, Finger and Essig, NACA Internet Report Database

[20] NACA Research Memorandum RM E52I10, May 25, 1953, "Effect of Rotor- and Stator-Blade Modification on Surge Performance of An 11-State Axial-Flow Compressor II – Redesigned Compressor for XJ40-WE-6 Engine", Conrad, Finger and Essig, NACA Internet Report Database

[21] WAGT memo, February 6, 1952, "Contracts NOa(s) 10385 and NOa(s) 10114 J40WE-6 and J40-WE-8 Engine Airflow Path Improvement for Altitude Operation" **B940VE2RG72.3.2NACP**

[22] WAGT memo, February 7, 1952, "Contract NOa(s) 10385FP, WEC Turbo-Jet Engine Model J40-WE-6. Rear Bearing Support Assembly – Corrosion of." **B940VE2RG72.3.2NACP**

[23] WAGT memo, February 15, 1952, "Contract NOa(s) 10385 FP, WEC Turbo-Jet Engine Model J40-WE-6. Engine WE030004 Hot Spot Condition on." **B940VE2RG72.3.2NACP**

[24] WAGT memo, October 20, 1952, "Contract NOa(s) 10385 – Report on Hot Spot Condition on J40-WE-6 Engine, Serial WE030004." **B942VE9RG72.3.2NACP**

[25] WAGT memo, February 11, 1052, "Contract NOa(s) 1035 – J40-WE-6 Compressor Axial Rubs" **B940VE4RG72.3.2NACP**

[26] WAGT Memo, April 4, 1942, "Contract NOa(s) – 10385 – Manufacture of Compressor Vane Assemblies" **B937VE8RG72.3.2NACP**

[27] WAGT Report A-1275, January 12, 1952, "Test of Grumman Aircraft Proposed Control to Prevent Stall on the J40-WE-6 Engine, Project No. 1521, S.O. 4A3838, Contract NOa(s) 10385" **B937VE9RG72.3.2NACP**

[28] BuAer Telegram, February 28, 1952, No subject. **B934VC1RG72.3.2NACP**

[29] WAGT Memo, March 18, 1952, "Contract NOa(s) 10385 J40-WE-6 Flight Limitation. Removal of." **B937VE9RG72.3.2NACP**

[30] BAR Memo, April 25, 1952, "Contract NOa(S) 10385 – J40-WE6 engine, serial WE-030014; information on" **B941VE8RG72.3.2NACP**

[31] BuAer Naval Speed Letter, May 6, 1952, "Contract NOa(s) 10385 – J40-WE-6 Flight Limitations" **B941VE8ARG72.3.2NACP**

[32] WAGT Memo, March 22, 1952, "Contract NOa(s) 10385 – Compressor Blade Tip and Seal Rubs." **B937VE9RG72.3.2NACP**

[33] WAGT Memo, March 31, 1952, "J40-WE-1, -6, -8, -22 Engines, No. 1 Thrust Bearing" **B937VE9RG72.3.2NACP**

[34] BuAer Memo, April 23, 1952, "Ignition systems for J40-WE-6 engines, request information on" **B937VE10RG72.3.2NACP**

[35] BuAer Memo, July 11, 1952, "Ignition Systems for J40-WE-6 engines – information on" **B942VE9RG72.3.2NACP**

[36] BuAer Memo, September 17, 1952, "J40-WE-6 engines, ignition system spark plug offset adapter; information on" **B942VE9RG72.3.2NACP**

[37] WAGT Memo, April 29, 1952, "Contract NOa(s) – 10385 – Physical Properties of Aluminum Compressor Discs in J40-WE-6 Engines" **B937VE8RG72.3.2NACP**

[38] WAGT Memo, June 28, 1952, "Contract NOa(s) -10385 – Physical Properties of Aluminum Compressor Discs in J40-WE-6 Engines" **B942VE10RG72.3.2NACP**

[39] BuAer Endorsement, May 12, 1952, "J40-WE-6 Engines – Radio Interference Testing of" **B942VE9RG72.3.2NACP**

[40] BuAer Memo, June 2, 1952, "J40-WE-6 engine Ser. No. 030009, compressor damage of" **B937VE10RG72.3.2NACP**

[41] BuAer Memo, June 10, 1952, "NACA information on effect of exhaust nozzle area on engine performance" **B937VE11RG72.3.2NACP**

[42] WAGT memo, May 14, 1952, "Model J40-WE-6, -8, -22 Engines, Automatic Ignition Features, Deactivation or Deletion of" **B937VE10RG72.3.2NACP**

[43] WAGT memo, June 13, 1952, "Contract NOa(s) 10385FP, WEC Turbo-Jet Engine Model J40WE6 Engine WE030014 Gearbox Failure on." **B937VE11RG72.3.2NACP**

[44] WAGT memo, June 20, 1952, "Contract NOa(s)-10385, Westinghouse Turbojet Engine Model J40-WE-6, High Speed Compressor Stall" **B942VE9RG72.3.2NACP**

[45] WAGT memo, July 10, 1952, "Contract NOa(s) 10385, J40-WE-6 Engine Serial WE030014 reported stall in flight, Information on" **B942VE10RG72.3.2NACP**

[46] BAR memo, August 22, 1952, "Contract NOa(s) 10385 – J40-WE-6 engine serial no. WE-030014; shipping instructions, request for" **B942VE10RG72.3.2NACP**

[47] WAGT Report A-1419, September 10, 1952, "Investigation of Damage Sustained by J40-WE-6 Engine Serial WE030014, During Flight on 21 May 1952" **B942VE12RG72.3.2NACP**

[48] BuAer Memo, December 1952, "Contract NOa(s) 10385, Westinghouse J40-WE-6 engine serial WE030014; Bureau of Aeronautics comments on report covering damage to" **B942VE11RG72.3.2NACP**

[49] WAGT memo, September 25, 1952, "J40-WE-6 engine, fuel system of" **B937VE11RG72.3.2NACP**

[50] WAGT Memo, December 8, 1952, "Contract NOa(s) 10385 J40-WE-6 Engine Fuel System" **B938VE8RG72.3.2NACP**

[51] BuAer Memo, December 17, 1952, "Contract NOa(s) 10385, Westinghouse J40-WE-6 engine serial WE030014; Bureau of Aeronautics comments on report covering damage to" **B942VE11RG72.3.2NACP**

[52] WAGT Memo, April 16, 1953, "Contract NOa(s) 10385, Revision to Westinghouse Restricted Report A-1419 Covering J40-WE-6 (WE030014) Engine Damage." **B942VE12RG72.3.2NACP**

[53] WAGT memo, June 24, 1952, "Contract NOa(s) 10385 – Use of Substitute Fuels in J30-WE-6 Engines." **B937VE11RG72.3.2NACP**

[54] BuAer memo, July 8, 1952, "Contract NOa(s) 10385; use of substitute fuels in J40-WE-6 engines" **B937VE11RG72.3.2NACP**

[55] WAGT memo, August 1, 1952, "Contract NOa(s) 10385 Use of Substitute Fuels in J40-WE-6 Engines" **B937VE11RG72.3.2NACP**

[56] BuAer memo, September 19, 1952, "Fuels for J40-WE-6 and -8 engines – request for information on." **B942VE9RG72.3.2NACP**

[57] BuAer memo, June 28, 1952, "J40-WE-6 Engine for Icing Test Under Project Summit" **B942VE9RG72.3.2NACP**

[58] WAGT memo, September 2, 1952, "J40-WE-6 Engine for Icing Test Under Project Summit" **B942VE10RG72.3.2NACP**

[59] BuAer memo, July 1, 1952, "J40 and J46 Compressor mechanical limits; request information on" **B937VE11RG72.3.2NACP**

[60] BAR memo, July 8, 1952, "Contract NOa(s) 10385, J40-WE-6, -22 Engines, Tool Engineering Changes in Design; Forwarding of" **B942VE9RG72.3.2NACP**

[61] WAGT memo, August 4, 1952, "Field replacement of turbine rotating parts and requirements for subsequent balancing – J40 engines." **B937VE11RG72.3.2NACP**

[62] BAR Naval Speed Letter, July 3, 1952, "Contract NOa(s) 10385; Douglas discussions on nozzle area settings" **B942VE9RG72.3.2NACP**

[63] WAGT memo, July 8, 1952, "Contract NOa(s) 10385 and 52-048, starting torque and speed requirements for J40-WE-1, -6, -8, -10, -12, and -22 engines." **B937VE11RG72.3.2NACP**

[64] WAGT memo, July 22, 1952, "J40-WE-6 Wind Tunnel Test Data; Transmittal of" **B937VE11RG72.3.2NACP**

[65] BuAer memo, July 31, 1952, "J40 gearboxes supplied by the Lycoming Spencer Division of AVCO, qualification of" **B937VE12RG72.3.2NACP**

[66] WAGT memo, August 6, 1952, "Contract NOa(s) 10385F.P., WEC Turbo-Jet Engine Model J40WE6. Dual Fuel Pump Diaphragm Failures. Contractor's action on." **B942VE10RG72.3.2NACP**

[67] WAGT memo, August 19, 1952, Contract NOa(s) 10385 J40-WE-6 Engine Exhaust Nozzle Seals, Information on" **B942VE10RG72.3.2NACP**

[68] WAGT memo, Sept 2, 1952, "Contract NOa(s), WEC Turbo Jet Engine Model J40WE-6 Exhaust Nozzle Assembly, Information on" **B942VE10RG72.3.2NACP**

[69] WAGT memo, September 2, 1952, "Contract NOa(s) 10385 F.P. , WEC Turbo-Jet Engine Model J40-WE-6. RUDM – Contactor's Action On." **B942VE10RG72.3.2NACP**

[70] WAGT memo, September 5, 1952, "Contract NOa(s) 10385 F.P., WEC Turbo-Jet Engine Model J40WE6. Control Cable Failure – Contractor's Action On." **B942VE10RG72.3.2NACP**

[71] BuAer Telex, September 26, 1952, Control Cable Failure **B942VE10RG72.3.2NACP**

[72] WAGT memo, October 9, 1952, "Contract NOa(s) 10385 – Pulsating Noise Associated with Operation of J40-WE-6 Engine Installed in the A3D Airframe." **B942VE10RG72.3.2NACP**

[73] WAGT Memo, October 23, 1952, "Contract NOa(s) 10385 F.P., WEC Turbo-Jet Engine Model J40WE6. Overhaul limits – Contractor's Recommendation on." **B942VE10RG72.3.2NACP**

[74] WAGT Memo, November 5, 1952, "Contract NOa(s) 10385; F.P., WEC Turbojet Engine Model J40WE6. Fuel Hose Assembly Interference; Contractor's Action on" **B942VE11RG72.3.2NACP**

[75] WAGT Memo, October 16, 1952, "Contract NOa(s) 10385, J40-WE-1, -6, -8 and -22 Engines, Elimination of Fuel and Oil Leakage" **B938VE12RG72.3.2NACP**

[76] BuAer Memo, November 6, 1952 , "Contract NOa(s) 10385, J40W-1, -6, -8 and -22 Engines, Elimination of Fuel and Oil Leakage" **B938VE12RG72.3.2NACP**

[77] WAGT Report A-1420, October 28, 1952, "Analysis of Gearbox Failure on J40-WE-6 Engine, Serial WE030006" **B942VE11RG72.3.2NACP**

[78] BuAer Memo, November 26, 1952, "Acceptable fuels for J40-WE-6, -8, and -22 engines, information on" **B942VE11RG72.3.2NACP**

[79] BuAer Memo, December 18, 1952, "Oil Hose Temperature Limits for J40-WE-6, -8, and -22 engines – approval of" **B942VE11RG72.3.2NACP**

[80] WAGT Memo, December 30, 1952, "J40 and J46 Engine Oil Temperature, Maximum Operating Limits, Verification of." **B938VE12RG72.3.2NACP**

[81] WAGT Memo, December 31, 1952, "Contract NOa(s) 10385; J40-WE-6/-8/-22 Engines, Incorporation of Anti-Icing Valve" **B942VE11RG72.3.2NACP**

[82] WAGT Memo, February 17, 1953, "Contract NOa(s) 10385, J40-WE-6 Second Stage Blade Root Cracking" **B938VE13RG72.3.2NACP**

[83] WAGT Memo, February 23, 1953, "Mechanical Limits of J40 and J46 Engines" **B938VE13RG72.3.2NACP**

[84] WAGT Memo, March 11, 1953, "Contract NOa(s) 10385, J40-WE-6 Exhaust Nozzle Actuator Bracket." **B938VE13RG72.3.2NACP**

[85] BuAer Speed Letter, March 8, 1953, "J40-WE-6 engines, military time for" **B942VE12RG72.3.2NACP**

[86] WAGT Memo, March 30, 1953, "Contract NOa(s) 10385, Westinghouse Turbo-Jet Engine Model J40 WE6 Serial 030024, protective hose cap in emergency fuel regulator – contractors comments on." **B942VE12RG72.3.2NACP**

[87] WAGT Memo, April 16, 1953, "Contract NOs(s) 10385. Westinghouse Turbo-Jet Engine model J40 WE6 Serial Nos. 030005 and 030012." **B942VE12RG72.3.2NACP**

[88] WAGT Memo, April 28, 1953, "Possible Compressor Disc Failures, J40-WE-1, -6, -8 Engines." **B938VE14RG72.3.2NACP**

[89] WAGT Memo attachment, June 25, 1953, "Contract NOa(s) 10385, Westinghouse Turbo-Jet Engine Models J40-WE-6 – Compressor Modification – Contractor's Recommendations on" **B938VE14RG72.3.2NACP**

[90] WAGT Memo, June 15, 1953, "Subject of Retirement of seven J40-WE6 engines." **B942VE13RG72.3.2NACP**

[91] BuAer Memo, July 31, 1953, "Contract NOa(s) 10385; J40-WE-6 Engines to be retained for aircraft usage" **B942VE13RG72.3.2NACP**

[92] WAGT Memo, August 18, 1953, "Contract NOa(s) 10385, WEC Turbo-Jet Engine Model J40-WE-6 – Gearbox Failure On." **B942VE13RG72.3.2NACP**

[93] WAGT Memo, January 13, 1954, "Contract NOa(s) 10385 F.P., WEC Turbo-Jet Engine Models J40-WE-1, J40-WE-6, and J40-WE-8 Engines. Operating Time Between Overhauls, Extension Of." **B943VE16RG72.3.2NACP**

Chapter 5
XJ40-WE-8 Procurement and Development

Procurement

The original specification for the WAGT model X40E2 was in response to requirements that required provision for an afterburner. The specification (WAGT-X40E2-2) submitted with the proposal made no mention of afterburning directly, although it was stated the design allowed for easy incorporation of a tailpipe extension. It is not clear if this statement was intended to address the afterburner requirement or was simply stating engine installation could include a tailpipe if needed.[1] The latter is likely, as tailpipes were used by some aircraft in early testing using the -6 model.

Actual development procurement of the -8 model was made under Contract NOa(s) 10114, dated November 5, 1948. The justification for the procurement was:

"Under reference (a) a development program was started which will result in a new turbo jet engine (XJ40-WE-6) of axial flow design to develop 7,500 pounds un-augmented thrust. This will be the largest axial flow engine available to the Navy and due to its relatively small diameter, is becoming more attractive from the standpoint of future airframe design. Nothing was included in that program to develop an afterburner which now appears very desirable and necessary. The development program covered herein is aimed at the development of a successful engine-afterburner combination which will be known as the Model XJ40-WE-8 engine. Further engine refinement and strengthening will be carried on as well as afterburner development which is expected to be very extensive in nature due to the high power rating of the basic engine. The result will be a basic 7,500 pound thrust engine which can be augmented by afterburning as compared to the engine to be developed under reference (b) which will be a basic 8100 pound thrust engine also with afterburning (XJ40-WE-10). The success of the XJ40-WE-10 engine is highly dependent on the results to be accomplished under the program covered by this clearance, since afterburner development covered herein will be utilized for the higher powered engine."[2]

From the above, it is clear that the development of an afterburner for the -6 was seen as a stepping stone to developing the afterburner for the J40-WE-12 engine, the afterburner model being J40-WE-10. The author believes references in the extant literature to "low power" and "high power" engines almost certainly refer to the J40-WE-6/-8 and J40-WE-10/12 respectively.

The Letter of Intent covering the initial design work on the -6 included work to design a tailpipe extension. The first thoughts on how that would appear become visible in a report with preliminary installation drawings in July of 1948. It is clear from these drawings that the initial tailpipe area control system proposed for the -6 had been changed at this point from an internal onion approach of the -2 to the clamshell design driven via a ram through a yoke assembly. It is not clear if this design change resulted from the A/B studies and was applied to the -6 retroactively or the -6 was changed first and the approach carried through to the exhaust nozzle of the -8. The approach allowed the A/B to be added simply via lengthening the actuation rods from the yoke to the clamshells.[3] This initial memo shows the A/B version of the -6 was still considered a -6 version, not a separate model. The contract called for conversion kits to modify -6's to -8's and back. Later development of the -8 resulted in modifications to the basic engine to such an extent that interchangeable kits became unfeasible and the items were canceled from the contract.

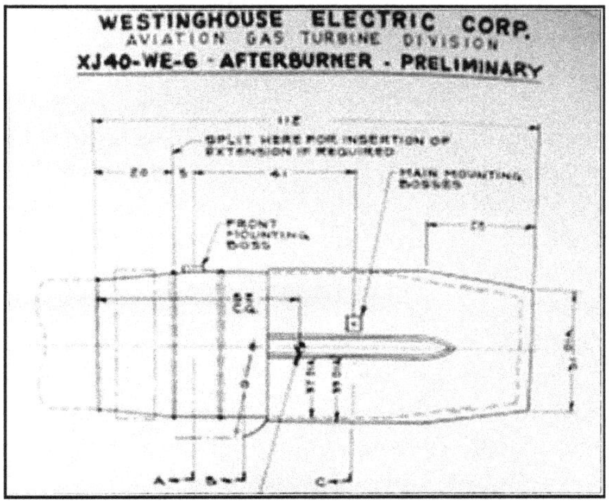

Figure 1 The preliminary installation outline of the A/B shows the engine is still considered a -6 even with an A/B.

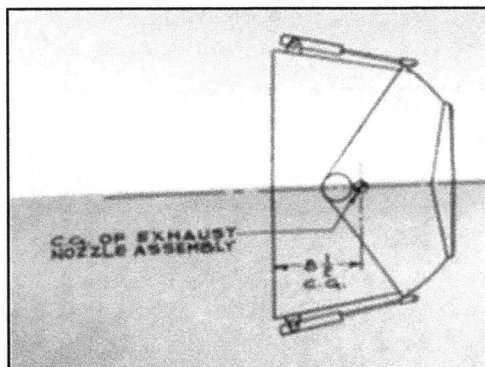

Figure 2 The exhaust nozzle is an eyelid design later changed to be driven by the same yoke and actuation rod system of the -6. Unnumbered drawing.

The total weight increase of the afterburner would be 570 pounds over the non-A/B -6. That would bring the total engine weight up to 2,922 pounds. Thrust at sea level would be 10,900 pounds, with a dry rating in military power setting of 7,300 pounds static thrust (due to duct loss), and an A/B SFC of 2.5.

The control system approval for the -6 included detailed design considerations that had to be met for final approval. While not specifically aimed at the A/B controls, since they would incorporate the same approach and the major components, the considerations were clearly applicable to the A/B as well. The concerns related to fail-safe asymmetric control in the event of a single eye-lid failure, engine lubrication loss if engine oil was used as the hydraulic fluid, field distortion of the eyelids leading to a lack of consistent power control power scheduling performance, and positioning of the eye-lids if the primary power control failed[4] The counter WAGT argument against a duplicate hydraulic system for the -6 was accepted (the single system would lock in its present position if the hydraulics failed) but a duplicate system had to be included in the design for the A/B. Engines could be delivered without it but it could be installed if service use proved it was needed.[5]

Engine weight was in flux, due to various changes from the mock-up board and design considerations. As of November 1948, the projected weight of the -6 engine had increased to 2,906.3 pounds, bringing the A/B -8 version to 3,562 lbs.[6]

Mock-up Board Report

The Mock-up Board was held on April 13 and 14, 1949 at the WAGT location in Essington, PA.[7] The detailed recommendations follow, but the primary discussion areas were:

A. Description of the engine – an improved -6 incorporating an afterburner and its associated controls and components. In addition it would have a new gearbox with external oil tank between the air inlet ducts and a lubrication system designed to provide continuous engine operation during inverted or negative-g accelerations. The static sea level thrust was 7,400 pounds dry with an SFC of 1.0 and a full afterburning thrust of 10,900 pounds with an SFC of 2.5. The engine was planned to be used in the XF3H-1, XF4D-1 and the XF10F-1 (until the -10 models were available) with other applications under consideration by BuAer.

B. During a prior mock-up review of another A/B engine, the responsibility for designing the cooling shrouds of the engine, both primary and secondary, had been deemed the responsibility of the airframe manufacturer. During the -8 review, this was reversed and the primary shroud responsibility shifted back to the manufacturer. The secondary shroud would remain the responsibility of the airframe manufacturer. The logic was that if the airframe manufacturers had to develop both shrouds, they would have to engage in extensive ground testing to develop proper cooling. (It was not noted in the review, but obviously any change in the primary shroud might have had an effect on engine performance beyond the engine manufacturer's control. The boundaries between the engineering interfaces related to turbojet installations were still in flux.)

C. The mock-up board generally agreed with the proposed fuel system and concurred with the desire to generally simplify the system. Areas of discussion were lowering the power level torque and fuel pump inlet pressure, but recommendations were deferred for further review by BuAer before final coordination of engine model specification.

D. Shaft bowing, resulting from unequal cooling of the shaft after engine shut down, resulted in engine vibration if the engine was restarted with the engine warm. The -8 would contain a small "shut-down" motor to rotate the shaft either continuously or intermittently in 180 degree increments to prevent bowing on the engine. This solution had not been proved yet or approved by BuAer on the -6.

E. Item changes that might have a possible effect on weight were: 1, 4, 12, 19, 21, 26, 28, 31, 32, 34, 37, 46, and 48.

F. Items requiring additional actions were listed in section A, paragraph 7. The mock-up was approved subject to ultimate completion and approval of those items.

J40-WE-8 Specifications at Time of Mock-up Board (April 13, 1949)				
Rating S/L	Thrust lbs	RPM	SFC	Gas Temp °F
Take-off (A/B)	10,900	7,260	2.50	1,240
Military (A/B)	10,900	7,260	2.50	1,240
Military	7,400	7,260	1.00	1,240
Normal	6,600	7,260	0.95	1,140
Cruise 1	5,000	7,260	0.86	840
Cruise 2	4,000	7,260	0.85	680
Idle	450	3,000		900
Weight	3,366 lbs			
Max Operating Altitude	50,000 ft			
Max Air Restart	40,000 ft / 1.2 ram pressure minimum 95% recovery efficiency			

The Mock-up Board Report, Section A, Paragraph 7 listed the following items:

A. Changes recommended without mockup and subject only to approval by BuAer release of drawings: 1, 4, 5, 7, 10, 12, 14, 18, 19, 21, 26

B. Recommended changes involving contract or specification changes requiring submission of proposals by the contractor: 48

C. Tentative recommended changes involving submission of design studies by the contractor: 31, 32, 34, 37, 45

D. Changes involving requests for additional information to be submitted to the Bureau of Aeronautics: 13, 24, 25, 27, 36, 41

The action items listed above in B, C and D are reproduced and discussed below in numerical order, each followed by the actual resolutions occurring over different time periods as development went forward:

Item 1 – Provide an extension and approved type drain valve for oil tank drain, not to protrude beyond the engine envelope, with sufficient clearance inside the envelope to permit addition of aircraft airframe fitting.

Resolution: A revised extension showing an approved drain valve was submitted and approved by BuAer.[8]

Item 2 – Provide a side removal oil filter instead of a bottom removal. (Disapproved – would have required the redesign of the oil pump.)

Item 3 – Provide for at least a 30 second negative G oil supply and an oil level indicator. (Disapproved – design provided indefinite (unlimited) negative-G and the indicator was not considered necessary.)

Item 4 - Install an oil filler cap and strainer with safety wiring on cap and a 10 mesh strainer with a blank cover plate on the top filler opening, along with scupper and drain connection for side filler not to increase engine envelope.

Resolution – Along with the relocation of the junction box proposed separately, the oil filler was moved aft approximately 6 inches from the original location. This allowed sufficient radial clearance to provide a satisfactory scupper within the engine frontal envelope. This would provide for 3 possible oil filling locations, left, right and center, and a survey letter would be sent (sent out July 22) to the airframe manufacturers for comments. The oil filler would be a 1 inch tube size with a removable 10 mesh strainer, sealed with a quick-disconnect self-locking type cap, chained to the adjacent scupper and would have lock wire provisions. The filler tube would be surrounded by a scupper with an overboard drain to handle spills or overfilling. A vent line from the top of the oil tank would be connected to the scupper for air release during filling. The brackets to mount the filler and scupper would be reversible to allow easy movement from the left to right side or vice versa. The weight increase would be 3.5 lbs.

Item 5 - Provide oil filler installation that can be made on either side of the engine in lieu of current right hand side only.

Resolution – See Item 4 above.

Item 6 - Revise shutdown motor arrangement so that the starter can be installed without removing the shutdown motor (Disapproved due to compromises of the basic starter design and gearbox.)

Item 7 – Provide better access to the power regulator blast cooling tube by moving the flange aft to improve clearance.

Resolution – Westinghouse redesigned the power regulator housing into a more compact form which improved the airflow. The power regulator was moved aft to remove it from the adjacent anti-icing hot gas line, changing the clearance from 1-5/8" to 3-1/8". They also changed the connection to the Boss type. They intended to make this change on the -6 as well.[9]

Item 8 – Provide adequate wrench clearance around the hydraulic pump pad for installation and removal of hydraulic pumps.

Resolution – Was demonstrated during the Mock-up Board review and took 16 minutes. Based on this, BuAer accepted the installation as acceptable.[10]

Item 9 – Provide alternate bosses on engine for spark plugs and alternate location of booster coils. (Disapproved – reason not stated but the issue was that the plugs would not be reachable in aircraft where the engine was installed over a fuel tank. Since no such aircraft was planned at that time, this might have been the reason for the disapproval.)

Item 10 – Eliminate the present fuel inlet connection on the afterburner and fuel pumps which had pads facing downward and replace them with pads on front of both pumps. Approved for the A/B pump, to be reviewed after final configuration of the simplified control system was determined.

Resolution – The fuel booster pump connection from the airframe was changed to be a single connection supplying fuel to both the engine dual fuel pump and the afterburner fuel pump along with reducing the A/B fuel pump inlet pressure requirements to be the same as the engine booster pump. (See also Item 36.)[11]

Item 11 - The engine manufacturer must provide the afterburner ejector (nozzle) that would be in use during the acceptance testing.

Resolution – WAGT forwarded a design to BuAer showing it would be within the existing nozzle dimensions of the -6 subject to minor changes resulting from development and add 1 pound to the engine.[12] The design was intended to be used on the -6 as well and WAGT asked that approval for the -8 also be the approval for the -6. The initial memo was followed with one of more detail showing the length of the nozzle would increase by 3 inches, increasing engine length to 277+/- 2 inches.[13]

Item 12 – Provide attachment holes for installing a seal flange between the A/B shroud and the aircraft structure to prevent reverse flow through the secondary ejector.

Resolution – Superseded by the redesign of the cooling system for the A/B.

Item 13 – Provide performance characteristics of afterburner ejector-flow ratio vs. pressure ratio for various ejector eyelid positions, needed for overall cooling solution design.

Resolution – Westinghouse produced an A/B cooling design with all the details in March 1950. The design used compressor air for cooling. The proposal included estimated skin temperatures, rate of heat rejection, jet wake patter (both temperature and velocity) and information on insulation blankets. This was very similar to the -10 model design, except that an iris type exhaust nozzle was planned for the -10 instead of the clam shell type on the -8. The design was sent out as Engine Change 70 for review.[14]

Item 14 – Provide more clearance on the jack pads on the rear flange combustion chamber casing for the oil lines on the right (starboard) side of the engine.

Resolution – The revised jack pads were sent out August 2, showing the new locations and the rerouting of the oil and air lines. The memo argued the new locations allowed ample clearance for all but the most adverse handling conditions. It was intended this change be applied to the -6 as well. There would be no weight increase.[15] A later memo shows the pads had to be moved again for the -6 and it was requested that the location be approved for the -8 as well.[16]

Item 15 – Eliminate all water traps in lines. (Disapproved)

Item 16 – Reduce the nozzle yoke size to ease design of the fairing on the non-A/B engine (-6). (Disapproved)

Item 17 – Make provision on the variable area nozzle to permit attachment of a position indicator. (Disapproved)

Item 18 – Relocate A/B spark plug to right side of engine provided there is no impact to A/B ignition.

Resolution – WAGT proposed eliminating the spark plug and related electrical system completely and switching the A/B to hot streak ignition based on successful tests on J34's with afterburners. The design change was accepted by BuAer in November.[17]

Item 19 – Add 3/16" holes in the forward fire seal flange spaced approximately 3" apart.

Item 20 – Minimize the number of engine instruments to simplify the pilot's duties. (Disapproved – reason given in the mock-up report is that the services would determine the required instrumentation for production aircraft. Given the push for a "simplified" power control system (See Item 28 below) this would require the pilot to be far more involved in engine management, which made it a curious board recommendation.)

Item 21 – Provide for retraction of the engine lifting eye at the top centerline in the plane of the aft mount points.

Resolution – WAGT sent their design for a retractable lifting eye to BuAer in September. The initial proposal had been sent out for a change in the XJ40-WE-6, but noted they intended to use the same design for the XJ-40-WE-8.[18]

Item 22 – Furnish an AN4139 adapter on the alternator pad, reducing the 10" bolt circle pad to a 5" bolt circle pad. (Disapproved – a specification item and the spec had to be changed so the action could be carried out.)

Item 23 – Add an (electromagnetic) radiation shield between the A/B and the shroud. (Disapproved)

Item 24 – Provide sufficient installation drawings to permit detail design based on drawings.

Resolution –Drawings including all Engine Changes (EC's) up through 76 were provided to close this item.

Item 25 – Establish eyelid configuration to determine if aft portion of aircraft is affected.

Resolution – Design showed no effect on rear of aircraft after the engine was lengthened to be interchangeable with the -10.

Item 26 – Relocate electrical junction box forward of plane of front engine mount.

Resolution – WAGT relocated the box and its connections as requested.[19,20]

Item 27 – Reduce power lever maximum operating torque below 50 inch pounds, current forces required were too high due to increased friction at low temperatures.

Resolution – WAGT suggested that their proposal for the Simplified Power Control System be approved. It provided for a 60% reduction in throttle torque with a maximum of 20 inch-pounds vs. the 50 inch-pounds of the -6 system with the extra A/B components presented at the Mock-up Board.[21]

Item 28 – Simplify the control system relative to the pamphlet distributed to the Mock-up Board, "Proposed Simplification of XJ40-WE-8 Power Control System, April 13, 1949". This would provide for less weight (26.5 lbs), fewer parts and the same reliability.

Resolution – Simplified system proposed for BuAer review (detailed elsewhere). Originally proposed by WAGT on April 13, 1949 and was in review.

Item 29 – Reduce throttle travel required from idle to shut off below the 26 degrees required. (Disapproved)

Item 30 – Reduce the time for obtainment of full military and A/B thrust from a 50% military power setting to less than the current 1.5 to 2.0 seconds from 50% to full military and then 3.0 to 5 additional seconds to full A/B (7 seconds total). Required for carrier wave-off condition. Sent to BuAer for specification coordination.

Item 31 – Make the quick disconnects in the fuel lines between engine and afterburner be self-sealing or if possible bring fuel lines into afterburner forward of quick disconnect flange. Required a study.

Resolution - Final design submitted with new fuel system lines shown.

Item 32 – Locate quick disconnect fittings for all three A/B fuel lines below the horizontal centerline to consolidate access to a single area. Required a study.

Resolution - Final design submitted with new fuel system lines shown.[22]

Item 33 – Include a flowmeter on the mock-up for proposed Air Force requirements. (Disapproved – requirements not finalized and flowmeter design was fluid at that time.) However, later the specifications were changed to require provisions for a flowmeter on the -8.[23]

Item 34 – Change anti-icing discharge connection to one from the current two on the mockup. Required a study.

Resolution – WAGT reduced the anti-icing discharge outlet to just one at a position of 45 degrees from the top of the engine on the left side.[24] This change was applied to the -6 as well.

Item 35 – Make anti-icing start and stop automatically, with a suggested constant small bleed for ice prevention. (Disapproved – no experience with system available)

Item 36 – Reduce the boost pump inlet pressure requirements to reduce excessive weight in boost pumping equipment along with the excessive power drain on the engine. Referred to BuAer.

Resolution – The A/B fuel pump was changed to accept fuel from the single feed of the airframe booster pump at the same pressure as the dual engine pump.[25] This was surveyed with Engine Change 49.

Item 37 – Provide more positive means than V type quick disconnect clamp for A/B attachment to engine and allow for small amount of initial misalignment between engine and A/B. (First part disapproved and the second required a study.)

Resolution – WAGT sent in a proposal in September to respond to part two, but later asked BuAer to ignore the proposal, which was superseded by a later proposal.[26] The later proposal was reviewed and approved, adding two pounds to the engine weight.[27]

Item 38 – Raise limiting sea level Mach number of .91 or provide override in control system to avoid excessive compressor outlet pressure. (Disapproved)

Item 39 – Request engine detail specification be revised to increase maximum air bleed of 20 lbs/min. (Disapproved – airframe manufacturer must submit requirements and justification to BuAer.)

Item 40 – Submit a study to make engine mountings system capable of withstanding 20g ultimate crash landing loads. Make the engine mounting system conform to airframe spec SSJC. (Disapproved)

Item 41 – Define the maximum permissible variation in pressure distribution within each duct inlet and between the two ducts. Required a study.

Resolution – In a memo dated April 12, WAGT established the positions of the intended pressure rakes in detail (7 individual probes, spaced 3-5/8", 6-1/4", 8", 9-1/2", 10-3/4", 12", and 13" from the horizontal centerline of the duct.) There were two rakes, one in the top and one in the bottom of the inlet duct along the centerline. The individual readings could be averaged arithmetically to obtain the average inlet pressure. The total variation (difference between the highest and lowest readings) should not exceed 5% of the average pressure under any normal operating conditions; static or flight; sea level or altitude. Yawed flight was not to be considered normal and no limit was set for such operation. (Note – the 5% pressure differential was the same established for the J34 and J46.)[28]

Item 42 – Investigate other means to reduce shaft bowing. Current system required the aircraft electrical system remain energized after stowing after a mission. (Disapproved)

Item 43 – Increase engine RPM to 8,000 to compensate for 10" generator pad. (Disapproved – pad design complied with current BuAer Electronics Division requirements.)

Item 44 – Provide clearance for variable volume pump. (Disapproved – but see the -22 model)

Item 45 – Study the ability to provide a special drive to reduce the 10" diameter pad to two pads for small alternator and drive for air compressor. (Disapproved – airframe manufacturer was responsible to provide.)

Item 46 – Recommend better access to components on aft side of gear box. Required a study.

Resolution – WAGT recommended that BuAer accept their report A-829 "A Study of the Application of Accessory Quick Disconnect Mountings for the XJ40, XJ46, and XJ34-WE-32 Engines" submitted April 25, 1949. There were several differences between it and the other engines as follows:

A. Quick disconnect fittings could not be used on the main oil pump because it had several cored oil passages in the mounting face, the pad face was not circular and the pump was driven by engagement of two gears in the pad face with a gear in the gear box instead of by a spline coupling. However, if the Governor Alternator were equipped with a quick disconnect interface, it could be removed in one minute solving the oil pump accessibility problem.

B. The Oil Cooler was a non-rotating accessory and was mounted by four lugs which picked up four studs on the back face of the gear box. A cored passage behind and between the studs supplied oil from a mating cored passage to the cooler from the oil pump. The two top lugs were readily accessible and the bottom two would be easily accessible if the Governor Alternator was removed.

It was recommended that the proposal be accepted for both the -8 and -6 engines.[29] It was approved as proposed.[30]

Item 47 – Request engine spark plugs be relocated for side accessibility. (Disapproved)

Item 48 – Remove tachometer alternator gearbox (RPM of pad was so close to the desired speed, deemed unnecessary in the circumstances).

Resolution – Was described and approved.[31]

Figure 3 XJ40-WE-8 Mock-up, right side, reviewed by the Mock-up Board. Note lack of cooling air piping or details of the exhaust nozzle assembly. Mock-Up Board Report, WAGT P40793, 4/19/1949. *Courtesy Hagley Museum and Library*

Figure 4 XJ40-WE-8 Mock-up of the left side of the -8 engine. Mock-Up Board Report, WAGT P40792, 4/19/1949. *Courtesy Hagley Museum and Library*

The mock-up board from February had requested (Item 45) that the A/B be designed as a cantilever extension of the -6 engine, attached at the rear of the engine and requiring no additional supports in the aircraft. WAGT produced a study, delivered in September (A-749) which evaluated the engineering changes of such an installation. It concluded that, based on the size of the A/B and the maneuvering and weight loads imposed on the engine that the basic engine weight would go up 108 pounds and the engine would be exposed to potentially damaging noise vibrations from the A/B. It recommended the A/B be attached with an extension and be separately supported by the rear of the aircraft. This would remove the loads from the engine itself and allow better vibration isolation as well.[32] The long delay in responding to the Mock-up Board's request indicates that the A/B design was on the back burner during much of this time period.

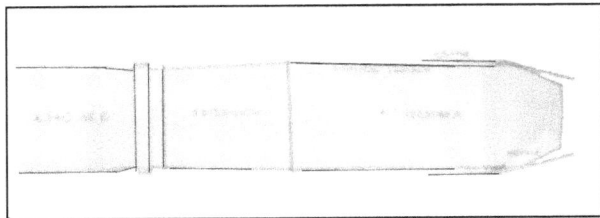

Figure 5 Basic simple drawing of the XJ40-WE-6 for a cantilevered A/B. *WAGT Report A-749.* **(Use of extensions with the various models is obscure.)**

Design of the A/B continued, but the remaining documentation only gives brief insights into the early design work. A memo of April 1949 found the design of the fuel pumps for the base engine and A/B to be unsatisfactory. They were of an interconnected nature that would allow some A/B function to remain if the A/B pump were to fail. BuAer found this to be an unsafe flight configuration as failure of either the aircraft boost pump or the A/B pump would cause complete failure of the A/B. BuAer required dual pumping systems due to the fact that the aircraft being

designed to use the A/B would require it to be in use during take-off.

At some point prior to June 1, 1949, the dedicated A/B version of the -6 began to be referred to as the -8 model and it will be referred to with that nomenclature from this point forward. No official documentation making this model change was located. It must have been realized at about that time that the A/B capability had to be designed into an engine from the start and required enough differences in design details to eliminate any chance the A/B function would be a "clip-on" capability of the basic -6 engine.

The August progress report showed good progress on the first version of the A/B, which would have a fixed nozzle for initial testing.[33] (These early nozzles and A/B construction were apparently tested on the -4 engines.)

In September, WAGT requested that BuAer consider allowing them to consolidate the primary and emergency power control units into one unit which would save eight pounds. They also asked permission to design the emergency nozzle control system to utilize the aircraft electrical power system. These changes should not be confused with the "simplified power control" request of BuAer. The request originated at the Mock-up Board for the XJ46-WE-2[34] and carried over to the -8 as well.

The issues related to the nozzle actuator system were the same as the -6, namely reducing the chance of total oil loss if a hydraulic line in the actuator system were to rupture. WAGT responded that an automatic pressure loss check valve (fuse) was investigated but it would not address slow leaks and was abandoned. They noted the hydraulic high pressure lines should be kept as short as possible (reducing risk of damage) by positioning the actuator cylinder on the rear bearing support and placing the high pressure oil pump as close to it as was practical. They rejected moving the actuator further to the rear and placing it on the nozzle for the reasons suggested by BuAer. One hand written note suggested WAGT should investigate internal integral oil lines (in the engine casing) but no evidence was found that this was ever formally suggested or investigated.[35] WAGT's design approach enabled reuse of much of their -6 design for the exhaust nozzle system, modified to handle the increased gas flow and temperatures expected with the A/B in use.

The specifications for the -8 first appear in specification WAGT-X40E2-2E of November 18, 1949 as part of Appendix D, newly added to the specification. This was revised in March 14, 1950 and was the earliest copy of this specification located. It now covered the -8 as well as the -6. This second revision only included the tables listing the performance curves that applied to the -8 but did not include any technical discussion of the engine beyond the basic -6 engine. The draft technical specifications (above and beyond the -6) would be added as Appendix D to the specification over time. They are summarized below as they appeared for the first time.

WAGT-X40E2-2F Appendix D J40-WE-8 Performance Specifications			
Total Weight of Dry Engine: 3,394 lbs (with up to +66 lb overage)			
Performance Guarantees			
Ratings	Jet Thrust lbs	RPM	SFC*
Take-off	10,900	7,260	2.48
Military	7,400	7,260	0.98
Normal	6,700	7,260	0.93
90% Normal	6,030	7,260	0.90
75% Normal	5,025	7,050	0.88
Idle	450	3,000	4.00

*Based on minimum heat of combustion of 18,800 BTU/LB, fuel at 77°F. (AN-F-58 fuel)

BuAer accepted the design of the actuating systems in August, although they noted they did not meet all the requirements. They found that the alternative designs considered were unacceptable due to complexity or weight. A later memo indicated the acceptance was based on time pressures to get engines into test.[36]

In this timeframe, WAGT asked BuAer to purchase additional equipment for the development of the engine and BuAer refused, citing they felt it was a Westinghouse responsibility and the equipment would be useful in time for future contracts, not just the current one.[37]

As part of the -8 design work, the gearbox was redesigned in detail to be symmetrical, moving to a dry sump configuration with an oil supply tank of larger capacity hung from the upper rear of the gearbox surrounding the stub shaft running from the front of the compressor into the gearbox. The true hour-glass shape of the gearbox allowed more room for larger gear sets in the upper portion of the gearbox, these being able to drive higher capacity generators. The design allowed for unlimited negative "G" or inverted operation of an aircraft and easier checking of oil levels during servicing. The design incorporated valves to block reverse flow of oil back into the gearbox during other than level flight to prevent flooding. It also allowed for the use of a three-toothed starter jaw adapter on the XIIF type starter drive (something desired but not supported on the -6 gearbox due to lack of space). Also, the -8 gearbox supported the required flange and spline adaptors for use of the Type XVI generator drive. The new gearbox added 44 pounds to the weight of the engine over the -6 asymmetrical design (not including the oil tank). This was offset by eliminating the tachometer generator drive adapter, saving one pound in weight.[38]

Engine Start Power Requirements vs. Temperatures				
Start time 30 Seconds on a Standard Day / 40 Seconds on a -55°C Day[39]				
	-45°C		15.6°C	
	Power (HP)	Speed (RPM)	Power (HP)	Speed (RPM)
Fuel Initiation (Est)	--	300	--	300
Firing Speed	70	860	35	860
Maximum Power	76	1,200	38	1,200
Cut Out	4	1,800	4	1,800

Preliminary performance curves for the design were showing that the engine would have slightly better SFC's at high thrust and higher SFC's at low thrust than the guarantees in the specification. WAGT requested in August that BuAer consider changing the specification, given that studies showed the current expected fuel consumption characteristics were more beneficial than the current specification.[40]

A discussion between Douglas and WAGT regarding the maximum fuel demand of the engine is enlightening insofar as showing the newness of designing airframes for A/B engines. The discussion entailed the maximum engine fuel demand vs. the ability of the airframe to supply the fuel. The airframe booster pumps, in parallel, could supply fuel at 48,000 lbs/hr and the engine fuel pumps (primary and backup) had a capacity of 16,000 lbs/hr. WAGT suggested limiting the vane type A/B fuel pump to 32,000 additional lbs/hr, which would allow the engine to perform at maximum design points and not exceed the airframe's ability to supply positive fuel pressure at the engine pumps. There would be a slight limitation on ram pressure attainable at sea level and -67°F. The engine performance would allow the aircraft to reach its design point at sea level and 60°F. If one of the parallel airframe pumps failed, maximum thrust could still be achieved up to 6,000 ft. The proposed volume limitation on the vane pump used for the Douglas XF4D-1 would be achieved by special shims inserted into the pump during test until the maximum volume was attained. Engines with such a volume limited pump would need special tracking, as they would not be interchangeable with other -8 engines unless the pump was replaced with a pump with no such limitation prior to alternative installation in another non-Douglas airframe.[41]

The September progress report showed that the fixed nozzle A/B was ready for testing but was waiting for an available test cell (the -4 50 hour test had precedence). The decision to make the -8 longer to allow it to be interchangeable with the -10 was expected to lengthen the development schedule of the -8.[42] The actual BuAer approval of the length change was received in October.[43]

In October, BuAer also approved a two pound weight increase to allow for the installation of a preflight emergency control check system on both the -6 and -8.[44]

During this period, Westinghouse realized that the clamshell approach to pressure control in an A/B engine was not the best approach from either physical strength or exit flow characteristics. They began to explore a multi-petal nozzle, which would provide either a conical nozzle or a converging-diverging nozzle with favorable characteristics at supersonic exit velocities. The following drawings, taken from a patent application, show the initial thinking for such a new feather nozzle and its construction. While this nozzle approach was not qualified for the -8 engine, it was planned to use such nozzles on later A/B versions of the J40 (-22B and -10).

Various changes being made to fundamental systems in the -6, such as a dump valve in the fuel system, scuppers, etc. were applied to the -8 as its design moved forward and were discussed in the -6 model development and production sections.

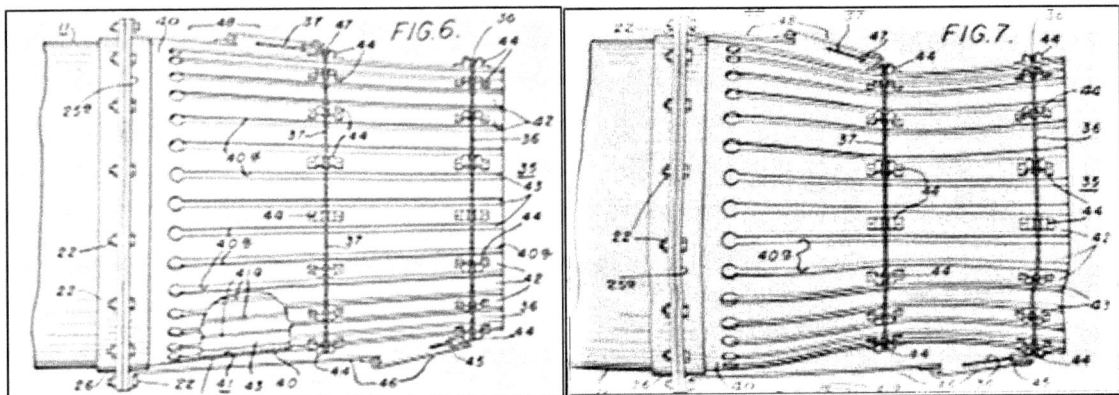

Figure 6 "Feathered Exhaust Nozzle" in place of clamshell, patent drawings. This might have been one of the early planned designs for the -10. See the -10 chapter. *WAGT Patent Application WM5124, September 29, 1949*

Weight concerns were back in November, front and center, as BuAer continued to pressure Westinghouse to reduce the projected weight of the -8 engine, citing feedback from the airframe manufacturers during weight surveys. The projected weight of the -8 engine was beginning to make it look less attractive and WAGT was strongly encouraged to re-engineer weight reductions into the engine. The requested WAGT weight of 3,584 lbs was rejected, being 177 lbs over the approved weight at that point. Every gain or loss of engine weight was subject to review at this point and, as will be seen, even while reliability issues with the -8 during testing were forcing strengthening of the engine, the Navy continued to press for the lower weights. WAGT was noted for not having a strong weight control process and was strongly encouraged to put a robust process in place. The increase in weight seen in the -4 and -6 was reoccurring in the -8 engine and BuAer placed as much pressure on WAGT to hold the line on weight as they possibly could.[45] Ironically, even as they pressed for lower weights, BuAer was authorizing weight increases for various design changes made to improve installation, such as the 5-point mounting change to the No.2 bearing housing (3 lbs), incorporated in WAGT-E40E2-2D.[46] Westinghouse's detailed response on their weight control process ultimately had no practical effect on reducing the weight of the engine, ultimately being driven by detail design changes for reliability or strength.[47]

The initial thoughts on the design of the A/B flame holder apparatus was captured in invention application WM 5268 dated December 30, 1949. The following drawings show a concentric three ring fuel feed and a two ring, V-shaped flame holder held in position via knife edge X-shaped supports. This design would undergo minor changes as detailed design progressed and testing started. From the very beginning, this design was integrated with the A/B power control unit to provide for three zones of A/B power, not just an on/off A/B.

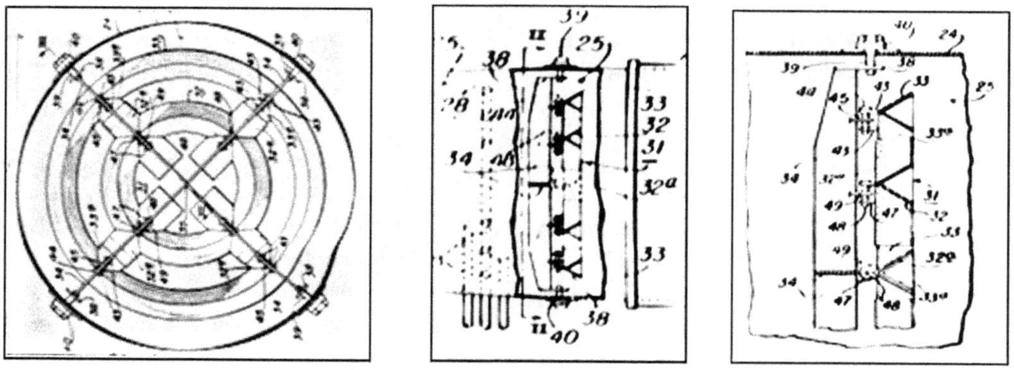

Figure 7 Early XJ40-WE-8 flame holder design elements sketches. Final design was not far from this, finalizing on a three ring cruciform supported arrangement. *WAGT Pat. App. WM5268, not filed, not dated*

The battle over weight continued in January with the negotiations to modify specification WAGT-X40E2-2E for release to the airframe manufacturers. A table in the memo summarizing the negotiations for the -8 showed WAGT's current position on the weight of the -8 based on approved or about to be approved changes:[48] In accepting the revised specification, BuAer required the following changes: Add 30 pounds for the increase in length of the engine to make it interchangeable with the J40-WE-10 (A/B version of the J40-WE-12 to be covered further on) and add three pounds for the addition of the 5 point engine mounting system. This brought the specification engine weight to 3,394 lbs.[49] Later evidence shows the three pound increase for the lugs on the -8 were never officially approved, bringing the specification weight back down to 3,391 lbs.[50] Much later BuAer came to understand that the -8 gearbox had strength changes in its design to accommodate the stresses related to the 5 lugs and that removing the lugs themselves only saved 1.5 lbs. That and a Grumman request that the -8 retain the 5 lugs caused BuAer to change the specification to include the lugs on the -8, bringing the specification weight back up to 3,394 lbs.[51]

A review of the need for a cockpit signal showing the A/B was in operation was sent out to the airframe manufacturers. It was found that none of them required such a signal and the requirement was withdrawn.[52]

Looking ahead (with the -6 about to undergo testing leading to a qualification run), BuAer accepted WAGT's request that qualification of the dual fuel pump for the -6 be considered qualified for the -8 as well.[53]

Features Affecting Weight of the J40-WE-8 in Changing the WAGT-X40E2-2E Specification	
Item	Weight (lbs)
WAGT-X40E2-2D Bid Specification	3,366.0
Modified Symmetrical Gearbox	8.0
Oil Filler Provisions	3.5
Oil Dipstick Provisions	1.0
Electric Junction Box Relocation	1.5
Firewall Seal Attachment Provisions	1.0
Emergency Control Preflight Check Provisions	2.0
Alignment Provisions for A/B	2.0
Removal of Tachometer Drive Gearbox	-0.8
Oil Drain Valve	3.0
Simplified Control System	-26.5
WAGT-X40E2-2F Weight	3,361.0

Development of the fuel controls for the -8 presented a challenge given the desire to limit the weight of such a system. The drawing included in Patent Application WM 5130 shows Westinghouse's approach to such a system while in development. It shows a three pump system with pressure relief and check valves to allow the pumping system to shift and balance fuel volume demands even if one of the pumps failed. At this point, the A/B pump was not intended to run except when A/B power was selected. The drawing shows the pumps being driven from the -2 type of gear take-off, not the -8 gearbox.

In July, WAGT increased the length of the compressor housing slightly to accommodate the strengthening needed to absorb the additional A/B loadings. This resulted in a .23" increase and had to be reviewed by the airframe manufacturers.[54]

The effects of supersonic flight and compression effects on the thrust curve calculations were beginning to be understood. In July of 1951, WAGT sent a revised set of performance curves to BuAer with an extensive

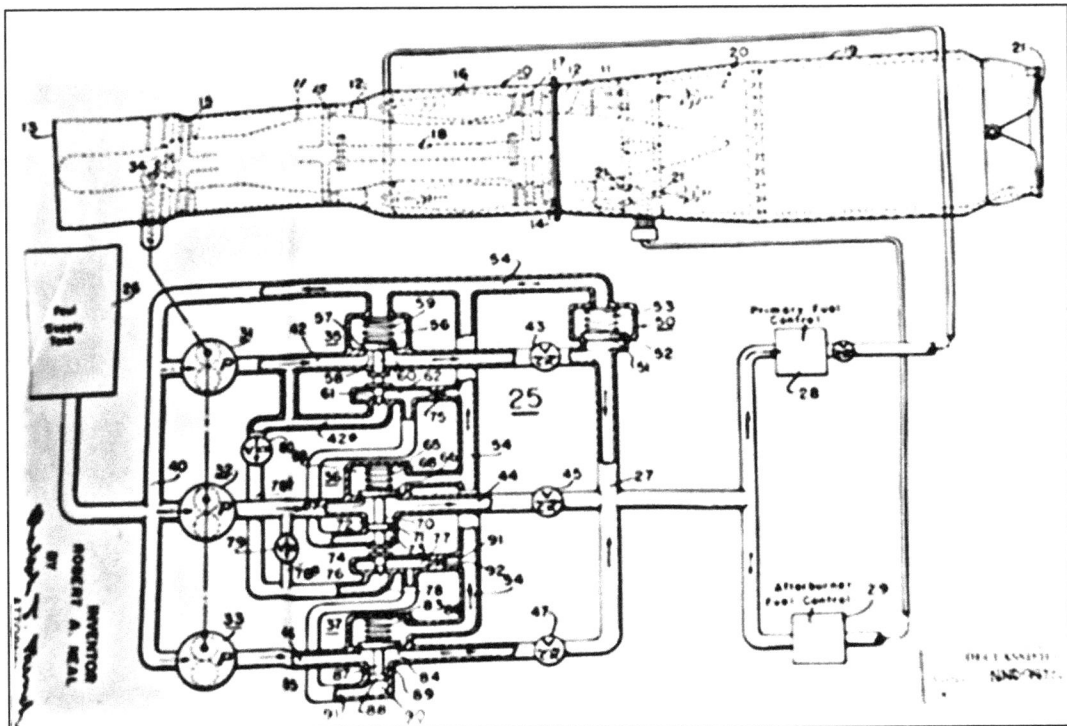

Figure 8 XJ40-WE-8 Fuel control system design to limit pressure on an A/B Engine. *WAGT Pat. Appl. WM 5130*[55]

explanation for the changes. Applied to the -6 and -8 curves, it resulted in minor changes at the upper limits of speed and altitude and changed the shape slightly. These findings would apply to all turbojet engine performance curves, not just the WAGT projects. Findings like these demonstrate the impact of increasing knowledge of how to apply compression effects and supersonic airflow characteristics to practical effect on airframe and engine development. The net of these changes were:

Changes in Thrust and SFC Curves Resulting in Applying Improved Calculation Methods[56]		
	Thrust Reduction	
	Minimum %	Maximum %
Maximum Thrust	0	19
Military Thrust	0	21
	SFC Increase	
Maximum Thrust	-2 (*)	20 (#)
Military Thrust	9	25

(*) 35,000 ft. low flight speed (a reduction in SFC)
(#) Sea Level high flight speed

The new performance and SFC curves were included in the draft specification WAGT-X40E2-2F sent to BuAer for review and approval on August 16, 1951. In their response, in questions relative to the -8, BuAer asked: 1) What accounted for the marked difference in the effect of pressure ratio on thrust for the J40-WE-6 and XJ40-WE-8 engines and 2) Help BuAer understand the change in SFC's using the new calculation method, as they did not understand it based on the data submitted.[57] The WAGT detailed response stated the curves in question now reflected the addition of the A/B extension and were adjusted for the new aerodynamic knowledge and reflected the A/B's presence whether in operation or not. The modified mathematical formulas were detailed for all of these calculations.[58]

In September, BuAer asked that the Appendices of specification WAGT-X40E2-2F covering the -8 (and the -22 production version of the -8) be resubmitted as a new separate specification and the appendices removed from the -2F specification under review at that time. The primary reason given was the change in the control system and different performance of the -8.[59] It is confusing that at this late date BuAer was asking for a separate specification for the -8 because as early as October 1948 both BuAer and WAGT were sending out preliminary performance curves to airframe manufacturers for the -8 and listed the specification number as WAGT-E40E3-2. (That much of the correspondence on the -8 can contain either the WAGT-X40E2-2x or WAGT-X40E3-2x reference has made tracking the development issues for this model a challenge. Apparently at times BuAer was just as confused.)

A/B design continued in detail. One invention Patent application (draft inset below) found in the files shows an approach to squeal suppression that allowed for thermal expansion of the components.

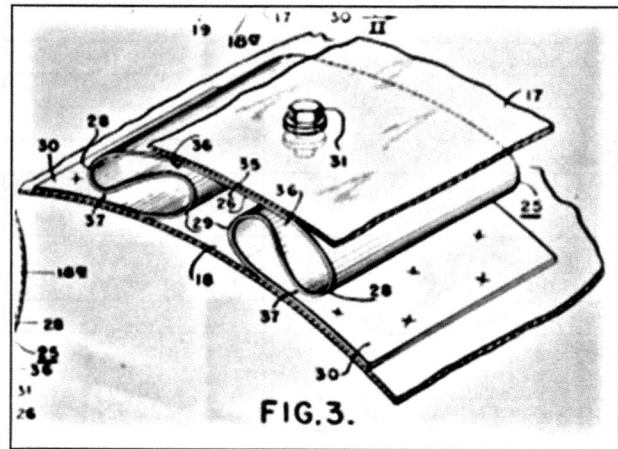

Figure 9 A/B Liner Attachment Approach. *WAGT Patent Appl. WM 269, October 5, 1953*

The October 1949 activity report showed the gearbox design and production falling behind schedule and the fixed nozzle prototype A/B ready for testing but waiting for the XJ40-WE-4 50 hour test to end.[60] The November report showed the same status, with the December report showing that two fixed nozzle A/B's were available and testing was expected to be complete by the end of January, 1951.[61]

The simplified fuel system was in dispute again in January. Several of the WAGT design assumptions were predicated on development and acceptance of the simplified fuel system previously proposed on Sept 9, 1949. No progress was being made due to the lack of Navy funding. In a memo on this subject, the first paragraph stated *"although Navy interest in the system has been expressed, no action for negotiation of the development has been initiated by the Bureau of Aeronautics due to budgetary considerations. Technical representatives of the contractor have been cautioned several times within the last several months that they should not assume that this bureau could or would negotiate the procurement."*[62] Development of the -8 continued assuming the electronic system would be used and modified for A/B functions.

Specification WAGT-X40E2-2E was accepted on January 7, 1950, with the following changes in Appendix D: *"1. The weight decrease of 26.5 pounds for the simplification of the engine control system will be accepted tentatively, pending receipt of a reply to and subsequent action on reference (a) (preceding paragraph above), 2. Add 30 pounds for the increase in length necessary to make the XJ40-WE-8 engine interchangeable with the XJ40-WE-10 engine, as authorized, and 3. Add 3 pounds for the addition of the 4 point engine mounting system."*[63]

BuAer eliminated the specification for minimum inlet pressure recovery by the airframe manufacturers in February 1950, reasoning that the airframe manufacturers made every attempt to design inlets with maximum recovery in the normal course of design.[64]

The February progress report was not encouraging. Only minor testing had been accomplished in January and problems with the hot-streak ignition pattern along with lack of test cell availability was holding up progress. The only test cell available was being held for XJ40-WE-6 testing. Some engine elements were still in design and others were undergoing weight reduction studies. The report warned that having an XJ40-WE-8 engine ready for test by July or August 1950 was doubtful.[65]

April brought a new requirement for crash landing strength. Specification SS-1C-2, which applied to aircraft structures, specified that acceleration on the pilot's seat of 40G's or less had to be survivable. All equipment aft of the pilot's seat had to be able to withstand such an impact without breaking loose and sliding into the cockpit area.

The three requirements resulting from this specification were:

1. The engine mounts had to be capable of withstanding a statically applied load in the aft direction equal to 40 times the engine weight, distributed in any proportion among the engine mounts, including those mounts not normally taking fore and aft loads.

2. The engine rotor, gear case, components and accessory supports or mountings had to have sufficient strength to prevent these items from breaking loose when a 40G acceleration was applied in the aft direction through the mounts as stated in requirement 1.

3. A weight estimate was needed for providing 40G capable engine mounts on the XJ40-WE-8 engine based on requirements 1 and 2.[66]

The March activity report showed a total of 44 minutes of actual test time on the Phase I (fixed nozzle) A/B. Failures of the squeal baffle cylinder and the secondary fuel injection igniter nozzle were attributed to A/B squeal vibration damage. Redesign was underway. Progress on other elements was satisfactory, but some parts were still in detail design. The possibility of getting an XJ40-WE-8 into test in August did not look good, but if the 150 hour test on the XJ40-WE-6 was to be completed by the end of June, the prospect would improve.[67]

In April, WAGT proposed dropping the "three pump" fuel system for the -8 fuel A/B emergency control system on the grounds that with the accepted proposal to move to a common connection to airframe booster pump from the engine and A/B, all of the sensors, valves and plumbing

associated with the "three pump" design were no longer needed, as logically, if the main fuel supply to the engine failed, the A/B would also cease to function. The original design guarded against a failure of the A/B fuel pump itself or the booster pump, switching automatically to the dual fuel engine pump if a failure was sensed. An estimated 8 pounds would be saved. This was proposed as Engine Change 68.[68]

In response to a WAGT memo stating that the throttle torque of the J34-WE-32 and -38 and the J40-WE-6 would be about 50 inch pounds[69], BuAer forwarded a review of a recent, quick study into both pilot preferences on throttle torque and the actual torque values as measured in turbojet aircraft at the test center. They confirmed that the specifications AN-N-30 "Engines: Aircraft Turbo-Jet General Specifications for, 14 June 1946" and MIL-E-5007 "Engines; General Specification for Aircraft Turbo-Jet, 19 July 1949" were validated, both requiring the throttle torque not to exceed 15 inch pounds. They required WAGT to respond to their plans to bring their fuel control design into compliance with the specifications.[70]

April's monthly report showed the total test time had only risen to one hour, with squeal still being experienced. The fixed nozzle would continue to be used until squeal had been eliminated. Some progress in that direction was being obtained. Again the status of the -6 completing its 150 hour test satisfactorily was cited as a hope in moving the -8 back towards its first test schedule.[71]

Douglas began design of an emergency air driven fuel pump and asked for design pressures and volumes to use from Westinghouse. WAGT responded at length through a letter to BuAer with the proper curves to use and pressure and volume assumptions that should be applied. Since the survey for eliminating the "three pump" fuel control system was still underway, several points were made assuming that it would be approved.[72]

The emergency fuel system dump valve proposed by WAGT back in May of 1949 was finally rejected as unnecessary by BuAer in June of 1950. They pointed out that since WAGT had stated in their proposal that the valve would not add weight to the engine, elimination of it should not reduce the weight either.[73]

The progress report for May showed total A/B testing time of only two hours, with squeal still occurring in every configuration tested to date. The basic design of the engine was complete, but some components had not been released to manufacturing because changes were expected based on testing of the -6. The qualification test of the -6 was expected in July and if successful, would ease progress problems on the -8.[74] Presumably, this meant the basic engine would cease to hold up -8 testing, but the report does not specify.

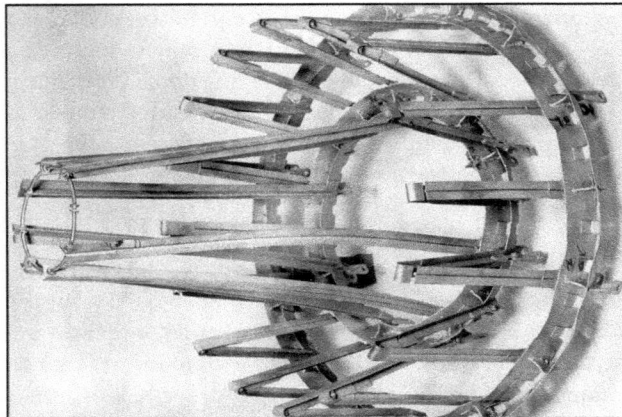

Figure 10 J40-WE-8 #3 Certification test engine A/B flameholder after test, WAGT P47604, 9/3/1952. *Courtesy Hagley Museum and Library*

On June, 22, 1950, the Letter of Intent contract was converted to a Cost Plus Fixed Fee contract with the intent of removing contractual issues that WAGT argued were inhibiting their ability to make better progress.[75] In August, the official delivery date of the crated A/B, having passed acceptance tests, was moved to November 30, 1950.

In August, citing the fact that difficulty was being experienced meeting the fuel consumption targets for the -6 and with the -8 being a development of the basic -6, WAGT offered BuAer two sets of alternative schedules for the -8 based on which set would be selected for the -6:

Alternative Milestone Dates for -8		
Milestone Description (Cont. #)	Schedule A Engine with Available Design and Uncertain SFC's	Schedule B Engine Meeting All Contract Guarantees
Mockup, Installation Drawings	10/31/1950	
150 hr Test and Report	03/31/1951	06/30/1951
150 hr Engine	04/30/1951	07/31/1951
Final Design Drawings	04/30/1951	07/31/1951
Final Summary Report	05/31/1951	08/31/1951

Monthly progress reports were not found for the -8 between June and September of 1950. It is possible the diversion of resources at WAGT to resolve problems with the -6 engine then being encountered could account for this

lapse. The October report reveals that little progress on the A/B design was being made. The Phase I fixed nozzle design was still the only design tested and it did not yet incorporate the integral afterburner cooling provisions. The main focus was on flame holder development at that point and tests had not achieved the required 10,900 lbs thrust. Two versions of the Phase II A/B had been ordered by WAGT, both with a variable area nozzle and integral cooling system. One version had a cross-corrugated internal cooling arrangement and the other had a concentric ring internal cooling arrangement. Both were expected to achieve the 10,900 lbs thrust specification. Neither would be available to meet the November delivery schedule. Weight reduction was satisfactory, with some elements of the redesign being installed on the No. 1 -6 engine for durability and performance testing, but assembly was not complete due to emphasis being put on the XJ40-WE-6 No. 2 engine for the 150 hour test under NOa(s) 9212. The BAR now estimated an XJ40-WE-8 not being delivered until January 31, 1951 or even later.[76]

In the latter half of October, WAGT surveyed the airframe companies about their desire to lengthen the -8 (and -1) by 27/64", widen the diameter across the nozzle push-rod cowls by 1/16" and allow the exhaust gases to exit in a new stream cross-section pattern. McDonnell rejected the request on the grounds that the aft section of the airframe cooling was already critical and they could not accept the change. They also complained about such a change being suggested so close to the qualification test with production engines expected quickly thereafter.[77]

November's report continued to show progress slippage. The combined control system (engine plus A/B) was being evaluated and initial design troubles resolved. One primary and one emergency fuel regulator were on hand but had not yet been tested. The weight reduction parts had still not been tested. XJ40-WE-6 #1 was to receive these parts but that engine was idle. The XJ40-WE-4 had suffered an exhaust nozzle failure and had to test a reinforced nozzle before it was available for A/B testing. Manufacturing problems were still being experienced on the Phase II A/B's but delivery was expected the second week of January. The outlook was now completion of the -8 150 hour qualification test by March 31, 1951.[78]

A minor internal design change in the engine reduced the distance between the engine mounts by 0.05" and just to be sure, WAGT sent out EC 74 requesting the airframe companies ensure there would be no problems. WAGT was assuming that the built-in tolerance of the airframe mounts to accommodate thermal expansion would allow the change without a problem, but was being cautious by sending out the survey.[79]

WAGT attempted to close out the Mock-up Board report recommendations in early January 1951 with the issue of a memo that addressed Item 12, and surveyed Engine Changes 75, 76, 77 and 78. Briefly they covered:[80]

1. EC 75 – Covered the incorporation of the "Simplified Control System", which was not shown at the time of the mock-up but was discussed. *(The BuAer study on this was not located.)*

2. EC 76 – Covered the final A/B cooling design, which incorporated EC Nos. 52 and 70. The new design used compressor air blown back through two 2.5" pipes to the A/B cooling shroud.

3. EC 77 – Covered the replacement of the high tension ignition system with the low tension high energy system.

4. EC 78 – To maintain common jack lug locations between the -6 and -8, the jack lugs on the -8 were relocated to be common to the -6.

BuAer reviewed the proposed A/B fuel control system (Report A-1041 [*Report not located.*]) and responded with the following comments:

1. The two external adjustments provided on the control system were found undesirable because it was expected that even though they allowed initial ground level adjustment to obtain optimum ground level performance, field operation would soon cause the adjustments to drift and resetting them in the field by limited experience service personnel was as likely to cause a gain as a loss in performance. It was requested that the external adjustment capability be removed. In addition, a new requirement was added: Submit a calibration of the A/B fuel control prior to qualification test of the component. The calibration should include the schedule of fuel flow, with associated production tolerances vs. compressor discharge pressure. The calibration should be made with Specification MIL-F-7024 Calibration Fluid.

2. Contamination of the control flow filter could adversely affect the accuracy of the fuel metering if it reduced the control flow below the regulating range on the constant flow valve. The regulating range of the constant flow valve should be made as wide as practicable and the housing for the filter should be designed so that the filter could be readily removed and cleaned.

3. Because it was expected that normal service use of the engine would cause a large shift in the calibration of the control, the pilot should be provided with a means in the cockpit to determine that full A/B thrust was available prior to take-off. A satisfactory operating procedure recommendation was requested to determine satisfactory afterburner operation.[81]

The progress report for December, the forwarding memo bringing progress up to date through January 15, reported progress on the -8 indicated initial testing including the Phase II cross-corrugated design A/B, would begin in early February. The combustion liner, exhaust collector and first stage turbine nozzle had been awaiting test results from the XJ40-WE-6 and were now available for the -8. No A/B testing had occurred during December through January 15. The cross-corrugated A/B was due from the supplier on January 22 (considerable trouble having been encountered during manufacture). The initial XJ40-WE-8 test engine would not have a complete set of accessories. To meet a March 31 qualification test date, considerable prototype testing needed to be accomplished. Key risk elements included the redesigned weight reduction components and the fact that the Phase II A/B had not been tested at all as yet. Also, the combined control system was new. Considerable doubt was expressed that the March 31 completion date could be met.[82]

BuAer once again reacted to the spiraling weight increases with a memo recounting that between reports from October through December, the -8 weight had increased from 3,390 to 3,423.7 lbs, a highly undesirable trend. The center of gravity was remaining constant at 69.6 inches, but the importance of weight and center of gravity to airframe designers was emphasized. The concern over weight management and BuAer's concerns was again stated.

January, 1951 showed the effort to get an engine into test had slipped the test date into late March. The new Phase II A/B would be tested first on a "house" engine (XJ40-WE-6 #1, Serial WE003002). A slotted seal design backup to the plain seal was in design for the nozzle. Only 9 total testing hours on the A/B components had been accomplished to date, most of which were used to evaluate flameholder design and elimination of the squeal, but nothing was definitively solved by the testing. Squeal was still present and parts durability was a problem. Testing of the first titanium compressor disks showed an excessive vibration of the eleventh stage. Machining of the 11th stage disks was stopped, but vibration testing continued on the 8th, 9th, 10th and 11th stages. All accessories were in hand and bench and engine test evaluation was underway with no holdups anticipated. The power regulator and combined fuel system had been though 400 hours of testing and more was planned.[83]

The acceleration control on the -6 was found to be totally unacceptable on the first few production engines being flown by McDonnell. Compressor stalls on slam accelerations were being experienced plus the acceleration time was 27 seconds. BuAer commented that since the simplified power control was to be used on the essentially similar -8 along with a new acceleration control, the focus should be to develop the simplified control and retrofit it to the -6 as the solution to the -6 acceleration problems rather than split the development effort on two separate systems.[84] WAGT proposed an override speed switch on the engine fuel control to force the nozzle to remain open while the engine accelerated from idle to military rpm, at which point the normal nozzle control would resume. This was sent out as EC 89.[85] A stall limiter comprised of an addition to the electronic control with a cockpit mounted manual adjustment for ambient temperature reduced acceleration times on the ground from 27 to 20 seconds on early -6 engines. It was later installed on all production -6 engines. It was not recommended for development as automatic since the -8 would not use the electronic control and the development effort would not be worthwhile.[86]

In March the lack of progress was acknowledged during a conference on February 20, 1951, resulting in the Qualification Test date deadline being moved to October 31, 1951 with the final summery report due November 30, 1951.[87]

The separate development contract NOa(s) 10114 was terminated in its entirety on December 28, 1951, however development and production of a limited number of -8 (and -1) engines continued through funding under contract NOa(s) 10385.[88]

The first "production" J40-WE-8 Serial WE031001 was shipped to McDonnell in January 1952 for ground testing only. The qualification test of the J40-WE-8 had not been run yet and the engine shipped differed in minor ways from the intended final configuration. The shipped engine would be brought up to final production standard with whatever modifications were necessary after the qualification test was accomplished successfully.[89] BuAer rejected the suggestion, saying the engine could not be shipped until a -8 had passed the 50 hour flight qualification test.[90] The production story will be covered in the next section.

The flow inlet connector on the F3H-1 to the fuel pump had been lengthened and testing showed it had improved delivery of fuel across the pump face. The tests indicated this was the third design of the connector to date.[91]

With very limited testing of the -8 A/B completed, but with significant experience accumulating from operating A/B versions of the J34 and J46, some details of the simplified control system (sometimes referred to as the "Simple Hydraulic Control System" or "Simple Control System") relative to engine performance and in particular A/B operation began to emerge. In a long memo addressing many different questions contained in different BuAer correspondence (some related to the J46 in parallel development), WAGT provided a consolidated response:[92]

A. A chart was provided showing a smooth increase of fuel flow during the starting cycle, but commented it

might be necessary for the pilot to manipulate the power lever between cut off and idle to prevent over temperature of the engine while accelerating the engine up to idle rpm.

B. The various acceleration times vs. pressures and temperatures of the -8 were presented in Curve No. 375900 as a series of plotted curves against time. The chart showed the engine achieving military RPM (7,260) a little over 7 seconds from a snap throttle movement (today termed a "throttle slam"). The data was taken from the first test XJ40-WE-8 #1 November 1, 1951.

C. The effects of the simplified control on SFC's were discussed in report A-1265. (Report not located.)

D. Experience with afterburner blowout on the early J34-WE-32 in the XF7U-1 after any two successive A/B light-off attempts at altitudes above 10,000 ft showed the blowouts were due to the A/B fuel flow dropping off during the period of from 2 to 5 seconds after advance of the power control level into A/B. Testing on the bench showed this only occurred when a compressor air purge of the A/B fuel manifolds had occurred. This had been incorporated to reduce A/B fuel ring clogging, but was later found to be ineffective and was deleted from the J34-WE-32 (and also not included on the J40-WE-8).

E. The importance of achieving full A/B fuel line filling was discovered in comparing J46 A/B testing with a Thompson air-turbine driven centrifugal A/B pump vs. the -8 gearbox driven centrifugal A/B pump, with the latter achieving full pressure with rapid filling of the lines, and the former causing A/B ignition problems due to surges in turbine outlet pressures. The differences showed the importance of withholding the initial buildup of A/B fuel pump pressure until full military RPM had been reached before allowing A/B fuel flow to commence. This feature was incorporated in the simplified power control by routing the signal to the A/B fuel pump valve via the military RPM switch. This change eliminated acceleration stall of the engines as well as premature actuation of the A/B igniter valve. The -8 could achieve full A/B fuel pressure in 0.8 seconds from A/B being selected with the power control level at military RPM.

F. A full failure analysis of the -8 simple control system had been completed and forwarded to BuAer earlier in response to a request for such a study.[93]

G. Air bleeds and possible problems with contamination were reviewed.

    a. The fuel regulator acceleration valve was controlled by compressor pressure rise across a bellows. No problems encountered in test.

    b. The afterburner fuel flow was controlled by compressor discharge pressure and no problems had been encountered in test.

    c. The A/B blowout switch used 3 pressures in the A/B section of the engine. Two of these communicated their pressures by an orifice to balance pressure ratios. This orifice tended to clog in early development and a dirt trap was incorporated. The development goal was to eliminate this orifice.

d. Further development added a manual trim control, to be operated from the cockpit by the pilot when operating either at military power or in A/B. This involved an adjustable knob to be used to avoid over-temp operation.[94] The knob was rotated counter clockwise to a "reset" position and then clockwise until a proper nozzle (T5) temperature was indicated.

The surging problem of the -6 at altitudes above 30,000 ft. became known at this point both from field operation of the -6 in the XF3H-1 and NACA testing. Analysis showed the compressor would have to operate at a pressure ratio just below 5.6 to eliminate the surging at altitude, but a compressor modified to give satisfactory results at altitude showed low airflow and insufficient stall margin at low speeds for acceleration. It was recommended the temporary fix to the -6 (open the first stage turbine nozzle by 16% and install the compressor outlet air mixer) be carried over to both the -8 for the qualification test engine, pre-qualification and early production J40-WE-8 engines. Investigation would continue.[95] The BAR recommended no decision on the -8 be made at that time pending receipt of requested information on the performance and exhaust nozzle area of the -8.[96]

The emergency control provisions of the simple hydraulic control for the -8 (as well as other planned engines) were the subject of a meeting on January 17 between WAGT and BuAer. The published minutes record BuAer's requirements for a fail-safe emergency control system for turbo-jets, listed here:[97]

1. The engine control system shall meet the following objectives insofar as practicable without special emergency provisions:
 a. The failure during take-off on a NACA standard day plus 40°F of any single functional part, except springs which operate over a low stress range, shall not reduce the engine thrust to below 95% of military thrust over the altitude range of sea level to 6,000 feet altitude.
 b. The failure of any single part shall not result in exceeding engine operating limits to the extent that engine failure cannot be prevented by corrective pilot action.
2. Any safety devices which are provided only for the purpose of meeting the objectives of paragraph 1 shall be the maximum practicable in simplicity and reliability. Only such manual and automatic safety features which, in the judgment of the Bureau of Aeronautics are of adequate

reliability, will be incorporated to provide emergency operation under the failure conditions of paragraph 1.
3. No devices, in addition to those required above, shall be incorporated for the purpose of protecting against the simultaneous failure of two control system parts except where the first failure can cause the second.

At the conference, there were three main topics of conversation:

1. Manual Temperature Control (Exhaust Nozzle Area Trim)
 a. Approved for use on early engines only. Deemed not acceptable for service engine use but allowed for initial flying for aircraft and engine performance evaluation.
 b. The Automatic Temperature Control (or equivalent) was a definite final requirement with WAGT indicating it would be available on November 1952 production engines and retroactively installable on prior engines shipped with the manual trim control.
 c. The automatic trim control would be incorporated in the engine model specification undergoing revision at that time. Manual Trim Control would be an accepted deviation.
2. Manual Override Emergency Exhaust Nozzle Provisions. Consisted of an emergency lever in the cockpit to permit pilot selection of either full open or full closed exhaust nozzle position in the event of aircraft electrical power failure. Variations to this feature were discussed and the following concluded:
 a. None of the features were entirely satisfactory for service engine use.
 b. As an interim measure, the available Manual Override was acceptable for early engines only. The primary objection to the manual feature was that pilot error could easily occur since the lever had to be moved in one direction in normal operation (non-A/B) and the other direction if using the A/B.
 c. Provision should be made as soon as practical to provide a locking arrangement to lock the nozzle in the position it was in at the time of electrical power failure. This feature would supplement the Manual Override for use on service engines until a more suitable arrangement could be provided.
 d. The use of electrical power for basic engine controls was discouraged for long range development. (It also violated BuAer's requirements that all engine power systems had

to be contained on the engine and not be located on the airframe.)

3. Fuel Control System. A review of the general operation of the system was presented and the emergency and fail-safe features described. A complete failure analysis was presented, the risk analysis supported by the background of extensive testing and service experience with critical elements of the fuel control.

    a. No special emergency provisions were required for the fuel control for either experimental or service use. Airframe companies could install interim equipment for limited emergency control for early experimental flying if they wished. WAGT would assist in such activities if asked.

    b. Two points in the fuel control system were identified as not entirely fail-safe:

        i. Acceleration control bellows – failure of the bellows would cause a loss of thrust to that corresponding to idle conditions. The current design was acceptable if it passed the qualification test. A study to determine the feasibility of a fail-safe design for the component would be done.

        ii. Over Speed Relay Drive Shaft – If the shaft sheared, the engine would start to over speed until the pilot retracted the throttle far enough to control the engine on the manual throttle valve. Again, the design was acceptable based on experience to date. Testing of extreme manufacturing tolerance conditions would be conducted to verify operation.

WAGT confirmed they would conduct the studies resulting from the above discussion.

An increase in the engine installation envelope to allow adequate clearance of the A/B fuel lines past the exhaust nozzle control yoke resulted in an additional .25″ in the radial diameter of the engine. This was issued as Engine Change 92 for the -8.[98]

Covers over the thermocouple leads to prevent shorting and damage from insulation blanket installation was surveyed as Engine Change 95. The thermocouple lead would be brought all the way to the junction box. This latter part of the change would apply to all but the first 3 -8 engines and be retroactively installed at overhaul for those.[99]

With the -6 now being used by airframe companies in their experimental aircraft to begin ground and flight testing, WAGT was under pressure to get the A/B version (-8) to them as quickly as possible. What was really needed was early delivery of the -10 and -12 models, which had the power required for the experimental airframes to achieve the desired flight performance, but those models were not even in view at this point of time in early 1952. Given the situation, WAGT proposed the following:[100]

1. Three "50 hour" engines for early flight testing, the ground running in test being charged to contract NOa(s) 52-403c. The control system of these engines would differ in minor ways, the qualification test engine having minor design refinements.

2. One pre-verification J40-WE-8 (Serial 031001) to be accepted under Contract NOa(s) 10385 for ground testing only, to be delivered at no charge, to be billed at the time of completion of modernization to flight configuration (i.e. the build standard of the 50 hour or 150 hour qualification at BuAer choice). If modernized at the time of overhaul, WAGT would furnish only the modernization parts.

3. Engine WE031001 differed in various ways from the engine (XJ40-WE-8 #2) in preparation for the 50 hour qualification test. WE031001 had originally been planned to be the 50 hour test engine.

4. A maximum of seven (7) J40-WE-8 engines verified by the 50 hour test would be delivered in advance of the initial qualified -8 engine deliveries. They could be built to any airframe standard including the XF3H-1 program. (Confusingly, at the point in time of WAGT's memo, the 5th and 6th 50 hour engines were to be built to the -1 standard for the Air Force program. Since no -8 was ever delivered to the Air Force, the requirement must have been dropped.)

5. At no increase in price, they would provide the parts required to convert the 50 hour test engines to the qualified configuration of the -8 or -1 engines.

6. Verification of any improved parts subsequent to the 50 hour test completion would be incorporated as agreed at that time.

The memo noted that two 50 hour J40-WE-12 engines were due at the same time under the contract and a proposal for those would be submitted later.

WAGT informed BuAer that they had reviewed their workload and priorities and rescheduled the estimated transmittal date for the specification to April 1.[101] That the specifications were still in flux and not agreed upon as yet is confirmed by a memo only a week later than the proposal above for early delivery of 50 hour engines for limited ground and flight testing. BuAer noted that the specification WAGT-X40E2-4 covering the -8 (and production version -22) was overdue and they requested it be expedited.[102] Late delivery of the specifications made it difficult for BuAer to decide if the engine build specifications for the early delivery engines would be acceptable.

Late February found WAGT changing the exhaust nozzle area of the -8 to be 726 in² via survey of Engine Change [96]. The change was necessary to achieve maximum thrust. Since the aircraft using the -8/-22 as an interim engine would actually be designed to have the -10 installed later, airframe companies were requested to design to the -10 nozzle area (new) of 962in². [103] The BAR reported the 50 hour unofficial verification test would be conducted on XJ40-WE-8 #2 and be underway by March 15. Delays from a

fuel pump damaging a gearbox resulted in changes to prevent the failure in the future. The engine had undergone 110 hours of running, of which 41 hours had been at military operation with 10 of those 41 hours in A/B. Inspection in preparation for the test resulted in the combustion liner and turbine nozzles being replaced. A new liner, an opened first stage turbine nozzle and new second stage nozzle were installed. [104]

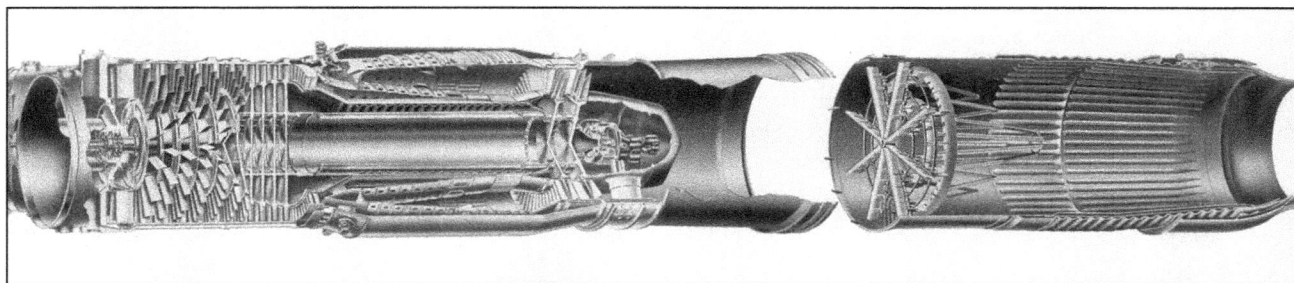

Figure 11 J40-WE-8 Cutaway of final configuration. Note shaft is now a one piece machined part. WAGT P47912, 10/17/1952. *Courtesy Hagley Museum and Library*

It was discovered through analysis that the afterburner yoke design needed an increase in stroke to fully utilize the area variations of the exhaust nozzle. The main change was to increase the stroke by approximately one inch, which brought the yoke within 7/32 inch of the rear bearing support. A survey was sent out to get feedback on the impact of such a change. [105]

February's discussions relative to the simplified control system and its emergency and failsafe provisions continued into March as WAGT continued to try to evolve a satisfactory solution both short and long-term. Continued BuAer review of the system based on more complete information from WAGT resulted in a WAGT response to a series of BuAer concerns and objections: [106]

BuAer's unacceptable features:

1. *The exhaust nozzle failure characteristics and emergency manual override.* These did not meet the fail safe requirements because aircraft electrical power was used for primary control purposes and the emergency mechanical override was unacceptable. WAGT agreed with the objection and stated they were studying an alternative hydro-mechanical control that would meet fail safe requirements and not require an emergency feature. A schematic arrangement had been tentatively agreed upon and a prototype schedule established with the production prototype available December 1, Qualification Test Start on February 1, 1953, Flight test start February 1, 1953, and first engine delivery June, 1953. (It is not clear if this early design was a minimum fail safe using mutually acceptable

alterations of the electrical systems consistent with meeting schedule requirements for interim service use and existing schedules, to be followed by a final design for full production engines, or if it was the final design.)

2. *Manual Temperature Trim.* This was not acceptable in a combat aircraft due to the requirement of the pilot to continually set the exhaust nozzle area relative to a cockpit indication of engine temperature. WAGT indicated an Automatic Control, developed to replace the manual temperature trim control, would be available for production engines in November 1952.

3. *Lack of a control to limit the compressor discharge pressure to a safe value.* On cold days at high speed, the temperature of the compressor outlet pressure could exceed the design values of the engine. The pilot had to manually reduce the power lever setting to lower the turbine inlet temperature to a safe value until either the aircraft slowed or gained altitude to fly out of the pressure limiting region. Pressure limiting ceased at any altitude above 16,000 feet regardless of outside air temperature. Afterburner use was not allowed while in a pressure limiting condition. WAGT was studying how to delete the requirement for a cockpit gauge for manual pilot control of pressure within the critical range of operations. Under consideration was a servo to operate in response to compressor outlet total absolute pressure which would drive a potentiometer to bias the electrical temperature trim signal to automatically reduce fuel flow. The goal was a definitive program by July 1, 1952 with incorporation into production engines by June 1953.

4. *No direct manual means for shutting off fuel flow to the afterburner.* WAGT noted the existing design already had an automatic fail safe shut off. If the engine electrical system

failed, or the power control was pulled back out of the A/B zone, the A/B pump shut-off valve automatically closed with a metal to metal contact and bypassed all the flow even with the pump operating at full volume. The distributor rotor was directly operated manually by the power control lever. The -8 also incorporated a resilient-seated shut-off valve in series with the distributor valve to eliminate any leakage past the metal to metal seal. This second valve operated from a servo driven by engine pump pressure at military power, making it independent of both the electrical power and afterburner pump operations. The net effect was that the pilot was actually manually shutting off the fuel flow to the A/B using the power control lever when it was moved from the A/B zone to Military Power or below.

5. *External fluid lines on the variable displacement nozzle actuating pump.* WAGT recognized the undesirability of having the external lines. They were using a Vickers pump and the lines could not be cast internally in time to meet the production schedules. The focus had been to verify the control features of the pump and secondarily study a pump rearrangement to eliminate the external lines of the control features. Current work would produce a prototype of the modified pump by December 1952 and have them available to incorporate into production engines starting in March 1953. Interim engines built with pumps incorporating external lines would have those lines held as close to the pump body as possible to lessen the possibility of handling damage. An alternative Bendix pump with internal lines, interchangeable with the Vickers design, was in test at WAGT at that time.

Other minor features needing correction:

1. *Lack of modulating afterburner thrust.* Modulation of the thrust of the -8 A/B was not recommended with the control. WAGT noted the A/B was built to be modulated and the recommendation was a conservative one until flight experience demonstrated modulation reliability. (In fact, on the -22 engine, A/B modulation was removed from the control unit.)

2. *Small thrust range at intermediate altitudes.* WAGT pointed out they had only been presenting idle thrust at low flight speeds, since it was more important than high flight speed thrust. WAGT enclosed a chart showing a more detailed picture of the altitude idle thrust. Below 35K feet, the J34-WE-36 and J40-WE-8 had very similar idle thrust values. Above 35K ft, the J40-WE-8 had considerably greater thrust. It was believed this additional data would resolve this design issue and it would be satisfactory to the Bureau.

3. *Externally adjustable fuel acceleration schedule.* It was too easy to be re-adjusted in service. WAGT responded that the original design was changed to be a jam nut against the

adjusting shaft of the aneroid bellows with a safety wired acorn nut over it to make such adjustments more difficult.

4. *Externally adjustable Microjet switch setting.* This was still in flight test evaluation and not released as yet to replace the current Aerotec switch being used. So far, it was considered necessary for the external adjustment to be kept for line maintenance checks as well as bench and engine adjustment. Sea level adjustment was not affecting altitude performance based on observations. Further evaluation to provide adequate locking and possible sealing against tampering would continue.

BuAer responded that WAGT had not addressed the data request for numerous items, including the now defined automatic temperature trim control that would be provided in lieu of the manual trim control. With the simplified control now designed, BuAer included in their response which data requests had been answered and which had not and included sample hand drawn chart examples of what they were looking for (included here).[107]

Even with continuing development of the simple control system, the very first pre-production J40-WE-8 (WE031001) was accepted for ground running only in an airframe on March 25, 1952. If the engine reached overhaul status based on any combination of run times, Westinghouse would only be responsible to replace parts improved or changed based on the qualification engine test, with BuAer supplying the balance of the overhaul components. If it required overhaul prior to meeting the agreed overhaul run limits, such overhaul/modernization would be entirely at Westinghouse expense.[108] WAGT requested a waiver on the performance requirements, as the engine as configured could not meet the specification thrust or SFC requirements. BuAer issued the waiver for this engine only.[109] The saga of this engine continued, as only ten days later, on the grounds of unspecified structural integrity of the engine, BuAer authorized it to be used for ground running on a test stand only. It could be shipped to McDonnell provided satisfactory liability arrangements were made between Westinghouse and McDonnell.[110]

In March, a change in design was issued for many O-ring seal grooves in the accessory gearbox to eliminate the chance that O-rings could ride over punched holes in the seal carrier. Doing so could introduce minor oil leaks as the O-ring would go out of tolerance. This change applied to all J40 models in production or in design.[111] Numerous CIDs were issued by WAGT during the life of the J40 program, only a few of which have survived in various files. Most addressed minor changes to ease production, clarify materials requirements or actually correct engine deficiencies as a result of testing or service use. In this case, oil leakage from the engine would be an ongoing concern during the life of the program. The greatly increased temperature environment of an installed turbojet made oil leaks a matter of even greater concern than those experienced with piston engines.

The number one thrust bearing design proved to be marginal, as testing with both the -8 and -6 engines discovered. The initial design was based on preliminary engine design studies which assumed dummy piston pressures in excess of those ultimately attained in the engine. The consequence was that the No. 1 bearing thrust load was larger than originally anticipated over the range of engine operating conditions. WAGT designed a new bearing that proved to have thirty times longer life and recommended that all -6 engines have their thrust bearing replaced at 150 hours or first overhaul after the new bearings became available. The -8 program would introduce the bearing with the 10th production engine and apply the 150 hour replacement rule to the first nine. (The -22 would have the new bearing starting with initial production). BuAer reacted to this saying a 150 hour life of a bearing was non-standard to industry practice or standards. While the BAR endorsed WAGT's recommendations, ultimately this was a price of replacement issue, as the qualification engines would pass their tests with the original design, making upgrades to the engine beyond the qualification build level at BuAer expense. The engine specifications did not cover or include individual component life expectancies.[112]

A curious situation became visible in April when Solar Aircraft Company, through their lawyers, notified Westinghouse that not all Solar patents on A/B designs and components were protected by government free license and that Solar would pursue the Government in the courts as their only avenue of redress if such patents were infringed upon. In such matters, Solar warned Westinghouse the Government was including provisions in their contracts with their suppliers that required assumption of defense of patent responsibilities for compensation due Solar. It is not known if any such action was taken relative to any of the A/B work on the J40.[113] It is not clear how Solar would have become aware of the Westinghouse A/B design details since all of the contracts and materials involved were classified as "Restricted" at the time.

The problems of the -6 entering service were now impacting WAGT's efforts to get at least a very limited number of 50 hour operation limited -8's to the airframe manufacturers. In a three page memo, BuAer spelled out their conditions for accepting any 50 hour engines:

1. Delivery limited to two J40-WE-8 and two J40-WE-1 engines with the following modifications:

a. A fix for inversion of radial distribution of turbine temperature exhibited in the -6 engine in NACA testing.

b. An exhaust nozzle area of not less than 675 in^2 and not more than 726 in^2.

c. A standard J40-WE-6 compressor and turbine section change to make the compressor stall-free (or an improved compressor).

d. An afterburner, the combustion performance of which was devoid of "squeal" or other obviously intolerable condition(s) of operation.

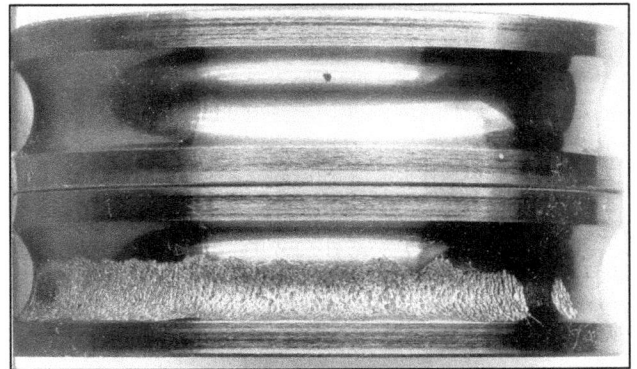

Figure 12 Condition of the original No. 1 bearing in XJ40-WE-8 after A/B testing, WAGT P45807, 12/5/1951. *Courtesy Hagley Museum and Library*

2. The -8 engines must:

a. Demonstrate sea level static performance commensurate with the performance of the XJ40-WE-8/-1 engine(s) which successfully completed the 50 hour test and to which they are identical (except perhaps for minor authorized deviations).

b. Demonstrate at least marginally satisfactory altitude-airspeed performance in service in accordance with the limits specified in the specifications.

c. Be delivered not later than: one -8 in June and one in August, with one -1 in September.

3. Not more than twelve -8 150 hour engines configured to the 50 hour engine will be accepted except:

a. The compressor must be of improved design over the -6 such that the standard -6 turbine is utilized.

4.   The recent 60 hour service test of the XJ40-WE-8 #2 did not meet the requirements listed above and the engine would not be accepted.[114]

Westinghouse wrote a long response in August to this memo, taking exception to some of the conditions. The most important were:[115]

1.   In agreement with the specifications, engines presented for acceptance would not have more than 16 hours total running time, with no more than 75% being above normal rated power. The specifications did not cover number of A/B starts or amount of A/B run time as BuAer was now requiring.

2.   The 150 hour qualification engine would have the 675 in² nozzle. Later testing would use the 726 in² nozzle.

3.   The contract only included guarantees for sea level performance, all other performance being estimated. BuAer inclusion of acceptance requirements for other than sea level performance was not in accordance with the contract.

By May, BuAer was struggling to keep up with the design differences between the various models, requesting that the automatic restart features of the control system on the -6/-8 engines be disconnected or eliminated. WAGT responded saying the latest control model for the -6 had the feature disabled and it was never present on the -8, as the -8 did not use the electronic control. The -8 had a switch in the cockpit to manually initiate an ignition restart.[116]

A survey was sent out to investigate the mounting of the automatic trim control device to replace the manual trim control that would be used on the first few engines. Airframe mounting of the device was a possibility.[117]

Late May had WAGT responding to a lot of requests for specific data on various engine models under development. Extracted from that memo relative to the -8 are:[118]

1.   Satisfactory starts had been made using the simple hydraulic control within the range of sea level ambient conditions occurring in the WAGT test house. Starting procedures for extreme ambient sea level static conditions still needed to be tested, something that could not be done at WAGT's facilities.

2.   The -6 both at AEL and NACA Cleveland had demonstrated satisfactory starting on JP-3 from 5K – 45K feet, giving both lean and some rich fuel flow ignition limit data. A report was in progress to be completed July 31.

3.   JP-4 testing was underway at the NACA wind tunnel to get the lean and rich limit data. Based on the limits of the facility and the time available, it was determined that 0°F was the lowest ambient start temperature achieved.

Additional data would be gathered once flight testing was initiated.

4.   A study using the procedures developed during the development of the J34 engine would be used for the -8 to determine the requested basic engine characteristics:
    a.   Sea level static 100°F ambient temperature.
    b.   Sea level maximum flight speed -65°F ambient temperature.
    c.   Maximum altitude minimum flight speed, standard ambient air temperature of +40°F.
    d.   Maximum altitude minimum flight speed, standard ambient air temperature of -40°F.
The study to complete the Item 4 work would be complete February 1, 1953.

5.   A report to be completed by March 15, 1953 responding to the request for data on transient and steady state limits of fuel flow as limited by compressor stall, main burner combustion blowout and A/B combustion blowout; provide data related to the rate of change of rpm with fuel flow rate; show the variation of turbine outlet temperature with exhaust nozzle area at military rpm; and characteristic engine time constraints. WAGT would utilize results from the -8 house engine to compile the data for the report.

6.   Failure analysis of the temperature control amplifier used in the simple hydraulic control and the transient response of the exhaust nozzle area to turbine discharge temperature when on automatic trim control would be conducted by Manning, Maxwell and Moore and be completed by September 5, 1952.

7.   A series of charts covering fuel flows, throttle settings, etc. had either already been or would be provided prior to the qualification test.

8.   The A/B blowout sensor was still in investigation and progress toward a satisfactory solution, not affected by dirt or debris, would be reported as part of the monthly progress reports.

During this period WAGT's progress in building engines was being hampered by materials shortages that were being addressed via the BRICKBAT program. The delays were similar to those experienced by the -6 program during this period.

WAGT issued the draft specification WAGT-X40E2-4 (replacing WAGT-X40E2-2F Appendix D), covering the XJ40-WE-8 and XJ40-WE-22 (formally called the XJ40-WE-8A). The -22 was the planned production model of the -8 to be built at the Kansas City plant.[119]

Concern continued over compressor pressure limiting and what signal would be present in the cockpit for all engines not equipped with automatic pressure limiting devices. In particular, how far below the limit would the signal be triggered?[120]

BuAer issued a request for the detailed data on engine performance prior to the WAGT committed February 1953 date noted in a WAGT memo of July 1. They noted that WAGT was soon to qualify the engine and that the data they are asking for must already exist since it was needed for the establishment of the control schedule. Without such data, BuAer expressed "grave" concern that the possibility would exist that the operation of the control as built would not be at all compatible with engine performance at other than static sea level conditions.[121]

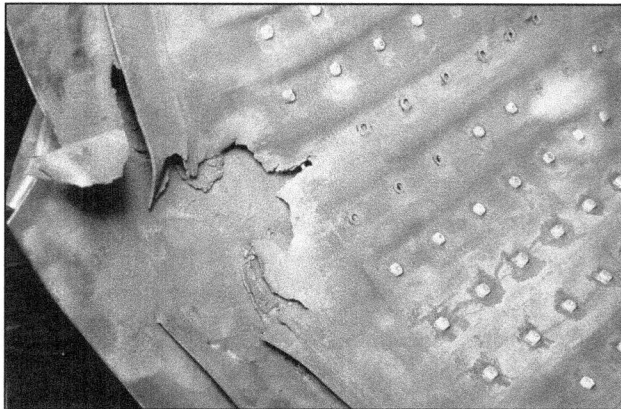

Figure 13 A/B Condition after first certification test attempt, WAGT P47913, 10/17/1952. *Courtesy Hagley Museum and Library*

Pressure now began to come from the Air Force MX770 missile program (X-10) for earlier delivery of the -1 engines. The AF was firm in requiring the full 726 in² exhaust nozzle size on their engines, but events forced the first (two) engines to be built with the smaller 675 in² nozzles. Since the qualification of the -1 was predicated on qualification of the -8 (due to their almost identical configuration and construction), the request made the scheduling of early engines to BuAer difficult, as all the early deliveries would come from Essington with its limited facilities.[122] The requested delivery schedule was changed and is reflected in the following table.

WAGT added a supplier of gearboxes for the -6 and -8 to the existing Westinghouse Nuttall Works, the Lycoming Spencer Division of AVCO, and BuAer agreed that since the gearbox itself was unchanged, a second 150 hr qualification test was not necessary. The first four boxes built by Lycoming Spencer would have both the housings and internal parts made by Nuttall. Thereafter boxes built for the -1, -6 and -8 would contain parts from both sources. They noted that the Lycoming Spencer gearbox for the -22 would have to undergo a separate 150 hr qualification, the entire box being entirely built by Lycoming Spencer and being of a modified design.[123]

Air Force J40-WE-8,-1 Production Schedule		
50 Hr Engines		
Date	Type	Number
Aug 1952	-8	1
Sep 1952	-8	1
Oct 1952	-1	2
150 Hr Engines, 675 in² nozzle		
Nov 1952	-8	3
Dec 1952	-8	3
Jan 1952	-8	3
150 Hr Engines, 726 in² nozzle		
Dec 1952	-8	1
Jan 1953	-8	1
Feb 1953	-8	4
Mar 1953	-8	2
Mar 1953	-1	2
Apr 1953	-8	2
Apr 1953	-1	2
May 1953	-8	2
May 1953	-1	2
One per month thereafter	-1	Until 16 under contract were completed

The use of alternative fuels (covered in the -6 development story) recommendations had one proviso related to the -8 only, in that leaded fuels (Avgas) had never been used in the -8 afterburner and no direct data of using such a fuel was available. However, based on burning such fuels in the J34 afterburners, some lead clogging of the fuel injection orifices was to be expected and if such fuels had to be used in service, regular inspections should be required to detect such clogging.[124]

The 150 Hour Qualification Test

Preparing for the 150 hr Qualification Test of the -8, it was planned to use the #3 house engine for the test. In addition to all the testing requirements listed in the specifications, an additional test requirement of the A/B nozzle had to be met for the failed aircraft electrical supply condition. In such an event, the nozzles were to slowly open to the full open condition, regardless of the aircraft power demand at the time of failure and not suffer serious damage while operating during that time period.

Specifically, WAGT proposed the following test using the manual override valve:

1. While operating at military power, shut off the D.C. power and allow the exhaust nozzle gates to drift full open. Restore D.C. power.

2. While operating at military power, close the exhaust nozzle with the override valve and retard the throttle to maintain maximum T_5 temperature.

3. Decelerate to 5,160 RPM and accelerate to maximum T_5 temperature.

4. Retard the engine to idle speed, open exhaust nozzle with manual override, jam accelerate to military RPM, close exhaust nozzle with manual override and retard the throttle to maintain maximum allowable T_5 temperature.

5. Return the manual override valve to the neutral position and accelerate the engine into afterburning.

6. Shut off the D.C. power supply and check for the opening of the exhaust nozzle.

WAGT noted that the 150 hr test engine demonstrated the above test at the start, 60 hr point and at the end of the 150 hr test without any problems. They also noted they understood the engine had to demonstrate the ability to withstand a minimum period of one minute at military RPM and twenty seconds in afterburning with the exhaust nozzle fully closed and still be operable for a short period under normal conditions. The engine would not have to withstand the repeated cycles of over-temperature, as set forth in the test procedure, without replacing possible damaged components (presumably this last after the "closed nozzle" test just prior).[125] The BAR reported he had observed the above test procedure on the qualification engine and it had passed satisfactorily.[126]

The engine passed its 150 hr Qualification Test in August but differed in some ways from the initial production engines, as will be seen in the next chapter. The -8 was now an interim engine until the planned

production -22 could be brought to production status at the Kansas City plant, but the service problems being encountered by the -6 would hinder the former being shipped until fixes were available.

Figure 14 Gearbox being installed on the stub shaft and intake of a J40-WE-8, WAGT P46550A, 7/3/1952. *Courtesy Hagley Museum and Library.*

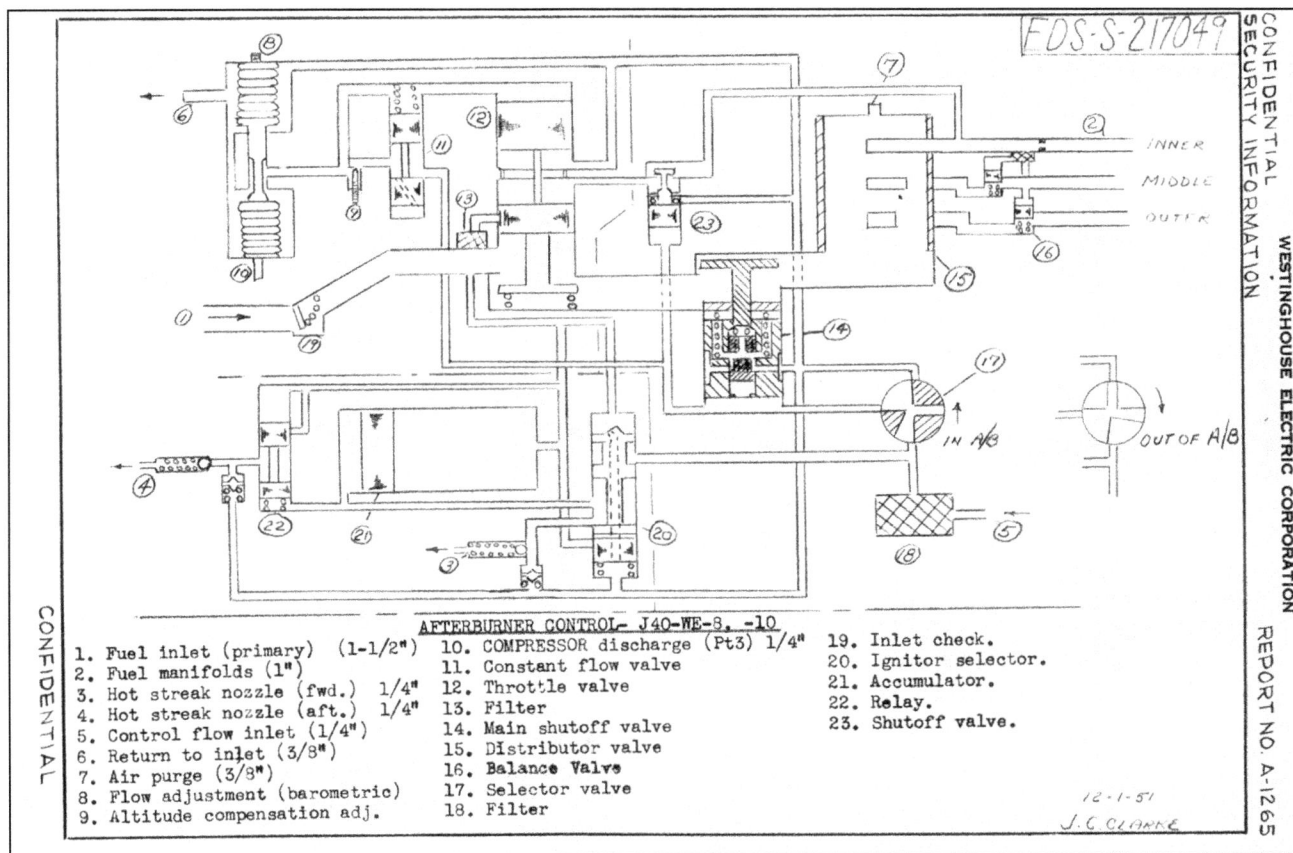

Figure 15 J40-WE-8 and -10 A/B Fuel Control Schematic, WAGT P47291, 7/21/1953. *Courtesy Hagley Museum and Library*

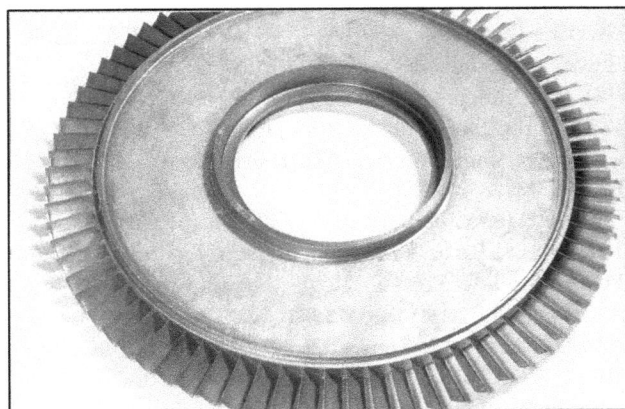

Figure 16 Experimental fabricated steel compressor disc for the -8, WAGT P49191, 4/15/1953. *Courtesy Hagley Museum and Library*

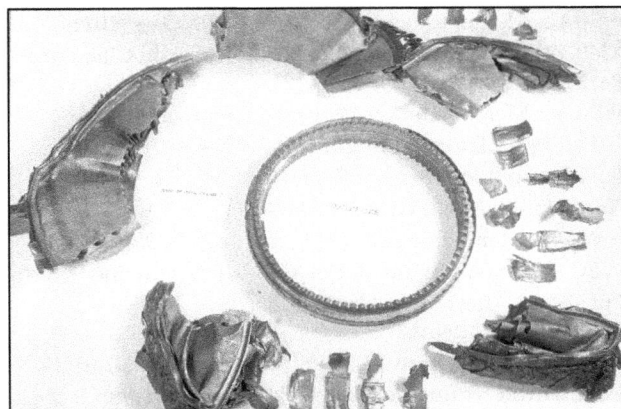

Figure 17 Experimental steel disc after testing, WAGT P50020, 8/26/1953. *Courtesy Hagley Museum and Library*

Chapter 5 - XJ40-WE-8 Procurement and Development Citations

[1] WAGT Proposal and E40E2 Specification, April 25, 1947 **B303V1RG72.3.2NACP**

[2] BuAer LOI Contract Proposal, November 5, 1948, "Letter of Intent for Contract NOa(s)-10114 Appn: 1791502.004 AN, 1949, Acct. 46823, Pro. 420A, Bu.Cont.No.20000" **B822V1RG72.3.2NACP**

[3] WAGT Memo, July 22, 1948, "Contract NOa(s) 9212; WEC Turbo-Jet Engine Model XJ40-WE-6; Installation Data; Transmittal of." **B303V4RG72.3.2NACP**

[4] BuAer Memo, July 28, 1948, "Contract NOa(s) 9212, XJ40WE-2 Engine, Design Study of Control System for." **B303V4RG72.3.2NACP**

[5] BuAer Memo, December 2, 1948, "Contract NOa(s) 9212, XJ40-WE-2 Engine; and Contract NOa(s) 9670, XJ46-We-2 Engine – Design Study of Control System for." **B303V4RG72.3.2NACP**

[6] WAGT Memo, November 15, 1948, "Contract NOa(s) 9212, Engine Weight Summary" **B304V5RG72.3.2NACP**

[7] BuAer Mock-up Board Report, May 27, 1949, "Mock-up Board Report For Model XJ40-WE-8 Engine Contract NOa(s) 10114" **B822VE1RG72.3.2NACP**

[8] BuAer Memo, October 31, 1949, "Contract NOa(s) 10114, XJ40-WE-8 turbo-jet engine mock-up board items 1 and 21; approval of"; References WAGT submission Conf. Ltr W1874E dated August 30, 1949. **B823VE3RG72.3.2NACP**

[9] WAGT Memo, June 2, 1949, "Contract NOa(s) 10114; WECE Turbo-Jet Engine Model XJ40-WE-8 Power Regulator Air Blast Inlet Connection Accessibility." **B822VE1RG72.3.2NACP**

[10] BuAer Memo, June 8, 1949, "Contract NOa(s) 10114, XJ40-WE-8 Engine, Hydraulic Pump Removal Clearance" **B822VE1RG72.3.2NACP**

[11] BuAer Memo, January 26, 1950, "Contract NOa(s) 10114; WEC Turbo Jet Engine Model XJ40-WE-8, Engine Fuel Pump Inlet Pressure Requirements and Accessibility." **B823VE4RG72.3.2NACP**

[12] WAGT Memo, June 16, 1949, "Contract NOa(s) 10114, WEC Turbo Jet Engine Model XJ40-WE-8 – Engine Furnished Afterburner Cooling Ejector." **B822VE1RG72.3.2NACP**

[13] WAGT Memo, June 29, 1949, "Contract NOa(s) 10114, WEC Turbo-Jet Engine Model XJ40-WE-8; Engine Furnished Afterburner Cooling Ejector." **B823VE2RG72.3.2NACP**

[14] WAGT Memo, March 5, 1950, "Letter on Intent for Contract NOa(s) 10114; WEC turbo-Jet Engine Model XJ40-WE-8 Engine Afterburner Cooling Provisions – Clarification of." **B823VE5RG72.3.2NACP**

[15] WAGT Memo, August 2, 1949, "Contract NOa(s) 10114, WEC Turbo-Jet Engine Model XJ40-WE-8 – Relocation of Jack Pads on #2 Bearing Housing" **B823VE2RG72.3.2NACP**

[16] WAGT Memo, September 16, 1949, "Contract NOa(s) 10114; WEC Turbo-Jet Engine Model XJ40-WE-8 – Relocation of Jack Pads on #2 Bearing Housing." **B823VE3RG72.3.2NACP**

[17] BuAer Memo, November 18, 1949, "Contract NOa(s) 10114 – Proposed afterburner fuel injection ignition system in lieu of electrical ignition for Model XJ40-WE-8 turbo-jet engine; approval of" **B823VE4RG72.3.2NACP**

[18] WAGT Memo, September 15, 1949, "Contract NOa(s) 10114; WEC Turbo-Jet Engine, Model XH40-WE-8 – Retractable Rear Engine Lifting Eye." **B823VE3RG72.3.2NACP**

[19] WAGT Confidential Letter, Serial W-1277, forwarded on July 7, 1949 to BuAer by the BAR. **B823VE2RG72.3.2NACP**

[20] WAGT Memo, December 8, 1949, "Contract NOa(s) 10114; WEC Turbo Jet Engine Model XJ40-WE-8; Electrical Junction Box Airframe Connections, Proposed Relocation of." **B823VE4RG72.3.2NACP**

[21] WAGT Memo, October 26, 1949, "Contract NOa(s) 10114; WEC Turbo-Jet Engine Model XJ40-WE-8 – Reduction of Power Lever maximum Operating Torque." **B823VE4RG72.3.2NACP**

[22] WAGT Memo, January 9, 1951, "Contract NOa(s) 10114 WEC Turbo Jet Engine Model XJ40WE-8. Design Changes Survey of. Engine Change Nos. 75, 76, 77 and 78." **B823VE6RG72.3.2NACP**

[23] BuAer Memo, January 13, 1950, "Contract NOa(s) 10114; Westinghouse Specification WAGT-X40E2-2D covering the models XJ40-WE-6, -8 and J40-WE-6 engine; flowmeter provisions for" **B823VE4RG72.3.2NACP**

[24] WAGT Memo, June 9, 1949, "Contract NOa(s) 10114; WEC Turbo-Jet Engine Model XJ40-WE-8; Proposed Change to Single Anti-Icing Discharge Port." **B823VE3RG72.3.2NACP**

[25] BuAer Memo, January 26, 1950, "Contract NOa(s) 10114; WEC Turbo Jet Engine Model XJ40-WE-8, Engine Fuel Pump Inlet Pressure Requirements and Accessibility." **B823VE4RG72.3.2NACP**

[26] WAGT Memo, September 26, 1949, "Contract NOa(s) 10114; WEC Turbo-Jet Engine Model XJ40-WE-8 – Alignment Provisions for Afterburner to Engine Quick Disconnect Flanges." **B823VE3RG72.3.2NACP**

[27] BuAer Memo, November 17, 1949, "Contract NOa(s) 10114; XJ40-=WE-8 engine; alignment provisions for afterburner to engine quick disconnect flanges" **B823VE4RG72.3.2NACP**

[28] WAGT Memo, April 12, 1949, "Contract NOa(s) 10014(*SIC*); WEC Turbo-Jet Engine Model J40; Allowable

Inlet total Pressure Variation; Survey of."
B822VE1RG72.3.2NACP

[29] WAGT Memo, October 20, 1949, "Contract NOa(s) 10114; WEC Turbo-Jet Engine Model XJ40-WE-8 – Improved Access to Components Mounted on Aft Side of Gear Box." **B823VE4RG72.3.2NACP**

[30] BuAer Memo, November 23, 1949, "Contract NOa(s) 10114; XJ40-WE-8 engine; improved access to compo9nents mounted on aft side of gear box" **B823VE4RG72.3.2NACP**

[31] WAGT Memo, August 8, 1949, Contract NOa(s) 10114; WEC Turbo-Jet Engine Model XJ40-WE-8 – Removal of Tachometer Gearbox." **B823VE2RG72.3.2NACP**

[32] WAGT Report, September 13, 1948, "Westinghouse Model XJ40-WE-6 Jet Engine Structural Requirements for Cantilevering an Afterburner and Extension." **B823VE3RG72.3.2NACP**

[33] BAR Memo, September 21, 1949, "Contract NOa(s) 10114 – Letter of Intent – Monthly Progress Report for August 1949 – Forwarding of." **B823VE3RG72.3.2NACP**

[34] WAGT Memo, September 9, 1949, "Contract NOa(s)-9670 and 10114 Proposal for the Simplification of the Emergency Fuel System on the XJ46-WE-2 and XJ40-WE-8 Engines, Reference WG-60999 and WG-62360" **B823VE2RG72.3.2NACP**

[35] WAGT Report, April 19, 1949, "Contract NOa(s) 9212, XJ40-WE-6 Engine, Daily Summary of Mock-Up Board Decisions, Study of Items 21 & 4" **B304VE6RG72.3.2NACP**

[36] BuAer Memo, August 31, 1949, "Contracts NOa(s) 9212, 9670, and 10114 – XJ40-WE-6, -8; XJ46-WE-8 and XJ34-WE-32 Engines – Exhaust Nozzle Actuating Systems" **B303VE9RG72.3.2NACP**

[37] BuAer Memo, June 21, 1949, "Naval Industrial Reserve Plant, Lester, Pa. – Acquisition of Additional Equipment" **B82VE1RG72.3.2NACP**

[38] BuAer Memo, August 5, 1949, "Contract NOa(s) 9212 Model Specification WAGT-X40E2-2B dated 18 May 1949", Proposed revisions. **B823VE2RG72.3.2NACP**

[39] WAGT Memo, August 11, 1949, "Contracts NOa(s) 10114 and 9212 – XJ40-WE-8 Accessory Gear Box Configuration." **B823VE2RG72.3.2NACP**

[40] WAGT Memo, August 17, 1949, "Contracts NOa(s) 9212 and 10114 XJ40-E-6 and -8 Performance" **B304VE9RG72.3.2NACP**

[41] WAGT Memo, August 31, 1949, "J40-WE-8 Engine Fuel System – Performance with Limited Capacity Airplane Fuel Systems" **B304VE9RG72.3.2NACP**

[42] BAR Memo, October 18, 1949, "Contract NOa(s) 10114 – Letter of Intent – Monthly Progress Report A-902 for September 1949 – Forwarding of." **B823VE3RG72.3.2NACP**

[43] BuAer Memo, October 25, 1949, "Contract NOa(s) 10114, lengthening of XJ40-WE-8 turbo jet engine; approval of" **B823VE3RG72.3.2NACP**

[44] BuAer Memo, October 6, 1949, "Contract NOa(s)-9212, Westinghouse Models XJ40E2A, XJ40-WE-6 and XJ40-WE-8 Engines, Preflight Check System for." **B304VE9RG72.3.2NACP**

[45] BuAer Memo, November 1, 1949, "Contracts NOa(s)-9212 and -10114; changes in specifications for XJ40-WE-4, -6 and -8 engine weights" **B305VE11RG72.3.2NACP**

[46] BuAer Memo, December 27, 1949, "Contract NOa(s)-9212, Westinghouse Specifications WAGT-X40E2-2D Dated 4 November 1949 covering the Models XJ40-WE-6, -8 and J40-WE-6 engines" **B305VE10RG72.3.2NACP**

[47] WAGT Memo, January 9, 1950, "Engine Weight Control System" **B305VE10RG72.3.2NACP**

[48] WAGT Memo, December 10, 1950, "Contract NOa(s) 9212 and NOa(s) 10114 Model Specification WAGT-X40E2-2E dated November 18, 1949." **B305VE10RG72.3.2NACP**

[49] BuAer Memo, February 7, 1950, "Contracts NOa(s)-9212 and NOa(s)-10114, Westinghouse Specification WAGT-X40E2-2E data 18 November 1949 covering the Models XJ40-WE-6, -8 and J40-WE-6" **B305VE11RG72.3.2NACP**

[50] BuAer Memo, April 25, 1950, "Contract NOa(s) 9212, Westinghouse specification WAGT-X40E2-2D of 4 November 49 covering the models J40-WE-6 and XJ40-WE-8." **B305VE12RG72.3.2NACP**

[51] BuAer Memo, August 25, 1950, "XJ40-WE-6/-8 engines; spare lugs on accessory gearbox housing" **B305VE13RG72.3.2NACP**

[52] BuAer Memo, February 24, 1950, "Contracts NOa(s) 9670, 9215, and 10114; XJ40-We-6 and -8, XJ34-WE-32 and XJ46-WE-2 engines; field survey; approval of" **B305VE11RG72.3.2NACP**

[53] BuAer Memo, May 2, 1950, "Contract NOa(s) 9212, WEC turbojet engine model J40-WE-6, laboratory qualification of the J40-WE-6 fuel pumping system" **B305VE12RG72.3.2NACP**

[54] WAGT Memo, July 12, 1949, "Contract NOa(s) 10114; WEC Turbo-Jet Engine Model XJ40-WE-8; Increase in Length of Compressor Inlet Housing." **B823VE2RG72.3.2NACP**

[55] WAGT Memo, March 16, 1950, "Navy Contract NOa(s) 9212-75 Patent Application WM 5130" **B505VE10RG72.3.2NACP**

[56] WAGT Memo, July 19, 1951, "Contract NOa(s) 9212 Westinghouse Model Specification WAGT-X40E2-2E dated 18 November 1949; Revised performance curves for" **B305VE14RG72.3.2NACP**

[57] BuAer Memo, October 17, 1951, "Westinghouse Specification WAGT-X40E2-2F dated 23 June 1951 covering the Models XJ40-WE-6 and J40-WE-6 engines; performance curves for" **B306VE15RG72.3.2NACP**

[58] WAGT Memo, December 27, 1951, "Contract NOa(s) 9212, Westinghouse Specification WAGT-X40E-2-2F dtd.

23 June 1951, covering model J40-WE-6 engine; performance correction curves for." **B305VE15RG72.3.2NACP**

[59] BuAer Memo, September 20, 1951, "Westinghouse Specification WAGT-X40E2-2F dated 23 June 1951 covering the XJ40-WE-6 and J40-WE-6 engines." **B306VE15RG72.3.2NACP**

[60] BAR Memo, November 23, 1949, "Contract NOa(s) 10114 – Letter to Intent for Monthly Progress Report, Month of October 1949 – Submittal of." **B305VE10RG72.3.2NACP**

[61] BAR Memo, January 19, 1950, "Contract NOa(s) 10114 – Letter of Intent for – Monthly Progress Report for December 1949." **B305VE10RG72.3.2NACP**

[62] BuAer Memo, January 20, 1950, "Westinghouse XJ46-We-2 and XJ40-WE-8 turbo-jet engines, simplified fuel systems for; status of action on Westinghouse proposal" **B823VE4RG72.3.2NACP**

[63] BuAer Memo, February 7, 1950, "Contracts NOa(s)–9212 and NOa(s)-10114, Westinghouse Specification WAGT-X40E2-2E dated 18 November 1949 covering the Models XJ40-WE-6, -8 and J40-WE-6 engines" **B823VE4RG72.3.2NACP**

[64] BuAer Memo, February 28, 1950, "Contract NOa(s) 10114; Model J40 turbo-jet engine; allowable inlet total pressure variation." **B823VE4RG72.3.2NACP**

[65] BuAer Memo, March 20, 1950, "Letter of Intent for contract NOa(s) 10114, Monthly Progress Report A-958 (month of February 1950); forwarding of." **B823VE5RG72.3.2NACP**

[66] BuAer Memo, April 11, 1950, "Contract NOa(s) 10114, XJ40-WE-8 engine, crash landing strength requirements" **B823VE5RG72.3.2NACP**

[67] BAR Memo, April 20, 1950, "Contract NOa(s) 10114, item 6, monthly progress report A-961 (month of March); forwarding of." **B823VE5RG72.3.2NACP**

[68] WAGT Memo, April 19, 1950, "Contract NOa(s) 10114; WEC Turbo-Jet Engine Model XJ40-WE-8; Elimination of the Emergency Afterburner Fuel Pumping Provisions, Survey of. Engine Change No. 68." **B823VE5RG72.3.2NACP**

[69] WAGT Memo, April 13, 1950, "Contract NOa(s) 9791, 10385, 10114, 10067, Throttle torque for" **B823VE5RG72.3.2NACP**

[70] BuAer Memo, May 5, 1950, "Contracts NOa(s) 9791, 10385, 10114, and 10067; Westinghouse Electric Corporation turbo-jet Models J34-WE-32, -38, J40-WE-6, XJ40-WE-8, XJ40-WE-10 and -12; throttle torque for." **B823VE5RG72.3.2NACP**

[71] BAR Memo, May 15, 1950, "Contract NOa(s) 10114, letter of intent for, monthly progress report A-973 (month of April 1950); forwarding of." **B823VE5RG72.3.2NACP**

[72] WAGT Memo, May 25, 1950, "Contract NOa(s) 10114; WEC Turbo-Jet Engine, Engine Fuel Pump, Inlet Pressure, Requirements for." **B823VE5RG72.3.2NACP**

[73] BuAer Memo, June 9, 1950, "Contract NOa(s) 9212 and 10114, XJ40-WE-6 and -8, Fuel System Dump Valve; approval of" **B823VE5RG72.3.2NACP**

[74] BAR Memo, June 15, 1950, "Contract NOa(s) 10114, letter of intent for, monthly progress report A-982 (month of May); forwarding of." **B823VE5RG72.3.2NACP**

[75] BuAer Contract NOa(s) 10114 (Superseding Letter of Intent for Contract NOa(s) 10114), June 22, 1950. **B823VE6RG72.3.2NACP**

[76] BAR Memo, November 20, 1950, "Contract NOa(s) 10114, item 6, monthly progress report A-1050 (October 1950) forwarding of." **B823VE6RG72.3.2NACP**

[77] BuAer Wire, October 31, 1951, referencing WAGT Survey of Engine Change 88 of October 12. **B824VE7RG72.3.2NACP**

[78] BAR Memo, December 19, 1950, "Contract NOa(s) 10114, item 6, monthly progress report A-1071 (November 1950) – forwarding of." **B823VE6RG72.3.2NACP**

[79] WAGT Memo, December 29, 1950, "Contract NOa(s) 10114 WEC Turbo-Jet Engine Model XJ40WE-8. Variation in Engine Mount Distance, Survey of. Engine Change No. 74." **B823VE6RG72.3.2NACP**

[80] WAGT Memo, January 9, 1951, "Contract NOa(s) 10114 WEC Turbo Jet Engine Model XJ40WE-8. Design Changes, Survey of. Engine Change Nos. 75, 76, 77, and 78." **B823VE6RG72.3.2NACP**

[81] BuAer Memo, January 16, 1951, "Contract NOa(s) 10114; J40-We-8 engine; afterburner fuel system" **B823VE6RG72.3.2NACP**

[82] BAR Memo, January 18, 1951, "Contract NOa(s) 10114, Item 6 – Monthly Progress Report A-1091 (December 1950) – forwarding of." **B823VE6RG72.3.2NACP**

[83] BAR Memo, February 23, 1951, "Contract NOa(s) 10114, item 6, monthly progress report, A-1106 (January 1951); forwarding of." **B823VE6RG72.3.2NACP**

[84] BuAer Memo, September 19, 1951, "Contract NOa(s) 10385, J40-We-6 engine, survey of acceleration control" **B940VE1RG72.3.2NACP**

[85] WAGT Memo, October 23, 1951, "Contract NOa(s) 10385 "EC Turbo Jet Engine Model J40WE-1 and -8. Outline of Speed Switch Control, Survey of." **B937VE8RG72.3.2NACP**

[86] WAGT Memo, November 16, 1951, "Contract NOa(s) 10385, J40WE-6 turbo jet engine, Acceleration control for." **B940VE1RG72.3.2NACP**

[87] WAGT Memo, March 13, 1951, " Contract NOa(s)-10114 Development Delivery Dates Our Reference WG-62360" **B822V1RG72.3.2NACP**

[88] BuAer Memo, December 28, 1951, Termination Notice, Document 019756 **B822V2RG72.3.2NACP**

[89] WAGT Memo, January 8, 1952, "J40-WE-8 Serial #031001 pre-Qualification Test engine for McDonnell XF3H airplane." **B937VE8RG72.3.2NACP**

[90] BuAer Speed Memo, January 14, 1952, "J40-WE-8 serial No. 031001 pre-qualification test engine" **B937VE8RG72.3.2NACP**

[91] WAGT Memo, January 18, 1952, "Model F3H Airplane – Engine Fuel Inlet connector for J40WE-8 engine; Performance." **B937VE8RG72.3.2NACP**

[92] WAGT Memo, February 1, 1952, "Contract NOa(s) 10385 and NOa(s) 10825; XJ40 and XJ46 Engines, Simplified Controls for." **B937VE9RG72.3.2NACP**

[93] WAGT Memo, September 24, 1951, "Contract NOa(s) 10385, Failure Analysis of J40-WE-8 Engine Simple Hydraulic Control", unnumbered/undated report attached. **B937VE9RG72.3.2NACP**

[94] WAGT Memo, February 4, 1954, "Contract NOa(s) 10385 WEC Turbo Jet Engines J40WE-1 and J40WE-8. Manual Trim Control for the Exhaust Nozzle, Survey of. Engine Change Nos. 26 for J40WE-1 engines and 93 for J40WE-8 engines." **B937VE9RG72.3.2NACP**

[95] WAGT Memo, February 6, 1952, "Contract NOa(s) 10385 and NOa(s) 10114 J40WE-6 and J40WE-8 Engine Airflow Path Improvement for Altitude Operation" **B940VE2RG72.3.2NACP**

[96] BAR Memo, February 8, 1952, "Contracts NOa(s) 10385 and NOa(s) 10114 J40WE-6 and J40WE-8 Engine Airflow Path Improvement for Altitude Operation" **B940VE2RG72.3.2NACP**

[97] WAGT Memo, February 14, 1952, "Emergency Features and Associated Operating Procedure of the Simple Hydraulic Control for the J40-WE-8, -22 and J46-WE-3, -8, -10 Turbojet Engines." Pages 2-3 **B937VE9RG72.3.2NACP**

[98] WAGT Memo, February 13, 1952, "Contract NOa(s) 10385, WEC Turbo-Jet Engine XJ40WE-8, XJ40WE-1, XJ40WE-22. Increased Clearance Envelope over Afterburner fuel lines, Survey of. Engine Change No. 92 for -8 engines, 24 for -1 engines and 3 for -22 engines." **B937VE9RG72.3.2NACP**

[99] WAGT Memo, February 15, 1952, "Contract NOa(s) 10385 WEC Turbo-Jet Engine Model J40WE-1, -6, -8, -22. Thermocouple Cover and Junction Box Connection, Survey of. Engine change Nos. 73 for J40WE-6, 95 for J40WE-8, 5 for J40WE-22 and 28 for J40WE-1." **B937VE9RG72.3.2NACP**

[100] WAGT Memo, February 15, 1952, "Contracts NOa(s)-10385 and NOa(s) 52-403c; J40-WE-8 Prequalification Engines" **B937VE9RG72.3.2NACP**

[101] WAGT Memo, February 5, 1952, "Contract NOa(s) 10385 Westinghouse Model Specification WAGT-X40E2 -4 describing the XJ40-WE-8 and -22 turbo-jet engines; transmittal date for" **B937VE9RG72.3.2NACP**

[102] BuAer Memo, February 26, 1952, "Contract NOa(s)10385, Westinghouse Specification WAGT-X40E2-4 covering the Models XJ40-WE-8 and -22 engines; transmittal of." **B937VE9RG72.3.2NACP**

[103] WAGT Memo, February 29, 1952, "Contract NOa(s) 10385 and 10067 WEC Turbo-Jet Engine Models J40WE-8, XJ40WE-10 and XJ40WE-22. Increase in Exhaust Nozzle Area, Survey of. Change No. 96 for -8 engines, 77 for -10 engines and 6 for -22 engines." **B937VE9RG72.3.2NACP**

[104] BAR Memo, February 29, 1952, "Contracts NOa(s) 10385 and NOa(s) 52-403c; J40WE-8 prequalification engines." **B937VE9RG72.3.2NACP**

[105] WAGT Memo, March 20, 1952, "Contract NOa(s) 10385 WEC Turbo-Jet Engine Model J40-WE-8 Afterburner Yoke increased stroke, Survey of." **B937VE9RG72.3.2NACP**

[106] WAGT Memo, March 24, 1952, "J40-WE-8, -22 and J46-WE-8, -10 engines, simplified controls for." **B937VE9RG72.3.2NACP**

[107] BuAer Memo, March 25, 1952, "J40-WE-8, -22, and J46-WE-8 engines, simplified controls, data regarding" **B937VE9RG72.3.2NACP**

[108] BuAer Naval Speed Letter, March 25, 1952, "Contract NOa(s) 10385; provisional acceptance of J40-WE-8 engine serial No. WE031001" **B937VE9RG72.3.2NACP**

[109] WAGT Naval Speed Letter, April 14, 1952, "Provisional acceptance of J40-WE-8 engine serial WE-031001; approval of" **B937VE9RG72.3.2NACP**

[110] BuAer Telegram to BAR, April 24, 1952, J40-WE-8 WE031001 structural integrity **B937VE10RG72.3.2NACP**

[111] WAGT CID 17049, March 31, 1952, "Change in the axial location of O-ring grooves in several seal housings" **B937VE9RG72.3.2NACP**

[112] WAFT Memo, March 31, 1952, "J40-WE-1, -6, -8, -22 Engines, No. 1 Thrust Bearing" **B937VE9RG72.3.2NACP**

[113] Correspondence, Law Office of Strauch, Nolan & Diggins, April 29, 1952 to Westinghouse. **B937VE9RG72.3.2NACP**

[114] BuAer Memo, May 2, 1952, "Contract NOa(s) 10385; acceptance of J40-WE-8/-1 engines" **B937VE10RG72.3.2NACP**

[115] WAGT Memo, August 8, 1952, "Contract NOa(s) 10385 Acceptance of J40-WE-8/-1 Engines Our Reference WG-63300" **B937VE11RG72.3.2NACP**

[116] WAGT Memo, May 14, 1952, "Model J40-WE-6, -8, -22 Engines, Automatic Ignition Features, Deactivation or Deletion of" **B937VE10RG72.3.2NACP**

[117] WAGT Memo, May 20, 1952, "Contract NOa(s) 10385, J40-WE-1, J40-WE-1, J40-WE-8, and J40-WE-22 engines. Engine mounting of Automatic Temperature Trim Control, and possible Airframe Mounting , Survey of. Engine Change Nos. 29 for J40-WE-1 engines, 97 for J40-

WE-8 engines and 7 for J40-WE-33 engines."
B937VE10RG72.3.2NACP

[118] WAGT Memo, May 27, 1952, "Request for Additional
Data on Westinghouse Turbo-Jet Engines."
B937VE10RG72.3.2NACP

[119] WAGT Memo, May 14, 1952, "Contract NOa(s) 10385
Westinghouse Model Specification WAGt-X40E2-4 dated
April 25, 1952, which describes the XJ40-WE-8 and -22
turbo-jet engines" **B937VE10RG72.3.2NACP**

[120] BuAer Memo, July 1, 1952, "J40 and J46 Compressor
mechanical limits; request for information on"
B937VE11RG72.3.2NACP

[121] BuAer Memo, July 23, 1952, "Effect of exhaust nozzle
area on engine performance" **B937VE11RG72.3.2NACP**

[122] BuAer Memo, July 26, 1952, "Contract NOa(s) 10385;
acceptance of J40-WE-8/-1 engines"
B937VE11RG72.3.2NACP

[123] BuAer Memo, July 31, 1952, "J40 gearboxes supplied by
the Lycoming Spencer Division of AVCO, qualification
of" **B937VE11RG72.3.2NACP**

[124] WAGT Memo, August 1, 1952, "Contract NOa(s) 10385
Use of Substitute Fuels in J40-WE-6 engines"
B937VE11RG72.3.2NACP

[125] WAGT Memo, August 22, 1952, "Contract NOa(s)
10385, J40-WE-8 Engine, testing with manual override
control." **B937VE11RG72.3.2NACP**

[126] BAR Memo, August 26, 1952, "Contract NOa(s) 10385,
J40WE-8 Engine, Testing with Manual Override Control"
B937VE11RG72.3.2NACP

Chapter 6

J40-WE-8 Mechanical Description

The J40-WE-8 was originally intended to be a -6 engine modified to accommodate an afterburner easily attached to the rear of a -6, enabling the two engine inventory types to be largely interchangeable. (The same approach was also attempted via conversion "kits" for the later -10/-12's). This was to be a minimal change engine, but additional testing showed that both basic components and those changed by the afterburner required engineering upgrades, most of those not purely afterburner-related being retroactively introduced to the -6 version while in development or later while in service use. Aside from the afterburner itself, the control system was the biggest change, the engine using a "simple hydraulic control". Only items unique to the -8 will be discussed.[1]

A.   External and Supporting Structure

This remained the same as the -6 with all the changes from -6 experience being incorporated. In addition, the No. 1 bearing was redesigned to a much higher life expectancy and, based on the testing results, the No. 1 bearing was also upgraded to allow unlimited A/B lights between overhauls. The afterburner assembly is covered separately below. The accessories gearbox was the hourglass shaped unit with dry sump lubrication. This design was also used on the -6 when development delays allowed it to be fitted to that model in lieu of the original wet sump design.

The afterburner combustion chamber, exhaust collector and nozzle assembly were attached to the diffuser by means of a quick disconnect clamp. Three manifold rings were followed by a flameholder assembly of Iconal steel. During engine operation, compressor discharge air was bled from the rear of the engine compressor through two external hoses to the shell of the afterburner combustion chamber. The afterburner construction was of the two wall shell type, with the corrugations in the layers lying at ninety degrees to each other. The balance of the unit was the exhaust collector and nozzle.

A NACA study of the afterburner showed the design of the manifolds and flameholder were not positioned in the optimum location. Moving them to the recommended location would have increased A/B thrust by approximately 8%. This change was never made as the recommendation came when the program was ending.

The variable area exhaust nozzle was comprised of two matched hemispherical gates hinged at top and bottom, positioned in unison by rods attached to a yoke mounted back under the No.2 bearing assembly. This design followed that of the -6 mechanism, incorporating changes necessary for the increased thrust forces and distance from the yoke to the nozzles. The gates were normally fully open up to 55% throttle position setting and fully closed from 80% throttle position on up.

B.   Flow Path Elements

The compressor remained an eleven stage axial, but used the C-20 design developed with the help of NACA feedback. It incorporated the flow mixers fitted to the -6 in the outlet guide vanes to resolve temperature inversion problems. The major change affecting this compressor was the finding that it had leakage paths in the curvic clutches that were allowing high temperature compressor gas from the later stages to leak forward and weaken the stage six disc. The curvic clutches were sealed with Bakelite and later a more sophisticated, less labor intensive method of using metal clips between the discs further sealed with silicon on the clip ends was approved. It was also determined that only the stages 6, 7 and 8 needed to be sealed and that the hotter temperatures in the last stages were well within their design limits and did not need to be sealed.

The turbine was later found to require the first stage guide vanes to be opened 16 degrees to address continuing compressor stalling above 35,000 feet.

C. Fuel System

The dual fuel pump delivered fuel through a micronic filter to the fuel regulator which served to meter the correct amount of fuel required by the engine. The regulator was designed to meter fuel for the steady state operation in direct response to the power lever position. Fuel for acceleration was regulated by the pressure rise through the engine compressor. Constant military speed regulation was provided by the isochronous governing action of the speed relay in conjunction with the fuel valve in the fuel regulator. The fuel regulator also provided a pump selector valve for switching the dual pumps manually by use of a switch in the cockpit. Either side of the dual pump could supply the full needs of the non-afterburning portion of the engine.

From the fuel regulator, the fuel flowed through a fuel-to-engine oil heat exchanger to a dump valve and distributor, and finally was injected as a fine spray from variable area fuel nozzles into the combustion chamber. The dump valve cleared the fuel distribution system of fuel during engine shutdown.

D.   Exhaust Nozzle Control System

This consisted of the exhaust nozzle control which contained the necessary control valves, an auxiliary electric control which contained the necessary control relays, the power scheduler, the air pressure switch, the temperature control, and the high pressure hydraulic fluid pumping element. The exhaust gate type of variable area exhaust nozzle operated through a linkage system. It positioned the gates, which were mechanically interlocked by means of the linkage to prevent asymmetric thrust in the event that one had become jammed.

a.   Non-Afterburning Operation

During non-afterburning operation of the engine, the exhaust nozzle control position was set in response to signals from the power scheduler, which was internally linked mechanically to the engine power lever. The auxiliary electric control contained a micropositioner which was a polarized sensitive relay. The micropositioner sensed the difference in voltage between the slider of the position setting potentiometer in the power scheduler and the slider of the position feedback potentiometer in the exhaust nozzle control and transmitted the appropriate commands to the control. These commands were received by the push-pull solenoid valve in the control which, in turn, caused hydraulic pressure to be applied to the proper side of the control piston. The resulting motion of the exhaust nozzle control piston continued until the desired position, as indicated by zero voltage between the sliders of the potentiometers, was attained.

The power scheduler action was over-ridden below military rpm by the speed switch which transferred the micropositioner coil to the open end of the scheduler, keeping the exhaust nozzle open until military rpm was reached. The speed switch was operated from the movement of the overspeed fuel valve. When that valve was in the wide open position, the speed switch was closed. When the overspeed fuel valve was in a controlling position,

the speed switch was open. The overspeed fuel valve remained in the wide open position until military rpm was reached. When military rpm was attained, a pressure signal was received by the overspeed fuel valve which moved it to a controlling position where it operated until the rpm was again less than military. At reduced engine rpm, the overspeed fuel valve again moved to a wide open position where it reactivated the speed switch.

b.   Afterburning Operation

When moving the power lever from military into afterburning, a switch in the power scheduler energized the circuit to open the afterburner fuel valve. Using hot streak ignition to ignite the afterburner, an air pressure switch energized additional circuits which shifted control of the exhaust nozzle from the power scheduler to the area position which would produce a military temperature at 100°F sea level conditions. In case of a flame out, the air pressure switch would operate the control circuits to close the exhaust nozzle to the scheduled 100°F military position and close the afterburner fuel valve. The circuits were then locked in this condition until the power lever was moved out of the afterburning range. Retracting the power lever out of A/B while the burner was operating de-energized the various afterburner control circuits, closing the fuel valve and returning the exhaust nozzle to its normal schedule.

The exhaust nozzle area was normally adjusted to produce military turbine temperatures on a 100°F day. Sea level ambient temperatures below 100°F prevented the engine from achieving maximum output. To compensate for this, a system was added to adjust the nozzle area to maintain a constant military temperature. The exhaust nozzle trim was in response to an electronic sensing of the difference between the maximum allowable turbine operating temperature and the actual measured turbine operating temperature.

The automatic temperature control consisted of a D.C. voltage amplifier for thermocouple signals (Microsen), an output stage and a two phase motor. The motor drove two potentiometers which were in the exhaust nozzle control bridge circuit to provide a trim action. Four vacuum tubes provided a temperature reference voltage supply for an oscillator and an A.C. amplifier. Operating power was drawn from the airframe 110 volts 400 cycle supply.

A manual override to the electric exhaust nozzle position circuit was provided. Failure of the electric supply would cause the automatic positioning control to close and the exhaust nozzle to slowly drift from its position at the time of failure to a closed position. The pilot was provided a mechanical linkage to move the servo valve either to the full opening position or closing position, using the temperature gauges to select the proper setting. This had to be changed by the pilot any time he changed the aircraft's altitude or speed and the engine was operating at military rpm (7,260). Below that rpm, the setting was full open.

E.   Afterburner Fuel Control System

This consisted of an engine driven centrifugal fuel pump and an A/B fuel control which throttled the discharge to schedule an A/B fuel flow proportional to mass air flow, as measured by engine compressor discharge total pressure. This pressure was sensed by an aneroid bellows in the A/B fuel control. Displacement of the bellows positioned a contoured metering pin which established the reference control pressure differential to be applied to the throttling valve of the A/B fuel regulator. By maintaining this same reference control pressure differential across the metering ports in the manually positioned distributor valve rotor, the throttling valve regulated the proper fuel flow to the A/B manifolds.

The A/B fuel control included provisions for automatic ignition of the A/B by means of an igniter valve of the accumulator type which discharged a measured quantity of fuel through an igniter nozzle upstream of the turbine as soon as the afterburner fuel pressure built up in the A/B fuel manifold lines. The ignition hot streak of fuel was relayed to the A/B fuel manifold rings by additional fuel sprayed through a relay igniter nozzle. When the pilot's control lever was advanced into the A/B position, fuel was admitted to the A/B fuel pump, which normally idled "dry" when not afterburning, by opening the gate-type fuel shut off valve at the A/B pump inlet. The initial advance of the control lever also opened the distributor valve ports to route fuel to the afterburner manifold rings. Retracting the power lever from A/B position closed the distributor ports, closed the A/B fuel valve and permitted the igniter valve to recharge with fuel. If an A/B flameout occurred, the pilot retracted the throttle out of A/B for a short period and then pushed it back into

A/B to begin the light-off sequence again. Afterburner light offs at full A/B were consistent up to 40,000 ft. Operation above 40,000 feet was possible on occasion. Above 50,000 feet, there was little additional power available and the A/B was not required to operate above that level. There were no operating limits on how long the A/B could be continuously operated.

The operating instructions for the engine noted that although the A/B had three zones of power depending on how far into A/B the control lever was advanced, ignition of the A/B in anything other than full A/B *(zone 3)* was inconsistent, particularly at higher altitudes. The three concentric rings in the A/B filled sequentially depending on how far into A/B the throttle was positioned. WAGT recommended that the throttle be advanced to full A/B if A/B was to be selected. Modulated A/B control was not an engine requirement and even though the basic A/B fuel control system had the capability to provide this, it was never developed to be reliable and in fact the fuel control for the production -22 and -22A had the intermediate settings eliminated.

The overspeed relay was mounted on the accessories gearbox and operated directly on the overspeed fuel valve. In the event the rpm of the engine was at military rpm and then either began to exceed or drop below military, this valve would begin to control the fuel to the engine to prevent an overspeed or an underspeed. At any rpm below military, this valve was held fully open. The speed switch was in operation as discussed above.

The dump valve opened when the engine fuel pressure was detected to fall to zero, dumping all fuel in the system to prevent fires on later starts.

G.   Afterburner Cooling

The afterburner was cooled between the two shells by compressor air taken from the rear of the compressor and using two pipes and was pumped to each side of the compressor on a continuous basis whether the afterburner was operating or not.

Chapter 6 - J40-WE-8 Mechanical Description Citation

[1] The mechanical differences between the -8 and the -6 were culled from numerous comments in hundreds of memos cited in the other chapters.

Figure 1 Mixer vanes used in the -6, -8 and -22 to eliminate temperature inversion problems in the turbine at altitude, WAGT P47979, 10/29/1952. *Courtesy Hagley Museum and Library*

Chapter 7

J40-WE-8 Production and Service Issues

◆━━━◆

Production was accomplished through contracts NOa(s) 10385 and NOa(s) 403c. The latter was used as an additional revenue vehicle to support the continuing development while the -22 was being prepared. Early production of the -8 was at Essington in the same environment simultaneously producing/supporting the -6. The production serial number block assigned was WE031001 to WE031999. As will be seen later, some of this block was reassigned to the -22 and -22A when they entered production.

The request for data on the transient response of the exhaust nozzle area to turbine discharge temperature when on automatic trim control was responded to with a memo saying the data was complete and the report would be delivered on September 26, not the earlier promised date of September 5. As it happened, auto trim would never be delivered on a production -8 engine.[1] The report was actually delivered October 2, 1952 as A-1456.

The shortage of parts due to alloy shortages continued at this time under the BRICKBAT program and affected the ability of Westinghouse to produce -8's. There would have been a delay in any event since the problems affecting the -6 had to be addressed and the corrections incorporated into the -8 engine, the turbine portion of which was essentially that of the -6.

The qualification engine, using a 675 in² nozzle was tested with a deviation from the specifications as follows:

Rating at Standard S/L	Thrust lbs	RPM	SFC – lbs/hr/lb	Turbine Inlet Temperature (TET)
Take-off	10,100	7,260	2.70	1,440°F

No new specification for the performance with the 675 in² nozzle engine was prepared, BuAer later waving the requirement.[2] However, each engine produced would be individually accepted until the standard specification -8 was being shipped. (There is no evidence that any -8 engine was ever shipped to an airframe manufacturer with a 675 in² nozzle. Delays for other reasons appear to have allowed the parts for the larger 726 in² nozzles to be fitted. However the first two -1 engines were shipped with the smaller nozzles.)

Cleanup of prior engine surveys related to various changes continued. WAGT was able to confirm that removal of the gearbox lugs and intake duct instrumentation bosses from the engines had no effect on the pressure readings from the new pressure probe locations. This was in answer to a concern of Grumman Corp. However, due to production pressure, it was not possible to incorporate this change in the -8 engine and the survey for this model was deleted.[3]

As related earlier, the thrust bearing was found to be marginal in design and unable to withstand the impact of large numbers of A/B light-offs. Data provided on test engine failures of the original bearing was:

Engine	Total Hours Of Operation	Hours of A/B Operation	No. of A/B Light-offs
XJ40-WE-6 #1	124.3	23.4	275 (est.)
XJ40-WE-8 #1	235.9	13.8	400 (est.)
XJ40-WE-8 #3	202.4	9.9	435 (Act.)
XJ40-WE-8 #7	134.3	11.4	210 (Min.)

BuAer rejected WAGT's suggestion that the bearings be replaced after the green run and that the engine not be inspected again after the Acceptance Test run. The engines with this bearing would then be limited to 100 A/B light-offs in service use. BuAer argued this approach would limit the engines to only about 50 hours of service use before overhaul. As an alternative, the BAR and WAGT proposed:

a. That the total number of afterburner light-offs for both green and final acceptance running and field operation be limited to 115.

b. That the maximum number of afterburner light-offs to be accumulated during green and final run be 30.

c. That any engine requiring more than 30 afterburner light-offs during green and final running be the subject of separate review.

This would allow the minimum number of A/B light-offs to be 85 after acceptance. Since BuAer expected to require 50 hour overhauls on these early engines, this change would not have any noticeable effect on service use. All engines after the first four would be retroactively upgraded with the new bearing at their first overhaul.[4]

The fuel control (58J651) used successfully in the 50 hr type test was to be replaced with a 58J851 improved model for all production -8's. The reasons were:

a. The throttle valve was mounted on the shutdown end of the throttle shaft to provide greater strength and to reduce the angular freedom between the two. This change had no effect on control operation beyond increasing strength.

b. The acceleration valve in the fuel control was redesigned to provide "fail-safe" operation. In the earlier valve, a failure such as clogging of the acceleration valve chamber bleed orifice would cause the valve to shut off all fuel flow to the engine. The new control failed safe by actually bypassing fuel through the control. Thus the only detrimental result of an acceleration valve failure was that the pilot must exercise care in accelerating the engine to prevent over-temps.

The new control had passed 400 hours of cyclic endurance testing plus a 50 hour hot test and a 50 hour cold test, with a 150 hour Engine Verification Test scheduled to be started in October. WAGT asked for an approval of the Change in Design (CID) requested so the new control could be used on all the -8 engines.[5] BuAer gave conditional approval of the CID, with the condition that the final acceptance test results were to be submitted along with a schematic drawing of the new control.[6]

The Operating Instructions (pre-acceptance) were issued for printing on October 24, 1952 in anticipation of pending shipments of engines. Actual availability for release of copies to the field was moved out into November.[7]

A new Engine Operating Limits chart was published in October to show the operating limits of the -8 (See Appendix B).

J40-WE-8 Serial WE031002 was authorized October 27 to be shipped to McDonnell in St. Louis for the start of air testing. The details of the configuration of early engines are covered below.[8] Actual shipment continued to be delayed by many factors still to be resolved.

BuAer turned its attention at this moment to McDonnell's proposed use of a 28 inch extension between the turbine and the afterburner in the XF3H-1 airplane. They admitted they had not paid much attention to the details of using dummy extensions on the -6, but now requested that WAGT submit any data they had above and beyond report A-1369 (not located – Auth.) submitted in May on the use of these extensions. They asked also that WAGT "hazard a guess" on the effects of such an extension and durability of the -8 engine. It was stated that unless WAGT confirmed the extension could definitely be unsatisfactory, BuAer would not require any pre-flight evaluation testing of the -8 with such an extension.[9]

Engine Change 101 was surveyed for the -8 in October, the change being to vent the booster pump seal to a drain to prevent damage from overpressure. This change also affected the first 60 -22 engines (as Engine Change 14). Specifically, the drain fitting for both the over-speed relay and the afterburner fuel pump check valve was changed from a "T" to a "Cross" fitting to add a connection for the seal drain.[10]

The subject of the maximum allowable exhaust gas temperature with the 675 in² exhaust nozzle was addressed in detail by WAGT in acknowledging to BuAer that engines would be individually accepted until standard configuration engines were in production. They noted that the Model Spec for the -8/-1 assumed a turbine inlet temperature of 1,425°F would be employed. The actual measure on the qualification engine was 1,440°F. The Maximum Allowable gas temperature in the Spec was 1,350°F and the qualification engine was qualified at about 90°F lower. The actual temperature of each engine was tested in acceptance test and the max allowable temperature recorded in the engine log book and on a decal which was attached to the engine, becoming the maximum allowable temperature for that specific engine. Given the above, they recommended the Model Specification remain unchanged.[11]

Figure 1 J40-WE-8 #1, Top (previous page) and bottom views. Projections around the end of the annular combustion chamber allowed pressure and temperature rakes to be obtained on the test bed. WAGT P45765 and P45766, 11/28/1951. *Courtesy Hagley Museum and Library*

In spite of all the continued development of items in preparation for production and determining how to introduce them into the production stream, one of the biggest challenges was the number six aluminum compressor disc, which, as it turns out, had developed a crack during the 50-hr Type Test. The engine (XJ40-WE-8 #3) had 202 hours of satisfactory operation when the crack was discovered. The disc failure was almost certainly related to the gas leakage paths in the compressor allowing the higher pressure and temperature gas from the last compressor stages to leak past the curvic clutches, through the bolt holes in a forward direction and to heat the 6th stage aluminum disc beyond the design temperatures. This problem was also being experienced in the -6 at this time and the solution was to seal the leakage paths with Bakelite. For the -8, the proposed solution was to replace the aluminum disc with one made of steel starting with the third engine produced and limit the operation time of the first two engines to a maximum of 65 hours total operation and 15 hours of military operation before first overhaul, with the same A/B light-off limits set previously. (These rules would apply to the first two -1 engines for the Air Force as well.)[12] The new disc would be tested on the XJ40-WE-8 #4 and the test results made available early in December 1952. The steel disc would add 35 pounds to the weight of the engine. A titanium disc would add nine pounds over the aluminum disc it replaced.[13] (Titanium was planned to be used in some of the compressor discs of the -10 and -12.) Steel discs (or another suitable solution to the aluminum disc problem) would be incorporated in engines starting with serial WE031004 and later. The first two engines (WE031002 and WE031003) were to be upgraded at their first overhaul.[14]

Yet another challenge to WAGT was the incorporation of an additional emergency fuel system at the request of McDonnell, to be used on the first flight of the XF3H-1 airplane using the -8 system, as a backup to the primary/emergency fuel control systems already installed on the engine. The control was described as:

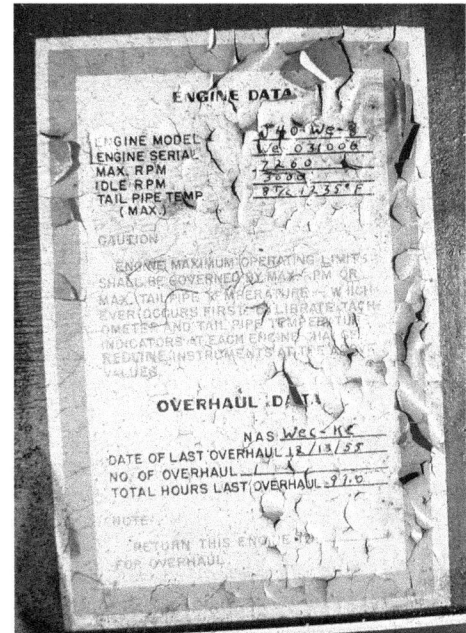

Figure 1 Example of engine decal on J40-WE-8 Serial WE031006 in the Air and Space Museum Collection. *NASM, Jeremy Kinney*

1. An 8 in² component installed on the airframe containing a fuel control valve.

2. An additional over-speed relay to be mounted on one of the airframe accessory drive pads on the engine gear box.

3. An adapter to the 4 bolt flange over-speed relay to the 6 bolt pad to be used.

4. Piping.

The new fuel control valve was placed in the by-pass line from the primary fuel regulator to the pump inlet. The discharge of the emergency fuel regulator was then routed into the fuel supply line from the primary fuel regulator to the engine. The higher backpressure occurring from this change in the primary fuel regulator required modifications of the acceleration valve, pump selector valve, and over-speed valve to re-reference these to the proper pressure.

McDonnell would supply a linkage system that in normal operation (the emergency fuel regulator power lever NOT switched on) would cause movement of the primary fuel regulator power lever only. A cockpit switch would allow the pilot to actuate a trim tab motor which would mechanically change the linkage system so that both the primary fuel regulator power lever and the emergency fuel regulator power lever would operate in parallel when the throttle was manipulated. Unless switched "On", the emergency fuel regulator power lever remained in the full "Off" position.

Pilot operating procedures were:

1. *Take-Off*

   a. Move the throttle to maximum (full afterburning) while on the primary system only.

   b. Throw the cockpit switch to activate the emergency power lever to parallel the primary throttle position. (NOTE: The throttle must be in maximum position first since once the emergency system was turned on, it is impossible to accelerate from military to maximum without detriment to the engine.)

   c. The emergency fuel regulator now supplies all fuel to the engine when the throttle is in either military or maximum power setting.

   d. After take-off, the pilot can throttle back. The emergency fuel regulator has no altitude compensation, so the pilot should move the emergency fuel control switch to the "Off" position. The primary fuel regulator immediately takes over.

   e. If during take-off the primary fuel system fails, there is no effect on the engine operation. If the emergency fuel system fails, operation is automatically returned to the primary system with no pilot action necessary. Since the emergency fuel system operates at slightly higher engine speeds than the primary, failure of the emergency fuel system would cause a drop of about 1% in engine RPM. If operating in afterburner at military RPM at the time of failure, a 3% RPM drop will be experienced momentarily.

2. *In Flight*

   a. If the primary fuel system fails (RPM or thrust drop noticed), the pilot has to throttle back to idle and then throw the switch actuating the emergency fuel system. (Throwing the switch at any other throttle position could cause over-temp damage to the engine or an engine stall, since the emergency fuel control had no acceleration schedule control.)

   b. With the emergency fuel control switched on, very slow manual accelerations can be accomplished with pilot

observation of the turbine outlet temperature to prevent over-temp damage.

3. *Landing*

   a. Advance the throttle to the military position and throw the switch paralleling the primary and emergency power levers.

   b. Throttle back as required for landing. Re-accelerations must be performed with caution not to exceed the allowable jet pipe temperature. Acceleration into A/B cannot be made in this condition.

4. *Remarks*

With the emergency fuel system actuated at military thrust, the emergency fuel regulator is feeding the engine. As the pilot throttles back, the emergency fuel regulator continues to supply all the fuel to the engine until the RPM corresponding to the top speed setting of the primary fuel regulator is reached. At this point the primary fuel regulator starts to supply fuel to the engine, still holding RPM at the primary fuel regulator top speed setting. As the throttle is reduced further, more and more of the fuel is supplied by the primary fuel regulator until the point is reached where all the fuel which can be passed by the primary and emergency fuel regulators that is required to maintain military RPM. At this point the primary and emergency fuel regulators are each supplying half of the fuel to the engine. This proportioning of fuel is maintained upon further retraction of the throttle. The military RPM flat spot with the emergency fuel system actuated will be longer than standard, military RPM being maintained in standard to throttle position approximately 55° and 35° with the emergency fuel regulator actuated.

A later memo in November noted that the modified primary controls could easily be returned to normal configuration, and that the first modified control shipped to McDonnell should only be used on WE031002. The modified control on WE031002 (and any engine using a modified control) was to be returned to normal configuration prior to the engine being returned to WAGT for overhaul. The second and third modified controls were shipped separately to McDonnell and were not associated with a specific engine.[15]

All -8 and -1 engines had the modified components capable of handling this emergency fuel regulator, since their basic functions were not affected by the changes if the emergency fuel regulator was not installed on the engine. At the time of the memo, WAGT was authorized by BuAer to run the qualification test using WE031002 and authorized the use of the modified parts on the engine, since no unmodified parts were available.[16]

On October 27, 1952, -8 Serial WE031002 was authorized to be shipped to McDonnell in St. Louis. The authorization noted the engine was needed for Flight Tests.[17] This engine had an aluminum sixth stage compressor disc, no anti-icing valve, a 675 in² nozzle and was limited to 65 hours operating time with no more than 15 hours at military thrust with afterburner light-offs limited to 85 before overhaul. It incorporated the modified fuel control to prevent un-commanded engine shut-downs (see below). Prior to shipment, it was further modified as described below, with shipment delayed.

The November progress report noted that the No. 4 development engine to be used in the 150 hr Qualification Test was being assembled as a J40-WE-22 engine. The engine incorporated the -22 gearbox with constant speed drive, a 726 in² exhaust nozzle, automatic temperature trim control, modified items from the -8 type test deficiencies and other minor changes. Test start was expected to be about December 21 and would be used to verify the constant speed drive for the -22. Successful completion of the test would qualify the -8 engine.

The second production -8 (WE031003) had revealed turbine hot spots which had to be cleared up and was being torn down for inspection. Delivery was expected in December. (It was configured the same as WE031002 above.) A second engine would not be delivered as scheduled in December.

Shortages of fuel controls and other minor items would delay testing of Serial WE031004 until January.[18]

WAGT had reported serious control system malfunctions during testing on the J40, J46 and models of the J34 that used the same fuel control and had proposed an improved acceleration control to prevent recurrence. BuAer accepted the verification testing proposed but noted if the failure analysis submitted earlier had been more thorough, the need for the improved provisions being incorporated would have been identified earlier and would have avoided much lost time and expense.[19] The problem uncovered was that surge pressures affecting the acceleration of the control valve could cause an un-commanded engine shut down. It was also discovered that clogging of an orifice in the acceleration control could also cause a shut down. The corrective action was to alter the fuel flow of fuel bypassed within the acceleration control to a stable low pressure. After doing this no further malfunctions had been experienced over 500 hours of operation on a J46 and 200 hours on a J40-WE-8. It was noted that the same correction had been applied to Chance-Vought J34-WE-32A engines which used the same controls. A fail-safe change was made to the control so that in the event a clogged orifice occurred, the control would fail to the wide open position. The control would be tested for 500 hours in a laboratory test and was to undergo an engine verification test on house engine XJ40-WE-8 No 4, due to start in October. It may have been these failures that triggered McDonnell's requirement for their emergency fuel control on first flight.

In early November, BuAer stated that in view of the decision to curtail the -8 program in favor of the J40-WE-22 engine, the AN engine handbooks and applicable accessories handbooks for the -8 would not be required. A

special version of the service instructions comparable to those previously furnished for the J40-WE-6 would be required.[20]

Figure 2 J40-WE-8 1st and 2nd stage turbine wheels being joined on assembly stand. WAGT P46561A, 7/3/1952. *Courtesy Hagley Museum and Library*

The fuel and oil leak fixes identified for the -6 were applied to the -8 in early November. Apparently the WE031002 engine had the modified gaskets, drains and scuppers installed prior to shipment.[21] It is not clear at what point further production engines incorporated the modified gaskets and scuppers prior to shipment to the field.

It was discovered that the existing tool (243681 S1 – T114) used to pull the No. 1 bearing could not pull the bearing without breaking it. It was determined that the bearing could only be removed undamaged by pressing on the outer race. The new tool (243681 S1 –T140) replaced the T114. Bearing replacements were occurring in the field at this point to incorporate the newly designed replacements in the -6 and -8.[22]

Figure 3 J40-WE-8 Burner basket being removed from the diffuser assembly, WAGT P46565, 4/2/1952.
Courtesy Hagley Museum and Library

McDonnell agreed to conduct certain flight test work on the J40-WE-8 for Westinghouse, although such testing would not be done on the first flights with the -8 installed. However, the desire to have the control system monitoring instrumentation available on the initial flights led McDonnell to request the instrumentation be installed on the WE031002 prior to shipment. WAGT requested permission to make the following changes to the engine prior to final acceptance testing and delivery (the BAR approved the changes):[23]

1. Compressor Inlet Pressure. Provide a "T" in the line which would transmit compressor inlet reference pressure to the engine control.

2. Afterburner Igniter Pressure. Provide a "T" in the line between the afterburner igniter valve and the afterburner igniter nozzles.

3. Compressor Outlet Total Pressure. Provide a "T" in the line which would transmit compressor outlet total pressure as a reference pressure to the engine control.

4. Turbine Outlet Total Pressure. Provide a total pressure rake in one of the four (4) boxes already available in the engine exhaust collector.

5. Aerotec Switch Pressure. Provide three "T's" in the Aerotec P_1, P_2, and P_3 lines located as close to the Aerotec switch as possible.

6. Four Thermocouple Harness. Provide a four thermocouple harness in the boxes already provided in the engine bearing support. McDonnell desired the turbine out temperature signal for a cockpit instrument, an oscillograph and a photo panel instrument. The cockpit instrument and oscillograph were connected to the regular eight-thermocouple harness on the engine and the photo panel to the auxiliary four thermocouple harness.

7. Afterburner Fuel Pressure. Provide a "T" in the bleed line between the afterburner fuel regulator and the engine boost pump outlet. During A/B operation, the tap would provide A/B pump discharge pressure and during non-A/B operation, would provide boost pump discharge pressure.

In November, WE031002 was formally accepted by BuAer and shipment was again authorized. This time it actually shipped on November 11, 1952 to McDonnell in St. Louis. Compared to the qualification engine, WE031002 deviated in the following ways (status of qualification items requiring redesign and test are included)[24]; target price for the -8 was $253,772.00:[25]

1. The seventh through the eleventh stage blades of the compressor had been ground off 0.010 inches on the radius because of a rub problem during acceptance testing.

2. The afterburner cooling valves had the rings removed. The valve configuration had been given a 20,000 cycle test and had accumulated a considerable amount of house running without incident. (The verification test engine would have one pinned ring valve and one ring-less valve configuration installed.)

3. The control flight test instrumentation provisions were incorporated.

4. The McDonnell emergency fuel control provisions were incorporated.

5. The flexible metal fuel line joining the booster pump with the afterburner and dual pump was the latest design.

6. The sixth stage compressor disc was an aluminum disc of the same design as that in the qualification engine. The engine usage limitations previously approved applied to the engine with this disc installed.

7. The A/B blow-out (Aerotec) switch arrangement was of the non-bleed type and included a ball-check valve mechanical time delay feature. The arrangement was to be verified on the house engine number four. Several hundred successful light-offs with that system had been reported. WAGT was required to immediately demonstrate 40

consecutive successful light-offs with this system on a house engine.

8. The thrust bearing was of the same design as the qualification test engine (this was the original -6 design). The new design bearing would be verification tested in house engine #4. The A/B light-off restriction limits agreed to earlier also applied to the engine.

9. The exhaust collector of the engine was of .050 inch material identical with the qualification test engine, but a redesigned collector of 0.080 inch material would be verified on house engine #4.

10. The engine had scuppers, drains and gaskets incorporated in accordance with the contractor's new design. A CID would be submitted for approval covering the scupper/drains/gasket changes on the -8.

11. The exhaust nozzle was a formed shroud in lieu of the spun design which had failed in the qualification test. The formed shroud would be verification tested on house engine number 4.

12. There was no anti-icing valve on the engine.

13. The engine weight was 3,548 pounds, as previously agreed for the first 50 hour engines. The maximum acceptable engine weight was requested from BuAer.

14. Performance was:

J40-WE-8 Serial WE031002 Performance		
Condition	Thrust (lb)	SFC
Maximum in A/B	10,600	2.18
Military	7,700	1.026
Normal	6,500	0.993
90% Normal	5,850	0.992
75% Normal	4,880	1.052

15. The A/B exhaust nozzle area was 675 in^2, approved for the first two -8 and -1 engines.

Two days after shipping the first -8 to McDonnell, WAGT had to decline a request for a J40-WE-8 test engine from Douglas on the grounds it would interrupt their ability to deliver flyable engines during that immediate time period.[26]

In response to the A/B extension question from BuAer, WAGT finally answered, stating that no problems with durability were expected, as the extension was designed to the same strength requirements of the engine and no problems with the extensions had occurred during testing. Some performance reduction was expected but they could not estimate how much this might be.[27]

The operating restrictions on the first two -8's (WE031002, WE031003) as well as the balance of the -8's to be produced were formally published to the airframe companies in a memo on November 18. These were listed as:[28]

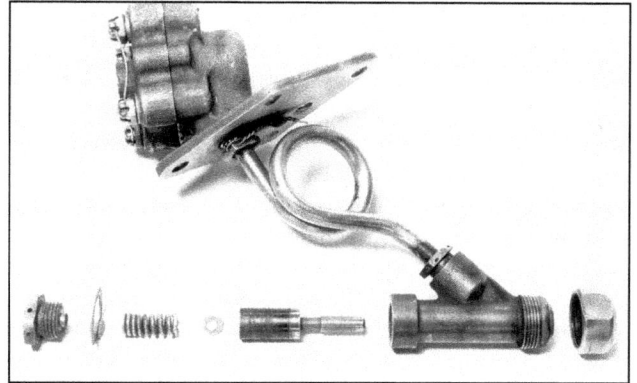

Figure 4 Fuel spray nozzle assembly, used in the -2, -4, -6, -8, -1, -22 and -22A models, WAGT P42095. *Courtesy Hagley Museum and Library*

The first two production engines should be returned for overhaul upon reaching any one of the listed limits:

1. 115 afterburner starts including those accumulated during initial and final acceptance testing at Westinghouse. (The total A/B starts at Westinghouse were limited to thirty.)
2. 60 hours of total time
3. 12.5 hours of military time
4. 7.5 hours of afterburner time

The balance of -8 engines should be returned for overhaul upon first reaching any one of the listed limits:

1. 60 hours of total time
2. 12.5 hours of military time
3. 7.5 hours of afterburner time.

Extensions of operating times for all J40-WE-8 engines would be contingent upon the condition of the engines at overhaul and also upon the operational characteristics during field use. The various handbooks would be forthcoming, but until they were available, the Preliminary Operating Recommendations were amended to include the needed information. Maintenance inspection instructions for engines restricted to less than 150 hours of total time would be forwarded after review with WAGT. This review was conducted and WAGT recommended (and BuAer accepted) "A" level checks at 30 and 90 hours and "B" level checks at 60 and 120 hours.[29]

In late November, BuAer forwarded a memo on acceptable fuels for the -8, noting that the first two engines did not have fuel lines compatible with JP4, but subsequent engines starting with WE031004 would have JP-4 compatible flex fuel lines. Use of fuels with a high lead content should be avoided, but if used, inspections of the afterburner fuel manifolds for lead residue clogging should be done frequently, based on experience.[30]

With flight engines now being shipped, WAGT notified BuAer in early December that there was a concern that an A/B squeal condition might exist in high density (high speed/low altitude) conditions. WAGT test information showed no squeal condition up to 0.83 Mach, but suggested flight tests be made at 10,000 feet at 0.80 Mach and further checks then extend at that altitude up to 1.2 Mach to verify no squeal condition existed beyond the WAGT end data point. The test procedure recommended was to operate in A/B for two minute intervals at 0.9, 1.0. 1.05, 1.1, 1.15 and 1.2 Mach (based on 100% duct recovery at the engine inlet), then the engine duct checked for squeal damage. Damage would appear as cracks in the inner shell of the A/B exhaust collector or around the bolts holding the two shells together. Also, particular attention should be paid to the diffuser brackets supporting the A/B cooling air lines. If the brackets failed and a large number of cracks developed in the inner shell, than squeal may have occurred. However, it was possible for the brackets or cracks to occur unrelated to squeal. A small number of cracks were not a concern and unbroken brackets were a reliable indication that squeal had not occurred.[31]

BuAer reacted by placarding the aircraft, requesting a detailed definition of the region of possible -8 A/B squeal within the Model Specification, and being apprised by the BAR as to what negotiations between WAGT and McDonnell were underway for engine tests in the XF3H-1 or F3H-1 airplanes.[32]

WAGT's response stated that the data limits on squeal were caused by their test facility's inability to test to higher limits. They noted that there were no agreements in place with McDonnell to conduct the recommended tests, stating they were interested in what would occur in such conditions. They further noted that in their opinion, the tests would pose almost no hazard and that should squeal be encountered it would be of a small magnitude, and damage incurred would be slight and repairable by routine welding operations in the field.[33]

Engine serial WE031003 was shipped from Essington to St. Louis and WE031005, the first engine to be delivered from Kansas City, was the fourth engine allocated by Speed Letter.[34] This was in response to a Speed Letter of December 10, which noted that the serial number block assigned to engines shipped from Kansas City was WE031005 to WE031024.

WE031002 was shipped February 11, 1953 from McDonnell in St. Louis to Kansas City for a blade change of some sort and then returned on February 15. The reason for the repairs was not stated but foreign object damage to engines was common at McDonnell during this period.[35]

The anti-icing valve was anticipated to be qualified and released for production during April, 1953. This valve

was to be used on the J40-WE-6/-8/-22 engines. Valves would be available for incorporation in June, 1953.[36]

The model specification WAGT-X40E2-4A, superseding WAGT-X40E2-4 dated April 25, 1952, was submitted to BuAer for approval on January 12, 1953. This only covered the -8. WAGT noted in their cover letter that specification WAGT-X40E2-6 would be submitted to cover the -22 and -22A, stating that the use of appendices to describe sub-types led to errors and confusion.[37]

A verbal request from BuAer to WAGT for determining the natural vibration frequencies of the -8 relative to the engine mounts was responded to with a test program proposed to be completed by February 27. WAGT requested that an airframe intake be available to attach to the engine, as this would change some of the vibration characteristics. Availability of such an intake might change WAGT's ability to conduct the test by the proposed schedule. Also, they pointed out the data provided would only apply to a test house engine installed on a test bed and it would be better to conduct such a test on an engine already installed in an airframe.[38]

In early February, WAGT wrote a long memo discussing the history of compressor limit philosophy and the various aspects of pressure limiting on early J40 engines. This provides excellent background to the emerging, more challenging flight conditions engines would have to endure. Portions of this memo are quoted here:

"Prior to 1950, all Westinghouse engines were designed to withstand the loads imposed by maximum power output operation at the rated operating conditions specified in the engine model specification. In general, it was found that the cold day rated operating conditions were the most severe from the mechanical design standpoint and required that the engine weight be increased to provide a satisfactory design. It was understood that the increased thrust output under cold day conditions could not be utilized efficiently and hence the engine weight increment could not be justified since the airplanes would require only the standard day engine performance for demonstrating airplane characteristics. It was apparent that a substantial weight improvement could be realized if a device could be incorporated as part of the engine control system which would eliminate the high loadings associated with a cold day as would be available on a standard day under the same conditions. An automatic pressure limiting device of this type was planned for the electronic engine control. When the simple hydraulic control was adopted there was not sufficient time to incorporate an automatic pressure limiting device on the early models. Therefore, the use of the less ideal system of a cockpit pressure indicator was planned. Studies are continuing on the various means of automatically limiting the engine loadings for use with the simple hydraulic or hydro-mechanical control systems.

"It has not been the intention of these limiting devices to permit safe operation beyond the ram conditions specified in the engine model specifications, the intention has been only to provide a lighter engine within the specified operating

limits. It has become apparent that, with the present state of aircraft development, there is the possibility that under extreme conditions the engine ram limits will be exceeded. An analysis of this phase of operation is included in the automatic pressure limiting study which is now being made."

The memo went on to say, in part:

"...the outer wall of the combustion chamber liner, the turbine shaft housing and turbine, in addition to the afterburner casing, is dependent on pressure limiting and, therefore, requires reduction of the power setting below the military condition.

"... the compressor pressure limiting system is not intended to protect the engine beyond the engine mechanical limits. The loadings which will be encountered under these conditions are being analyzed at the present time. It has been found after a survey of the field, that the maximum compartment pressure will not exceed the ambient pressure by more than 2 psi.

"... the engine variable which should be sensed in order to limit operation below the engine mechanical limit is the pressure difference between compressor outlet total pressure and aft compartment static pressure......it is the intention that this pressure limit will completely protect the engine below the engine mechanical limits.

"... the J40-WE-6 engine was designed for the maximum loads encountered under the ram conditions specified in the engine model specification while operating at military conditions. For this reason, no pressure limiting is required on this configuration.

"The exact date of when an automatic pressure limiting device can be incorporated and the details of such a device is not completely defined at this time.....it is estimated that this study will be completed by March 15."[39]

From this memo we can discern the emerging need for more sophisticated intake systems that could automatically restrict airflow volume, pressure and flow velocity at the engine face to that which the engine could safely accommodate. WAGT was trying to solve this problem independently on the engine only and this would continue to be an area of concern to BuAer. It also shows that WAGT had to continually educate BuAer about the differences between the models (-6 and -8/-22 in particular) as they worked to support the -6 in the field and the -8, which was close to flying. In a later memo, WAGT committed to complete the study by June 1.[40]

In approving new performance curves for the -8, BuAer noted that definite discontinuities were present in the power lever versus thrust schedules at altitudes above sea level and that they were obviously undesirable. WAGT was asked if the control schedule could be improved to lessen or eliminate the discontinuities.[41] This request was addressed in a cover memo to some updated engine performance curves. WAGT noted that the characteristics on the curve in question were inherent in the fuel regulator used on the -8, -22 and -22A engines and was caused by an uncompensated fixed fuel flow schedule below 100 percent RPM. They noted the discontinuities were not present at sea

level over the range of ambient temperature nor were they present between approximately 50 and 100 percent thrust at any altitude. In their experience, the region below 50 percent thrust was not normally used at altitude. A typical flight pattern applicable to the J40 engine from a performance standpoint indicated that the lower thrust requirements associated with altitude loiter conditions could be met with the control as it was. As an example, sea level loiter required approximately 18 percent thrust and could be realized with the current control schedule, whereas 35,000 feet loiter required a thrust greater than 50 percent which was a condition above the range of the control schedule discontinuity. To provide a straight line thrust relationship at all altitudes and flight conditions would require an altitude and flight speed compensated manual fuel schedule between idle and approximately 55 degrees of throttle position. This would require considerable redesign of the fuel regulator and would be a departure from the simplicity and reliability intended in changing to the present hydraulic control. As no complaints or examples of undesirable power lever versus thrust conditions had been received from aircraft manufacturers, WAGT felt the fuel regulator was satisfactory as it stood. No plans were in place to incorporate any changes at that time relative to discontinuities.[42]

Grumman notified WAGT on March 10 that an intake duct was available for use in testing.[43]

Early April 1953 found BuAer writing regarding how engine logbooks were maintained relative to zero time after overhaul. They pointed out that standard practice was to zero time the engine at the end of the green run, assuming the teardown of the engine after the green run was the same level of "completeness" of an overhaul teardown and that the green time would have been a small part of the total operating time of the engine on balance across numerous engines. In the specific case of WE031003, they found that the acceptance run time after the green time run had exceeded 26 hours of testing, yet the engine was zero timed at the end of these tests. Since the engine had one part in particular, the 6th stage aluminum disc that was limited to 60 hours of running after green time running, this left the engine with less than 40 total service hours before mandatory overhaul. They requested that a procedure for zero-timing engines be discussed.[44]

The engine allocation schedule as of April 1 showed that -8's, WE031002-WE031003 had been delivered. Even numbered engine buildups (not their serial numbers) from No. 8 through No. 36 would be built as -8's, the balance to be -1's.[45]

On April 28 WAGT recommended the grounding of all -1, -6 and -8 engines because of the failure of two house engine compressors due to disc failure. Excessive heat was occurring but WAGT did not understand the problem yet.

(This was the air path leakage problem discussed earlier.)[46] A house engine with Bakelite seals between the disc clutches and around the through bolt holes was sent to the test cell on May 29.[47] This test was successful in lowering the number 6 compressor disc temperature to specification levels. In the wire updating the test engine status, it was noted that WE031002 had shipped without the A/B nozzle aft fairing, which was to be upgraded to 0.80 inch material from the original design of 0.50 inch material.

A component qualification test was conducted on XJ40-WE-22A #3 for the Bakelite sealing, an Automatic Temperature control was modified to incorporate improved shock mounting of the voltage and amplifier tube and the (726 in^2) Exhaust Nozzle Fairing made of 0.50 inch material. The results were, respectively:

A. The Bakelite effectively sealed the leaks in the compressor stack and the number 6 disc remained within its thermal design limits throughout the test.

B. The Automatic Temperature Control operated satisfactorily during the entire test, the engine running within a 15 degree band of turbine outlet temperature and for 90% of the test indicating exactly 1180°F, the original setting.

C. At 39 hours of qualification testing and 45 hours of total running, a crack was noted in the fairing located on the aft flange of the channel along the weld. The crack was stop drilled and the test continued satisfactorily, the fairing accumulating a total of 68 hours. Inspection of the fairing at the end of the test revealed rubs in the area of the channels which were caused by the afterburner eyelids. Further inspection revealed that the aft circumferential stiffeners had separated from the basic part at the welds. Prior to the test the welds had been found to not be ideal, but since no other A/B's were available, the test proceeded. The eyelids had also distorted and the axial hat-shaped stiffeners had rubbed on the fairing. The A/B used in the test had accumulated 247 hours of running prior to the test and now had 315 hours and had not been overhauled. It was concluded the cracking would not have occurred with a new or overhauled A/B being used in lieu of the one actually used.

WAGT asked that the Automatic Temperature Control and Exhaust Nozzle Fairing be qualified based on the test results.[48] This was accepted by BuAer on June 11.[49] In the acceptance letter, they released WE031002 and WE031003 to return to flight status. WE031004 and WE031005 were accepted with Bakelite compressor sealant but noted all the other acceptance criteria for WE031004 and subsequent engines had to be met, including weight. It stated the last weight limit set for -8's was 3615 pounds and asked for the actual weights of WE031004 and WE031005 to be forwarded to BuAer.

The preliminary results of the compressor loading limit study showed that the limiting control schedule selected previously was unsatisfactory from the standpoint of compressor protection. Determination of a modified pressure limiting control schedule and a study of the engine parts loadings encountered with this modified schedule would delay the completion of the study until July 1.[50]

In July, at BuAer's request, WAGT wrote a memo detailing their review of the Marquardt Corporation's work relative to their proposed control system (the PC-1) for jet engines. After complimenting Marquardt for the work done to date, they noted it was a novel approach to the problem. Their issues with it relative to WAGT turbojets at that point and design approach issues in general were:[51]

A. The control system proposed represented a considerable departure from the control arrangements then in use or contemplated. To bring the system to a point of satisfactory operations would require 5 to 10 man years of WAGT effort on top of any Marquardt required effort. It did not appear justified by the benefits to be gained.

B. The design presented the following issues:

a. The design used "referenced" RPM to produce maximum thrust in all flight conditions. The complexity of choice of engine design RPM and compressor stall margins was a difficult problem. WAGT's existing engines all were designed to operate at essentially constant actual RPM and a referenced RPM system would not be expected to provide optimum performance.

b. The use of oil as the operating medium for the fuel flow computing mechanism was not a good choice as in cold conditions it became too thick to provide proper operation of the mechanism. This was a safety issue requiring heaters or their equivalent. Secondarily, the mixing of oil and fuel was difficult to prevent. Also, using a separate hydraulic system had definite disadvantages relative to maintenance in service and the ability of the Power Control System to operate satisfactorily during acrobatics.

C. The existing development schedule at WAGT did not allow sufficient support to the extent necessary to do a creditable job. WAGT would continue to provide to BuAer engine performance and envelope information where available.

D. WAGT regretted not being able to support the development of the subject power control.

Allocation of production engines began in earnest in July, with WE031005, WE031006, and WE031007 previously allocated. It is believed WE031006 went directly to Douglas, as this engine was in the F4D that set the new world speed record. Further allocations were:

- McDonnell - WE031008*#, WE031010 and WE03010
- Douglas – WE031009*, WE031011 and WE03013
  * Completed in August and ready to ship from Kansas City.
  # Configured with the same special instrumentation fittings as WE031002 at the request of McDonnell and approved by BuAer.[52]

A "Naval Speed Letter" indicated WE031014 and WE031015 would be allocated at a later date.[53]

The engine mechanical limits study was completed on July 10 and forwarded to BuAer on August 10 by the BAR. The study report (A-1626) again summarized the earlier design philosophy of the -8 to save weight. A survey of the aircraft manufacturers had revealed that all of them expected their aircraft to be operated in the regions that could expose the engine to damage as currently designed. The report recommended an automated pressure limiting device combined with an automated temperature control. Other solutions requiring pilot action and/or performance limiting of the engine in certain circumstances were rejected as undesirable. The former control reduced the maximum RPM setting 15% as the difference between the compressor outlet pressure and the afterburner compartment pressure, and the temperature control was reduced 30% over the same control range. This gave a control relationship between RPM and temperature which held referred RPM and temperature constant for a given Mach number over the range of ambient temperatures. This approach allowed full use of military and A/B power settings and eliminated the need for pilot action to protect the engine. The report noted the A/B outer pressure casing would require strengthening to accommodate the overpressure of 114.5% over the current design limits of the -8 (and it is presumed, -22/-22A as well at this point). It noted such redesign was being undertaken. All subsequent WAGT engines would incorporate such a control.[54]

An earlier BuAer request that a second wire in parallel to the first in the automatic temperature control be added as a fail-safe should one of the wires fail, leaving the engine with no temperature control except that of the pilot, who might be otherwise occupied. WAGT agreed to this and said it would incorporate it at the "earliest possible convenience".[55]

A survey had been sent out regarding reinstallation of Afterburner Squeal Detection Instrumentation per CID 19421. It covered the -1, -8, -22 and -22A and it was approved for -1 engines by the Air Force on August 11, 1953[56]

August showed two -8's were accepted, serials not recorded, but likely were in the group of WE031004-WE031007.[57]

In September a CID was issued to cover a new A/B fuel shut off valve to be placed as close as possible to the entry point of the fuel lines into the A/B. The purpose was to allow the lines to retain fuel but not drain into the A/B on either non-A/B being selected or after engine shutdown.[58] This would allow faster A/B light-offs.

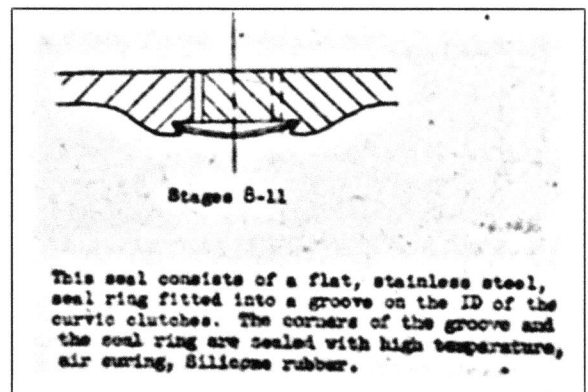

Figure 5 Compressor stage curvic clutch sealing solution for the -6, -8, -1 and -22. *WAGT Drawing, 9/10/1953* [59]

The solution to compressor air leakage involved a WAGT proposal (August 19) to replace the 1st and 5th stage compressor discs with new designs, replace the 7th stage disc with a steel one of the same design, and apply a "silastic" silicon based rubber ring to the curvic clutches off all the stages of the compressor, even the first 5 stages that did not need it. The latter was because WAGT believed that leakage might be occurring there as well and compressor efficiency might be improved by sealing them. On September 10, based on test results using -22A engines, the BAR endorsed this solution for all -1 and -8 engines (including retrofitting all -8's already built and shipped up to WE031007), all new -8's subsequent and the first block of -22A's. (Apparently no sealing of the actual through bolt holes was done, as the seal at the curvic clutch interface would preclude leakage inside the clutch.)[59] BuAer approved the new mechanical seals on October 12 with all repairs at no additional charge to the Navy.[60]

The layout of the new exhaust nozzle shroud and fairing design for the -1 and -8 were released to the airframe manufacturers on August 17. It applied to WE031003 and up, but only if fitted with the 726 in² nozzle.[61]

BuAer modified the service restrictions on -8 engines regarding allowable operating times on September 21 based on WAGT input to 100 hours of all condition time prior to overhaul with no limit on military or afterburner time.[62]

WAGT's reasoning behind their recommendation was as follows:[63]

A.   WE031002 had used up its allowable time and was returned to Kansas City back on February 16 for first modification and overhaul. The specific modifications were to replace the No. 1 bearing and the sixth stage aluminum disc with a steel disc, in both cases in advance of their respective CID's (19364 and 19380). Permission was given to bring the engine up to the latest configuration as of that date and return the engine to the field quickly, as it was urgently needed. On teardown the following discrepancies were noted and corrected:

    a.   Metal particles were found in the oil system caused by seal rubs, inspected the lube system for foreign particles; replaced main and rear oil pumps because of possible damage; disassembled the exhaust nozzle actuator under engineering guidance for replacement of damaged parts and then modified to incorporate new design elements; oil reservoir cleaned; all oil filter elements replaced; gearbox torn down to allow inspection for damage and cleaning plus incorporation of the overboard drain on the A/B pump gear assembly.

    b.   Blade tip rubs on the last six compressor stages and slight galling in the first stage segmented liners – These were likely due to high vibration caused by seal rub and distortion of the rear bearing support. Minor rework corrected the condition and then was followed by corrosion protection.

    c.   Nicks on 29 compressor blades – none severe, they were blended out.

B.   In addition, the following modifications were made to the latest configuration:

    a.   New housing assembly on the front bearing installed.

    b.   Changed the bolts, torque, lubrication, and locating plate on the first stage compressor disc.

    c.   Installed a new steel six stage compressor disc.

    d.   New fuel distributor 58J41-4 installed.

    e.   New fuel booster pump 58J146-6 installed.

    f.   New check and surge valves in the A/B lines installed.

    g.   Tie-wire flexible hose connections to the A/B rings installed.

    d.   Removed the A/B bleed solenoid.

    e.   Modified the A/B fairing to EDS-S228238 (part from Essington).

    f.   Reworked or replaced all hardware damaged or missing.

C.   The engine was shipped back to the field in March.

D.   When WE031002 and WE031003 were last returned from the field, they were equipped with opened rear bearing seals. Since that time, the opened seal configuration had been tested for over 300 hours on two -22 engines with completely satisfactory operation. This removed the last restriction to longer overhaul periods, WAGT argued. The latter plus the condition of WE031002 and WE031003 at their last overhauls supported their argument to increase the overhaul period to 100 hours for WE031003 and subsequent with no power operating limits, recording A/B time for record purposes only. Inspections of WE031002 and WE031003 would continue at overhauls with the goal of increasing the operating time between overhauls to 150 hours.

Shortly after the above memo recommending extended periods between overhauls, WAGT also recommended extending Bakelite sealed compressor rotor inspections from 15 to 30 hours based on field inspections showing the sealing was remaining in excellent condition.[64] The BAR concurred with the recommendation in his cover memo to BuAer, making the 30 hour inspections coincide with the regular 30 hour inspections.[65]

The Kansas City BAR accepted 3 J40-WE-8 engines in September.[66]

Delays occurred, making anti-icing valve availability for -8 engines move to February 1, 1954 from the earlier date of January 1 because various components were rescheduled with the supplier, delaying qualification testing.[67]

A field problem of A/B pump drive shaft ball bearing failures was reported, with both WE031003 having such a failure at 90 hours of operation and WE031004 at 53 hours. All -8 engines in the field were grounded. Engine WE031011 had also had such a failure at 6.5 hours during green testing. The failures were caused by clogging of the oil jets to the bearing, causing overheating and subsequent failure of the micarta ball retainer. A bronze retainer design was substituted and the oil jets enlarged from .025 to .032 inches. Three hundred and fifty hours of running on XJ40-WE-22A Nos. 1 and 3 incorporating the new bearing and enlarged jets had been accomplished without problems. All engines in the field had been immediately modified to the new standard and WAGT recommended the grounding of the engines be lifted.[68] No cause for the jet clogging was mentioned. Engines in the field modified were: McDonnell – WE031002, WE031006; Douglas – WE031005, WE031007.

In October WE031013 and WE031014 were shipped to Douglas.

BuAer informally requested information on cost and weight reduction if the anti-icing provisions were removed from the -6, -8, -22 and -22A engines. Westinghouse said their prices did not include costs for such provisions (it is not clear if that meant there would be additional charges or if they never priced the feature separately) and the weight reduction of eliminating the valve would be 12.5 pounds (including a cover plate to blank off the boss). An additional 3.5 pounds might be saved if the boss on the combustion can were removed, conditional on where the particular part was in production planning and tooling. They asked that any request for deletion of anti-icing be made via the specifications.[69]

WE031003 was returned from Patuxent River, MD to Kansas City for unknown repairs on October 27.[70]

In November, WE031012 was allocated to Douglas, the engine to be specially calibrated [71]at BuAer's request, such additional calibration not to exceed one hour of running time. WE031014 and WE031015 were allocated to McDonnell.

A revised Operating Recommendations manual dated November 1, 1953 was issued November 6, 1953 by the Kansas City BAR with his endorsement, replacing and updating the December 1, 1952 version of this manual. The manual covered the -1, -8 and -22A engines. Interestingly, the manual was marked "Confidential" by WAGT but the classification was raised to "Restricted" after it was issued.

A production summary memo of November 11, 1953 reported 12 J40-WE-8 engines produced and accepted through October 31, with two more -8's to be built under the contract (one from Kansas City and one from Essington.) This would bring total production of -8's to 14.[72]

An issue surfaced between WAGT and BuAer over the inclusion of changes in engines before the necessary CID's had been issued and approved. The subject engines were actually being shipped to the end users after acceptance testing without the necessary CID approval(s). The emergence of this issue in the September time frame came when WAGT was under very heavy pressure to deliver flyable -8 (and -1) engines at the same time that fixes were emerging to the many problems both the -6 and test -8 engines were uncovering in use. In addition WAGT was trying to transfer production engine building and testing to the new Kansas City plant. WAGT responded to a BuAer wire on this subject in September with a long memo, stating:[73]

A. There was no deliberate attempt to force BuAer to accept CID's because of exigencies without following the system provided in the contract.

B. Only the "sincere" desire to deliver the best quality engine possible made WAGT feel "it necessary to circumvent the CID system" by incorporating engine improvements prior to a given CID being fully coordinated.

C. Even with the changes made to the CID process, there would always be changes that, due to their importance, needed to be expedited into engines in advance of complete CID coordination.

D. Additional changes to the process were in the works, but not discussed in their memo, but the process was long and BuAer's personnel were not yet fully conversant with the new process following up on surveys and getting an answer within two weeks.

E. The new engine models were generating many CID's from initial operation in the field due to lack of flight test activity (before service or airframe manufacturer use). Flight testing of later models, just initiated, should reduce the burden on the CID system.

F. Additional manpower had been employed to follow and expedite the processing of CID's.

G. The change of moving J40 design control to the Kansas City plant did not affect CID processing in the time period in question with one exception, the costing determination of a CID, which remained an Essington responsibility. This latter activity was being analyzed for a possible future move to Kansas City.

H. The first two -8's (and first -1) were assembled in Essington but then shipped to Kansas City for acceptance. They contained changes not yet approved through the CID process, which had been the Essington BAR's responsibility. Without WAGT's knowledge, BAR Essington decided it was the BAR Kansas City's responsibility to get the CIDs on the engine changes approved. After this instance, a process was put in place to avoid confusion on any future engines built and/or tested at two different locations.

So, having stated basically that speed had been of the essence and that they were only trying to help in spite of knowing the changes weren't approved, and given that the Kansas City BAR had dropped the ball in the final analysis, WAGT felt they had demonstrated their support for the CID process and that the violations were necessary, and probably would happen again!

BuAer's Kansas City BAR basically agreed with the letter, but his endorsement shows that he really felt that WAGT's CID process was cumbersome and normally incurred delays. He argued that although the -8 was accepted as a production engine, it was still basically a development engine. Having accepted it at a lower level of development than normal, a greater number of CID's could be expected. Finally he noted that WAGT Kansas City was improving in CID processing timeliness.[74]

The BuAer General Representative Central District did not agree however, finding that WAGT should never ship an engine unless BuAer was aware of any changes in the engine and that they had been cleared by BuAer in some way. He noted the BAR at Essington had been delegated to issue "Engineering Advices" in advance of full coordination of CID's to obtain BuAer approval prior to an engine being shipped. He recommended that the Kansas City BAR be given the same delegated authority.

He also argued that there was confusion and poor coordination between WAGT and the two BARs incidental to the change in design control. He stated that in his opinion the transfer of design responsibility for the engine from Essington to Kansas City was in trouble. Even though WAGT had stated that design responsibility for the J40 program had been shifted to the Kansas City plant, what was actually established was a split responsibility - split authority operation. He supported his argument by pointing out that the chief engineer, principle engineering offices, laboratory facilities and accounting personnel were all in South Philadelphia, not in Kansas City. Delays due to this fact were inevitable. He recommended higher level BuAer offices express their concern over the handling of CIDs and direct WAGT to take immediate and positive steps toward improvement of the situation.[75]

The Chief of BuAer concurred with the recommendations and issued orders to the various parties in November.[76] This situation has been covered here to demonstrate the fraying relationships between WAGT and BuAer that were adding to the pressures the whole program was under as the actual production version of the -8 (the -22 and -22A) was being readied for approval testing.

Procurement and production of accessory handbooks on the various components of the J40 were much delayed, with a schedule of availability for most (but not all) of these items published in December, the schedule showing most would start becoming available in the first six months of 1954 in no certain order. Four items were still in negotiation with the vendor.[77]

The 1953 target price for the -8 was $309,842.[78]

Specification WAGT-X40E-4D covering the -8, -22 and -22A engines was approved on January 29, 1954 and distribution approved.[79] Many changes were made: clarification of cold/hot cycle accessory testing protocols, deletion of the anti-icing valve, electrical component connection clarification, and updated performance curves were the prime changes along with many minor clarification changes/corrections to the text.

On January 13, 1954, WAGT recommended that the overhaul period for -8 engines be increased to 150 hours based on the latest six 150 hour test cycle tests on the J40-WE-22A (which were basically -8 engines) uncovering no problems. The BuAer program manager accepted WAGT's recommendation for the -8 during the Chief of BuAer's absence and a formal letter of approval was sent over the Chief's signature on February 2, 1954. The extension was conditioned on an acceptable combustion liner and complete turbine inspection at 90 hours.[80]

WAGT surveyed removal of the four holes located at the top section of the engine mount ring, since they were not used by the interested airframe contractors. The change was requested to simplify manufacturing and fabrication.[81] It

appears this applied mostly to -22/,-22A engines, but some of the -8's may have been shipped with no holes in the engine mount ring.

The initial copy of the service instructions for the -8 and -1 sent to BuAer in January were returned to WAGT for corrections to remove material BuAer determined were security concerns. After removal, the manual was resubmitted to BuAer.[82]

Excessive oil temperatures on the -8 resulted in a survey of a new oil cooler design and relocation as used on the -22 and -22A in March. This was EC No 118.[83]

May found WAGT submitting supporting data for installing a 16% opened first stage turbine on the -8 to reduce compressor stalling, which was still occurring. A similar change had been made to the -6. The use of the C-20 compressor on the -8 was supposed to have eliminated stalling without the change to the first stage turbine, but this had proved not wholly effective. Two test engines incorporating the change showed a loss of 75 lbs of thrust in A/B and 150 pounds at military power, but the engines still met the specification. Four flights (with a -8 on a modified B-45 test bed) had been made. Fuel consumption increased both in A/B and in military power, but still met the specification. Charts showing the minimum airspeed required to raise the compressor inlet temperature above -32°F and/or the maximum turbine outlet temperature tolerated to keep the engine out of the stall range were provided and placards suggested for the pilot to follow in the cockpit.[84]

In August, WAGT recommended that curvic clutch air sealing of the 9th, 10th and 11th compressor discs be eliminated for several reasons, the primary one being that the steel discs were designed to run at the higher temperatures encountered at those positions in the compressor stack and that sealing them did not add to performance or increase their reliability. Secondarily, removing the seals would lower costs and decrease compressor assembly difficulty. Engine serials WE031006, WE031007 and WE031012 were used in tests with and without the seals and showed no significant impacts on thrust or fuel consumption. Such tests were also run on -6's, -1, -22 and -22A engines with similar results. Removal of these seals was also recommended for all but the -6 engines, presumably because that model engine was no longer being used, or use would shortly end and no further uses for the -6 had been identified.[85]

Removal of the top oil filler and the oil reservoir overflow prevention float valve was surveyed in September. The change would facilitate manufacture. The side filler and its dipstick would be retained for oil filling and oil level determination. The change also applied to the -1, -22 and -22A.[86]

Engine WE03015, while in use at El Segundo and Edwards Air Force Base, was found to have cracks in the exhaust collector in excess of the allowable rework limits. This was an older design that had been superseded in the later -22A engine by a design that eliminated this type of cracking. The newer part was forwarded and the engine repaired in the field. Interestingly, the BAR cover letter does not mention or request special instructions for other -8's still in the field, possibly implying there was minimal usage of -8's by this point. All -22/-22A engines used the new collector.[87]

WAGT reported on the results of a magnetization test on the dual fuel pump which contained a steel liner in the emergency checkout spring in the fuel regulator. The tests show that magnetization of the pump had no effect on switchover of the pump to emergency, but would prevent the pilot from rechecking the emergency pump for approximately five minutes after release of the pump test switch.[88] It is not known why this test was run, but a query must have been made for some reason. The subject line of the memo is a good demonstration of how many engine models shared this component, in this case eight.

A series of Engine Notices were sent out in 1955 concerning the -8 for the reasons listed. It is unlikely that many -8 engines were still being used, only the -22A being left in production at that point.

Notice No.	Purpose
KN-98	The inspection of the lubricating oil filter cap is necessary to locate the oil filter caps which have been damaged by excessive torque during installation.
KN-97	To provide instructions for the replacement of the internal oil tube assemblies in the rear bearing support. The tube assemblies being replaced were damaged during manufacture by excessive grinding of the weld surfaces. (NOTE – Replaced KN-86 which was never issued.)
KN-76	The addition of a Teflon back-up ring to the rear oil pump relief valve liner will prevent the "O" ring from being damaged by pressure pulsations.
KN-89	To provide instructions for inspecting bolts used in flanged fuel connections to assure proper length and thread engagement.

The contract files indicate that there were a total of 15 J40-WE-8 engines produced. However, at the government hearings (see Appendix B), the Navy indicated that 14 had been built. Since serial numbers up to WE0301015 were assigned, the higher number is likely the correct number constructed and accepted. One engine, WE031006, used in the world speed record breaking XF4D-1, was preserved and is in the possession of the Smithsonian Air and Space Museum (below)

Figure 6 NASM J40-WE-8 Serial WE031006, World airspeed record engine used in the Douglas XF4D-1. Also used in the Grumman XF10F-1 for limited testing. Afterburner is stored in a separate engine container. *NASM, Jeremy Kinney*

Chapter 7 - J40-WE-8 Production and Service Issues Citations

[1] WAGT Memo, September 10, 1952, "Transient Response of Exhaust Nozzle Area to Turbine Discharge Temperature when on Automatic Trim Control." **B937VE11RG72.3.2NACP**

[2] BuAer Naval Speed Letter, October 7, 1952, "J40-WE-1/-8 Engine Performance Guarantees" **B937VE11RG72.3.2NACP**

[3] WAGT Memo, October 10, 1952, "Contract NOa(s) 10385; 52-403c, 52-1171, 52-048. WEC Turbo-Jet Engine Models J40-WE-1, -8, -22, -10. Removal of Gearbox Lugs and Inlet Duct Instrumentation Bosses, Information on." **B937VE11RG72.3.2NACP**

[4] BuAer Memo, October 13, 1952, "J40-WE-8 and -1 engines, restrictions due to bearings" **B937VE11RG72.3.2NACP**

[5] WAGT Memo, October 8, 1952, "Contract NOa(s) 10385, Change in Design 19243 Supplement A, New fuel control for the J40-WE-1, -8 and -22 Engines, Verification of" **B937VE11RG72.3.2NACP**

[6] BuAer Memo, October 20, 1952, "Contract NOa(s) 10385, Change in Design 19243, supplement A, new fuel control for J40-WE-1, -8 and -22 engines" **B937VE11RG72.3.2NACP**

[7] WAGT Memo, October 24, 1952, "Operating Recommendations for J40-WE-1, -8, -22 Turbojet Engines" **B937VE11RG72.3.2NACP**

[8] BuAer Telegram, October 27, 1952, Telegram 272234 **B937VE11RG72.3.2NACP**

[9] BuAer Memo, October 28, 1952, "Contract NOa(s) 10385; use of extension in J40-WE-8 engine" **B937VE11RG72.3.2NACP**

[10] WAGT Memo, October 29, 1952, "Contract NOa(s) 10385; 52-403c, 52-408. WEC Turbojet Engine Models J40-WE-8, J40-WE-22. Booster Pump Seal Drain, Survey of. Engine Change Numbers 101 and 14 for the J40-WE-8 and -22 respectively." **B937VE11RG72.3.2NACP**

[11] WAGT Memo, October 30, 1952, "Contract NOa(s) 10385 J40-WE-8/-1 Engines Performance Guarantees" **B938VE12RG72.3.2NACP**

[12] WAGT Memo, October 24, 1952, "Contract NOa(s) 10385, Recommendation for Engine Operating Time for early J40-WE-8 and -1 engines." **B937VE11RG72.3.2NACP**

[13] BAR Memo, October 30, 1952, "Contract NOa(s) 10385, Recommendation for Engine Operating Time for early J40-WE-8 and -1 engines." **B937VE11RG72.3.2NACP**

[14] WAGT Memo, October 24, 1952, "Contract NOa(s) 10385, Recommendation for Engine Operating Time for early J40-WE-8 and -1 engines." **B937VE11RG72.3.2NACP**

[15] WAGT Memo, November 21, 1952, "J40-WE-8 Engine Emergency Fuel Control, P/N EDS-L-222277, for McDonnell Aircraft Corporation, Information on" **B942VE9RG72.3.2NACP**

[16] BuAer Memo, October 31, 1952, "Contract NOa(s) 10385, J40WE-8 Engine; McDonnell Emergency Fuel System for" **B942VE9RG72.3.2NACP**

[17] BuAer Telegram, October 27, 1952, Authorization of Shipment of WE-031002. **B942VE9RG72.3.2NACP**

[18] BuAer Monthly Progress Report for November, 1952; Dated December 10, 1952 **B93VE12RG72.3.2NACP**

[19] BuAer Memo, November 5, 1952, "J40-WE-8 and J46-WE-8 engines, improved acceleration control for" **B938VE12RG72.3.2NACP**

[20] BuAer Memo, November 6, 1952, "Supplemental Contractor-Furnished Equipment List for J40-WE-8 and -22 Turbo-Jet Engines dated 1 October 1952" **B942VE11RG72.3.2NACP**

[21] Bar Memo, November 6, 1952, "Contract NOa(s) 10385, J40WE-1, -6, -8, and -22 Engines, Elimination of Fuel and Oil Leakage" **B938VE12RG72.3.2NACP**

[22] CID 9028, October 2, 1952, sent out in November by BuAer **B942VE11RG72.3.2NACP**

[23] BAR Memo, October 31, 1952, "Contract NOa(s) 10385, J40WE-8 Engine No. 031002; Flight Test Instrumentation on" **B942VE11RG72.3.2NACP**

[24] BAR Memo, November 12, 1952, "J40WE-8 Engine Ser. No. 031002; Acceptance of" **B938VE12RG72.3.2NACP**

[25] WAGT Memo, December 8, 1952, Target prices in negotiation with BuAer for 1952 fiscal year **B942VE11RG72.3.2NACP**

[26] WAGT Memo, November 14, 1952, "Contract NOa(s)-10385, Request for J40 Test Engine" **B938VE12RG72.3.2NACP**

[27] WAGT Memo, November 14, 1952, "Contract NOa(s) 10385, Use of Extension in J40-WE-8 Engine" **B938VE12RG72.3.2NACP**

[28] BuAer Memo, November 18, 1952, "J40-WE-8 engine operating restrictions; information on" **B938VE12RG72.3.2NACP**

[29] BuAer Memo, December 15, 1952, "J40-WE-8 operating restrictions, information on" **B938VE12RG72.3.2NACP**

[30] BuAer Memo, November 26, 1952, "Acceptable fuels for J40-WE_6, -8, and -22 engines, information on" **B942VE11RG72.3.2NACP**

[31] WAGT Memo, December 4, 1952, "J40-WE-8 Afterburner Squeal" **B938VE12RG72.3.2NACP**

[32] BuAer Memo, December 17, 1952, "Contract NOa(s) 10385; J40-WE-8 afterburner squeal" **B938VE12RG72.3.2NACP**

[33] WAGT Memo, February 8, 1953, "Contract NOa(s) 10385; J40-WE-8 Afterburner Squeal" **B938VE12RG72.3.2NACP**

[34] BuAer Speed Letter, December 17, 1952, "J40-WE-8 Allocation" **B938VE12RG72.3.2NACP**

[35] BuAer Wires, February 12 and February 15, 1953 regarding shipment of WE031002. **B938VE12RG72.3.2NACP**

[36] WAGT Memo, December 31, 1952, "Contract NOa(s) 10385; J40-WE-6/-8/-22 Engines, Incorporation of Anti-Icing Valve" **B942VE11RG72.3.2NACP**

[37] WAGT Memo, January 12, 1953, "Contract NOa(s) 10385 Westinghouse Model Specification WAGT-X40E2-4A dtd 1 December 1952 which describes the XJ40-WE-8 turbojet engine." **B938VE12RG72.3.2NACP**

[38] WAGT Memo, February 3, 1953, "Determination of the natural frequencies of the J40-WE-8 engine vibrating with respect to the engine mounts." **B942VE12RG72.3.2NACP**

[39] WAGT Memo, February 10, 1953, "Mechanical Limits of J40 and J46 Engines" **B942VE12RG72.3.2NACP**

[40] WAGT Memo, April 15, 1953, "Mechanical Limits of J40 and J46 Engines." **B938VE12RG72.3.2NACP**

[41] BuAer Memo, February 26, 1953, "Contracts NOa(s) 10385 and NOa(s) 52-403c; revised performance curves for Westinghouse Models J40-WE-8, -22, and -22A engines, approval of" **B938VE12RG72.3.2NACP**

[42] WAGT Memo, April 10, 1953, "Contracts NOa(s) 10385 and NOa(s) 52-403c; Revised Performance Curves for J40-WE-8, -22, and -22A Engines." **B938VE12RG72.3.2NACP**

[43] BAR Memo, March 10, 1953, Grumman intake duct available **B938VE12RG72.3.2NACP**

[44] BAR Memo, April 3, 1953, "Recording of engine operating time in log book" **B938VE12RG72.3.2NACP**

[45] BuAer Memo, April 17, 1953, "Contract NOa(s), J40-WE-8/-1 engines; delivery of" **B938VE12RG72.3.2NACP**

[46] WAGT Memo, April 28, 1953, "Possible Compressor Disc Failures, J40-WE-1, -6 and -8 Engines." **B938VE12RG72.3.2NACP**

[47] BAR Telegram, June 2, 1953. **B938VE15RG72.3.2NACP**

[48] WAGT Memo, June 3, 1953, "J40 Component Qualification Test, Information on" **B942VE15RG72.3.2NACP**

[49] BuAer Naval Speed Letter, June 11, 1953, "Contract NOa(s) 10385; acceptance of J40-WE-8 engines" **B942VE15RG72.3.2NACP**

[50] WAGT Memo to BAR, June 16, 1953, "Mechanical Limits of J40 and J46 Engines." **B937VE11RG72.3.2NACP**

[51] WAGT Memo, July 9, 1953, "Marquardt Model PC-1 Power Control System" **B939VE16RG72.3.2NACP**

[52] BuAer Memo, July 23, 1953, "Contract NOa(s) 10385; provisions for special instrumentation on one additional J40-WE-8 engine." **B942VE13RG72.3.2NACP**

[53] BuAer Speed Letter, July 23, 1953, "Contract NOa(s) 10385, J40-WE-8 Engine Allocation" **B942VE13RG72.3.2NACP**

[54] WAFT Report A-1626, July 10, 1953, "Analysis of Engine Protection Methods in Flight Regions of High Gas Loadings" **B942VE13RG72.3.2NACP**

[55] WAGT Memo, August 13, 1953, "Modification of the Automatic Temperature Control" **B942VE13RG72.3.2NACP**

[56] BuAer Memo, August 17, 1953, "Contract NOa(s) 52-403C, 10385, 52-048. WEC Turbojet Engine Models J40-WE-1, 8, 22A and 22. Survey ltr W356J reinstallation of Afterburner Squeal Detection Instrumentation Per CID 19421" **B949VE16RG72.3.2NACP**

[57] BAR replacement wire, Covering August 1953 Shipments of -8's and -1. **B949VE16RG72.3.2NACP**

[58] WAGT Memo, September 29, 1953, "Contract Noa(s) 10385, 10825. WEC Turbojet Engine Models J40-WE-1, J40-WE-8, J40-WE-22A, YJ46-WE-8A, J46-WE-8A, J46-WE8B, J46-WE-8, J46-WE-12A, J46-WE-12B, J46-WE-18, J46-WE-20. Engine Information Transmittal of." **B949VE18RG72.3.2NACP**

[59] BAR Memo, September 10, 1953, "Contract NOa(s) 10385 – Westinghouse Turbo-Jet Engine Models J40-WE-6, -8 and -22A Compressor Disc Sealing Program, Information on." **B949VE17RG72.3.2NACP**

[60] BuAer Memo, October 12, 1953, "Mechanical Seals for J40-WE-8, -22A and -22 engines; approval of" **B949VE18RG72.3.2NACP**

[61] WAGT Memo, September 17, 1953, "Contract NOa(s) 10385, 52-40c, WEC Turbojet Engines XJ40-WE-8 and J40-WE-1, -8. Exhaust Nozzle Shroud and Fairing Design, Information on." **B942VE13RG72.3.2NACP**

[62] BuAer Memo, September 21, 1953, "J40-WE-8 engines, allowable operating times between overhauls" **B942VE13RG72.3.2NACP**

[63] WAGT Memo, August 12, 1953, "Contract NOa(s) 10385, J40-WE-8 and J40-WE-1 Engines, Operating Time Between Overhauls – Extension Of." **B942VE13RG72.3.2NACP**

[64] WAGT Memo, September 2, 1953, "Contract NOa(s) 10385-FF, WEC Turbo-Jet Engine Models J40-WE-1, J40-WE-6, and J40-WE-8. Inspection Periods for Bakelite on Compressor Rotors – Extension Of." **B942VE13RG72.3.2NACP**

[65] BAR Memo, September 21, 1953, "Contract NOa(s) 10385-FF, WEC Turbo-Jet Engine Models J40-WE-1, J40-WE-6, and J40-WE-8. Inspection Periods for Bakelite on Compressor Rotors – Extension Of." **B942VE13RG72.3.2NACP**

[66] BAR Wire, October 1, 1953, Engine Acceptances in September **B943VE15RG72.3.2NACP**

[67] WAGT Memo, September 16, 1953, "Contract NOa(s) 10385 – Westinghouse Turbo-Jet Engine Models J40-WE-1, -8, -22A and -22. Contractor's Schedule for Incorporation of Anti-Icing Provisions." **B942VE13RG72.3.2NACP**

[68] WAGT Memo, September 25, 1953, "Contract NOa(s) 10385 – Failure of Ball Bearing on Afterburner Pump Drive Shaft, Information on." **B943VE18RG72.3.2NACP**

[69] WAGT Memo, October 26, 1953, "Contract NOa(s) – 10385 and 52-048, Removal of Anti-Icing Provisions from J40-WE-6, -8, -22 and -22A Engines, Our References WG-63300 and WG-67750" **B943VE14RG72.3.2NACP**

[70] BuAer Telegram, October 27, 1953, "For Repair to Support Scheduled Flight Tests" **B943VE14RG72.3.2NACP**

[71] BuAer Wire, November 5, 1953, J40-WE-8 Engine Allocation Revision **B943VE14RG72.3.2NACP**

[72] BAR Memo, November 17, 1953, Contract NOa(s) 10385 engine models J40-WE-1, -6 and -8; Contract NOa(s) 10825 engine models J46-WE-8A and 8B; engines accepted." **B943VE14RG72.3.2NACP**

[73] WAGT Memo, September 25, 1953, "Contract NOa(s) 10385 – Westinghouse Turbo-Jet Engine Models J40-WE-1 and -8, = Changes in Design Incorporated on Engines Prior to Customer Approval." **B942VE13RG72.3.2NACP**

[74] BAR Endorsement Memo, September 25, 1953, "Contract NOa(s) 10385 – Westinghouse Turbo-Jet Engine Models J40-WE-1 and -8, = Changes in Design Incorporated on Engines Prior to Customer Approval." **B942VE13RG72.3.2NACP**

[75] BAGR-CD Memo, September 27, 1953, "Contract NOa(s) 10385 – Westinghouse Turbo-Jet Engine Models J40-WE-1 and -8, = Changes in Design Incorporated on Engines Prior to Customer Approval." **B942VE13RG72.3.2NACP**

[76] BuAer Memo, November 25, 1953, "Contract NOa(s) 10385, Westinghouse Model J40 jet engines, changes in design for" **B943VE14RG72.3.2NACP**

[77] WAGT Memo, December 21, 1953, "Estimated Delivery Dates for J40 and J46 Vendor Accessory Handbooks" **B943VE15RG72.3.2NACP**

[78] BuAer Memo, January 4, 1954, "Contracts NOa(s) 10385, 10653, and 10825; CY 1953 Target Prices, Information on" **B943VE15RG72.3.2NACP**

[79] BuAer Memo, January 29, 1954, "Contract NOa(s) 10385 and 52-403c; Westinghouse Specification WAGT=X40E2-4D dated 2 December 1953 covering the Models J40-WE-8, -22 and -22A Engines, approval of" **B949VE20RG72.3.2NACP**

[80] BuAer Memo, February 2, 1954, "Contract NOa(s) 10385 F.P. – WECO Turbo-Jet Engine Models J40-WE-1, J40-WE-6, and J40-WE-8 Engines-Operating Time Between Overhauls – Extension of" **B949VE20RG72.3.2NACP**

[81] WAGT Memo, January 28, 1954, "Contract NOa(s) 10385, 52-408, WEC Turbojet Engine Models J40-WE-8, -22A and -22. Alteration to Engine Mount Ring, Survey of. Engine Change No. 117, 53 for the J40-WE-8 and the J40-WE-22 Series Respectively" **B949VE20RG72.3.2NACP**

[82] WAGT Memo, March 24, 1954, "Special Version of Service Instructions for Westinghouse Turbojet Engines Model J40-WE-1 and –WE-8, dated 20 January 1954." **B943VE18RG72.3.2NACP**

[83] WAGT Memo, March 17, 1954, "Contract NOa(s) 10385. WEC Turbo-jet Engine Model J40-WE-8. Oil Cooler Substitution and Relocation, Survey of. Engine Change No. 118." **B943VE18RG72.3.2NACP**

[84] WAGT Memo, May 7, 1954, "Contract NOa(s) 10385: Incorporation of 16% Open First Stage Turbine Nozzle per CID 43386 on J40-WE-8, -22, -22A Engines and per CID 43386A on J40-WE-1 Engines; Substantiating Data for and Effect on Performance of" **B949VE20RG72.3.2NACP**

[85] WAGT Memo, August 24, 1954, "Contract NOa(s) 10385 – Removal of Mechanical Seals in Compressor Stages 9, 10, 11 for J40-WE-1, -8, -22A, and -22 Engines" **B949VE20RG72.3.2NACP**

[86] WAGT Memo, September 15, 1954, "Contract NOa(s) 10385, 52-048. WECO Turbojet Engine Models J40-WE-1, -8, -22A, -22. Removal of Oil Reservoir Float Valve Assembly – Survey of" **B944VE20RG72.3.2NACP**

[87] WAGT Memo, September 3, 1954, "Contract NOa(s) 10385, Westinghouse Turbo-Jet Engine Model J40-WE-8. Rear Bearing Support Failure – Contractor's Comments On." **B944VE20RG72.3.2NACP**

[88] WAGT Memo, September 27, 1954, "Contract NOa(s) 10385 and 10825. Westinghouse Turbojet Engine Models J40-WE-1, -8, -22A, -22 and J46-WE-8, -8A, -12 and -18. Magnetization Effect During Ground Checkout of Dual Fuel Pump, Information Concerning." **B944VE20RG72.3.2NACP**

Chapter 8

J40-WE-1 Procurement, Development, Production and Service

Procurement

This model of the J40 was procured by the Air Force for use in the North American MX770 pilotless cruise missile project. This was accomplished through interdepartmental purchase orders AFMIPR 33(038) R-51-80N (with two amendments) and MIPR (33-038)R-49-235N for 16 engines. These engines were added to BuAer contract NOa(s) 10385 with WAGT in January 1951.[1] The initial proposal was to substitute 18 (later reduced to 16) -1 engines for planned -6 production engines. BuAer cancelled two J40-WE-6 engines on order to make additional funds available to pay for the higher price afterburner engine. A later interdepartmental purchase order (MIPR 33-600-4-02R-319 dated January 19, 1954) was placed for 14 additional engines, bringing the total order to 30. Each X-10 airframe, of which 13 were built, required 2 engines.

The Air Force justified the additional purchase expense as being cost competitive with other WAGT engines at $16.70 per pound of military thrust. Initial price for each engine was $182,034. The engines offered the thrust required to drive the test vehicle to almost Mach 2 speeds for long enough periods for the airframe to thermally "soak" and stabilize.

The -1 engine was a J40-WE-8 engine modified as follows:

1. The fuel regulator was relocated.
2. Added a new mounting bracket for fuel regulator.
3. North American Aviation (NAA) supplied a power lever actuator motor mounted on Westinghouse bracket.
4. A new fuel inlet fitting on the fuel regulators.
5. Mounting lugs on gearbox removed for fuel regulator and aircraft equipment to clear aircraft ducting.
6. Completely new piping forward of the compressor as required to accommodate the above and to meet the following NAA installation requirements:

   a. Inlet duct seal flange
   b. Starter air duct
   c. Inlet duct "Y"
   d. Landing gear mechanism
   e. Oil cooler inlet fitting
   f. Afterburner pump inlet fitting

7. Booster pump removed to provide drive for NAA hydraulic pump.
8. Afterburner fuel shutoff valve rotated 90 degrees.
9. New linkage between fuel regulator and afterburner fuel control.
10. New electrical harness including junction box.
11. Oil drain valve removed and plug in line provided.
12. Provision made for installation of NAA provided flow meters in engine and afterburner fuel lines. Engine would have two jumpers.
13. Afterburner fuel lines rerouted to clear NAA lower left hand compressor air bleed ducts.
14. All anti-icing provisions and cover ports removed.
15. Tee fitting added in the compressor inlet pressure line.
16. A new firewall and associated piping to provide more installation clearance at the bottom of the engine.
17. The oil cooler for J46-WE-6 was used to facilitate piping.
18. Weight 3,580 lbs vs. the J40-WE-8 at 3,585 lbs.

Development

The Air Force -1 version suffered from the same development issues as described for the -8. The X-10 airframe it was to power had limited capacity for fuel, so shortfalls in fuel consumption guarantees and increases in weight had the potential to have a serious impact on the MX770 program and received significant attention during the development period.

The basic program management plan had BuAer driving development as well as assuming responsibilities for ensuring that Air Force requirements were met. The Air Force agreed to accept the qualification test results for the -8 engine as qualification of the -1 as well. Administratively, it added a layer of communication with attendant delays in keeping the Air Force informed as to design changes that would potentially affect their program. Air Force Air Material Command at Wright-Patterson Air Force Base, Dayton, Ohio was the coordinating and approval authority for the program.

Development of the -1 was done to comply with specification WAGT-X40E3A-1.

NACA had been doing analysis of afterburners and presented to BuAer two curves showing their calculations of afterburner performance. BuAer asked for WAGT's comments on the NACA letter and curves before forwarding them to North American Aviation. WAGT's response was clear:

"This contractor considers that the afterburning data presented in Enclosure (1) of Reference (a) are for the most part theoretical, i.e. not based on the J40 test data deemed necessary to provide North American reliable advice concerning the J40-WE-1 engine for the MX770 missile. It is not possible to substantiate the curves presented because (a) constant afterburner combustion efficiency was assumed, (b) little or no calculated detail afterburner test data was available at the time from wind tunnel tests, (c) no figures are quoted for the effect of afterburner cooling air bleed on the values presented, (d) the assumption of an area coefficient of 1.0 is not realistic and has been found to be an important item when considering maximum exhaust limitations on a hot day with appreciable inlet duct loss, (e) these data do not describe the effect of the non-symmetrical flow distribution of the MX770 inlet duct, an effect which has gained increasing importance in view of subsequent testing at NACA simulating this distribution.

"It is the contractor's opinion that these data will serve North American no useful purpose and should be withheld in order to avoid confusion and misunderstandings. Therefore, it is requested that Enclosure (1) of Reference (a) not be forwarded to North American Aviation"[2]

The memo went on to mention they were about to start testing with a mock-up of the MX770 inlet duct which had been provided by NAA.

Development proceeded and the next major snag was that BuAer had assumed, based on some earlier correspondence, that engines with the 675 in² nozzles would be acceptable to the Air Force. The Air Force discovered this and informed BuAer they definitely required the 726 in² nozzles on the -1's in order to be acceptable.

Figure 1 North American Aviation X-10 port side intake mockup, WAGT P48171 and P48173, 12/1/1952. *Courtesy Hagley Museum and Library*

The Air Force issued a letter in August stating they would accept the first 8 engines with a weight penalty of 147 pounds, with a not-to-exceed weight of 3,580 lbs. They also noted that based on prior correspondence, the specification guaranteed weight should be lowered to 3,408 lbs due to the weight savings of moving to the simplified control system and the specification updated accordingly. They noted the weight reduction program to get the engine down to the guaranteed weight should not change the price of the engine. They urged that although BuAer was reconsidering the specification weight for the -8/-10 engines, the Air Force needed the lighter engines to allow the X-10 to meet its program objectives and urged BuAer to aggressively pursue their weight reduction program.[3]

August also found the Air Force trying to get on top of the installation configuration of the -1 engine in anticipation of upcoming deliveries. In a comprehensive memo, they noted that all items from the -1 mockup meeting of May 21, 1951 had been addressed. They asked that all the surveys of installation changes since that time be included in the latest drawings. They noted that since the closure of the mock-up items, the installation surveys that followed had been sent out and incorporated in the XJ40-WE-1 and they wanted the surveyed items included on the qualification engine to be listed. WAGT responded with a complete listing of surveyed items for the -1, the complete list follows.[4]

XJ40-WE-1 Installation Survey Items*		
Survey No.	Description	Disposition
1	Rerouting of A/B Fuel Lines within J40-WE-8 engine envelope	Complete. Item 24 revised this.
2	Change fuel inlet fitting	Complete
3	Change routing of small fuel lines	Complete. Rescinded by removal of governor alternator in connection with the incorporation of the simple hydraulic control.
4	Change fuel inlet line to oil cooler	Complete
5	Installation of NAA control actuator motor	Dummy motor mount to be incorporated in the engine to provide a bearing point for the control linkage in the same relative position as the bearing points on the NAA servomotor. Guaranteed throttle torque not to exceed 40 inch pounds.
6	Moisture Drain in inlet duct.	Complete. CID 14856
7	Remove accessory mounting bosses from the accessory drive gearbox	Complete
8	Compressor pressure limiter	See Survey 169
9	Study to define special engine instrumentation	Cancelled
10	Reword paragraph E4B-10 to the XJ40-WE-1 model specification	Cancelled. Paragraph to be deleted.
11	Study operational requirements in relation to one hour continuous afterburning as required by MX-770 application	Requirement to be included in final XJ40-WE-1 specification.
12	Change firewall outer curve as per NAA layout 9606-941029	Complete. CID-14928
13	Request for photographs of mock-up	A J40-WE-1 metal shell mock-up was supplied to NAA under Call 32 of Contract AF 33(038) 8543.
14	Addition in inspection parts or methods to inspect hot end of engine	See Item 17.
15	Withdrawn	
16	Compressor Pressure Limiter	Cancelled by item 20.
17	Combustion Chamber Housing Inspection Ports	Cancelled by BuAer ltr 015935 of 5 November 1951.
18	Relocation of No. 2 Bearing Thermocouple	Complete. CID-14747
19	Increase in Exhaust Nozzle Area & Engine Length	Complete. CID-19123
21	Outline of Speed Switch Control	Complete. CID-14815F
22	Additional Thermocouple Circuit	Complete. CID-14754 on J40-WE-8/-22 engines. Not required for missile application and will not be incorporated on XJ40-WE-1 engines.

23	Spark Plug Removal Clearance	Complete. CID-14531E
24	Increased Clearance Envelope Over A/B Fuel Lines	Approved. No change to engine.
27	Radial Axis Tolerance Between Engine Mounts	Complete. CID 14912
28	Tailpipe Thermocouple Cover & Junction Box Connection	Not required for missile application. Will not be incorporated on J40-WE-1 engines.
29	Engine or Airframe Mounting of Automatic Temperature Trim Control	The automatic temperature trim will be engine mounted for shipment but would be airframe mounted upon engine installation.

* Items 20, 25, and 26 were omitted in original report.

As qualification testing approached for the -8, along with the agreed acceptance of the proposed Westinghouse test program for qualification, the AF had these comments relative to the testing of the -1 and the components to be used:

1. They suggested that twenty-five (25) jam accelerations be added to the test.

2. They felt that 50 hours of successful testing including the minimum of maximum military and A/B time as specified in MIL-E-5156 would be acceptable for the first two engines for delivery in October. Such tests would include an official inspection by the BAR, Essington.

3. The turbine discs which were to be used on the first 16 engines would come from a batch that included a manufacturing error (the nature of which was not detailed), but these would be acceptable following satisfactory completion of the qualification test on the first stage turbine assembly. It was assumed the lead time for standard -8/-22 disc assemblies was adequate for them to be incorporated into the following 14 engines on order. In the event of a delay due to testing failure or otherwise, they requested information on when standard turbine disc assemblies could begin to be incorporated in manufacture.

4. They noted that the Westinghouse request for permission to have first stage turbine blades produced for these defective discs was entirely Westinghouse's responsibility.

WAGT noted in a December memo that the 25 jam accelerations would be accomplished during the 150 hour verification run on the J40-WE-8 and that the first stage turbine discs would be subjected to at least 150 hours of accumulated running "at maximum and military operation at least equal to that required in qualification testing". They clarified that 19 discs were involved, one of which was to be used in testing, one for a spare part for the Air Force, 16 to be used in the first batch of engines and one unit to be scrapped.[5]

In October, the new fuel control (58J851) passed its 150 hour qualification test and was to be used on the 150 hour verification test of the -8. The actual fuel control functions were unchanged. This modified control had the throttle valve moved to the shutdown end of the throttle shaft to provide greater strength and reduce the angular freedom between it and the throttle shaft. It also had a redesigned acceleration valve incorporating a failsafe feature to deal with potential clogging of the chamber bleed valve should it occur.[6]

In the middle of October, an early Photostat copy of the operating recommendations for the -1 were forwarded to the Air Force.

Acceptance authority was again discussed in a BuAer memo at the end of October, with BuAer proposing the Essington BAR deal directly with the Wright Air Development Center regarding any deviations on -1 engines, the specifications for which were being directly coordinated between WAGT and the Air Force. If that approach was too slow, the AF was asked to consider granting the BAR the authority to accept the first two -1 engines under some pre-agreed to set of minimums. BuAer asked the current specifications for the -1 be forwarded and also to be kept up to date with any specifications which pertained to Air Force versions of Westinghouse engines.[7] WAGT responded to this request in November, forwarding an uncoordinated copy of WAGT-X40E3-2B (dated October 31, 1952) to BuAer.[8]

The Air Force considered the normal deviation acceptance process too lengthy and likely to introduce delays. They were willing to authorize in advance the acceptance of the first two XJ40-WE-1 engines on the basis that their performance exceeded the understood guarantees. The maximum allowable gas temperature values for each engine would be considered individually for each engine submitted for acceptance. They proposed that the BAR should prepare a list of all deviations from the qualification test configuration and installation survey requirements for each engine submitted for acceptance.[9]

At that point, the specification still called for the use of MIL-F-5624A Grade JP-3 fuel. Weight was not to exceed 3,580 lbs. Continuous A/B operation had to be possible for one hour. Other data:

J40-WE-1 Dimensions

	Room Temperature	Max. Operating Temperature
Length	287.0 in	288.8 in
Width	41.5 in	41.8 in
Height	43.0 in	43.3 in

The end of October found WAGT agreeing to include calibration data in their 150-hour Qualification Test report for the -8 (run with the 675 in² nozzle). The performance guarantees for the -8/-1 would only be deviated from at one point, that of thrust. (The guarantee was 10,900 lbs static thrust at standard day conditions and the qualification engine produced 10,100 lbs due to the smaller nozzle.) It was noted that BuAer did not intend to write separate specifications for the small nozzle engines. As noted in the -8 development story, the subject of the maximum allowable measured gas temperatures was addressed, the individual engine limits being noted on a decal at the end of acceptance testing.[10] An example of such a decal is on page 141.

Production

The delays in development of the -6 and -8 brought the Air Force schedule into conflict with BuAer's. The initial schedules for engines were coordinated and agreed to in March, 1951. Immediately slippage began as follows:

March 1951 Production Allocation for J40-WE-1 Engines		May 1951 Production Allocation for J40-WE-1 Engines	
1	January 1952		January 1952
1	February 1952		February 1952
2	March 1952	1	March 1952
4	April 1952	4	April 1952
4	May 1952	5	May 1952
4	June 1952	4	June 1952
2	July 1952	3	July 1952
		1	August 1952

Based on X-10 airframe construction progress and engine development progress a new production schedule was proposed:[11]

50 Hr Engines		
Date	Type	Number
Aug 1952	-8	1
Sep 1952	-8	1
Oct 1952	-1	2
150 Hr Engines, 675 in² nozzle		
Nov 1952	-8	3
Dec 1952	-8	3
Jan 1952	-8	3
150 Hr Engines, 726 in² nozzle		
Dec 1952	-8	1
Jan 1953	-8	1
Feb 1953	-8	4
Mar 1953	-8	2
Mar 1953	-1	2
Apr 1953	-8	2
Apr 1953	-1	2
May 1953	-8	2
May 1953	-1	2
One per month thereafter	-1	Until 16 under contract were completed

It would turn out that the first two -1 engines produced as 50 hour engines (between overhauls) would have the smaller nozzle in spite of the Air Force requiring the larger nozzle. Acceptance of the smaller nozzles and not waiting for engines with the larger nozzles allowed North American to begin learning to mate the engine to the airframe and begin ground running and taxiing tests.

The BuAer memo to the Air Force regarding the new "best efforts" Westinghouse schedule referenced the follow-on order for 14 more -1 engines and noted that the earliest date they could start to be delivered would be July 1953 and suggested that the suitably modified -10 engine (some documents refer to the Air Force version of the -10 as the -5) might be more readily available by then in lieu of -1 engines.[12]

As a consequence of the slipping schedules, the constant speed drive on the gearbox was anticipated as now being available. This gearbox was originally developed for the -10 engine, was to be moved into the -8 and -1 program, but ultimately was only being delivered with the -22/-22A models.

Westinghouse originally planned to assemble the -1 engines in Essington and then ship them to their new Kansas City plant for green runs, green testing, teardown, final build-up and final test. To make best use of their testing facilities, they requested in November that authority be given to both the Essington and Kansas City locations for testing and acceptance. They noted they had received similar authority for the -6 and -8.[13]

The Air Force in December noted that the weight of 3,580 pounds was acceptable for the first 16 engines, noting that a summary of the correspondence relating to weight of the engines indicated the original design objectives could not be met in time to meet engine delivery commitments due to reasons under control of the contractor. They further requested a proposal from Westinghouse for the price of incorporating titanium components in the last 14 engines (as indicated by Westinghouse in a letter dated October 30) and the average cost per pound of weight saving in the proposal.[14]

With engines about to enter production, the oil temperature limits were discussed again, with Westinghouse asking for permission to operate the engines with the following temperature limits:

Maximum Oil Cooler Discharge Temperature	240°F
Max. Scavenge Temp	300°F
Lubricant	MIL-0-6081-1010

In support of the proposed limits, they noted that operating a test -6 engine configured as a -8 engine for 5 hours and 51 minutes of operation with the oil discharge temperature above 240°F showed the controlled rubber parts and rubber hoses to be in satisfactory condition. The thrust bearing showed damage due to similar testing of the -8, attributable to the need for the new larger bearing, but the damage was not attributable to oil temperature. The new bearing had been tested under high load conditions in 150 hour tests at 300°F oil inlet temperature (at the pressure pump) using MIL-L-7808 as a lubricant. Ten hours of operation had been accomplished with oil inlet temperatures above 250°F using MIL-0-6081 Grade 1010 oil.[15]

Two J40-WE-1 engines were assembled in Essington in December but were not shipped to Kansas City for green testing, tear down and acceptance as planned. These activities for these two engines occurred at Essington.[16] These would have been serial numbers WE030101 and WE030102. The number sequence for -1 engines was from WE030101 to WE030130 (taken from the original -6 block of serials).

The AF accepted the two engines with 675 in² nozzles, aluminum 6th stage compressor discs, and a 65 hour overhaul limit with 15 hours of military operation. Afterburner light offs were initially limited to 115, then reduced to 85 until the new thrust bearing could be installed. Oil hose temperature limits were set at 300°F with the AF suggesting that XJ40-WE-1 engines with tight fitting

exhaust nozzle seals had to be exceptional regardless of whether the oil lines were rubber or stainless steel. All later upgrades to bring these first two engines into compliance with the qualified 150-hour Qualification Test configuration were to be at no additional cost to the Air Force.[17]

The engine log for engine WE030102 is extant and shows that the engine was accepted by the BAR Essington on March 31, 1953. This engine actually was run-in and finally tested at Essington, not Kansas City as was WAGT's original intention. Although this engine was signed off by the BAR Essington with a 726 in² nozzle on April 1, 1953 prior to shipment to NAA, it is known that the first two -6 engines shipped with 675 in² nozzles. (The 726 in² nozzle was not qualified by the Navy until June 11. The shrouds and fairings were qualified for the larger nozzle on or about June 29.) Westinghouse upgraded many components on this engine to -8 Qualification Test level prior to shipping it to NAA in Downey, CA. The most interesting change was the removal of the automatic temperature control and replacing it with the manual control. All changes were completed on April 1, 1953. This engine was de-preserved and installed in missile GM-19307 for the first time on May 14, 1953. It was removed on May 23 for more rework. It flew for the first time October 15, 1954, including 32 minutes in the air during the first flight of the X-10 GM-19307.[18]

A rough draft of the Service Instructions for the -1 was sent to the Air Force on April 1, 1952.

Later in April, BuAer instructed that all even numbered J40's built after No. 8 and through No. 36 should be built as -8's and the odd numbered engines as -1's. The -1's would be shipped to NAA in Downey, CA.[19] In late April the situation regarding overstressed compressor discs due to overheating developed and it was recommended the engine be grounded until a solution was found. Covered in the -8 production chapter, the grounding applied to -1's and the early solution of applying Bakelite sealing between the curvic clutches was applied to the early -1 engines at the Westinghouse support center on the West Coast and the engines put back in service by the end of October. As soon as this problem became known, the AF sent a memo asking for confirmation that the fix would not cost them anything.

July found the Air Force asking for prices and development schedules to add retractable inlet screens and a compressor inlet temperature limiter. WAGT reminded WADC that their bulletin No. 80-24 relieved engine manufacturers from the responsibility for engine intake screens in the case of a guided missile aircraft. WAGT stated their current engineering work precluded them from undertaking development of a compressor inlet temperature limiter. They further stated that their current approach was to limit compressor discharge pressure, which effectively limited temperatures to a predetermined value. The device in question modulated exhaust nozzle area and reduced engine speed as needed. They would investigate combining the compressor discharge pressure limiter and temperature limiter in one integral unit.[20]

BuAer's clarification on cost responsibility for fixes could not have been encouraging to the Air Force. The memo stated that "NOa(s) 52-403 covers research and development work on the XJ40-WE-8 of which the XJ40-WE-1 is an adaptation. This research and development work, however, does not include test and field fixes nor service problem cost for engines delivered under NOa(s) 10385. NOa(s) 10385 is a re-determinable incentive-type contract to which are attributable all costs of obtaining acceptability for the engines produced thereunder. The costs of work accomplished by the contractor on any particular test, fix or servicing may or may not result in a revision to target cost depending on the circumstances." In other words, fixing the compressor might end up costing the Air Force more money.[21]

In answer to a request in August for substantiation test data on the automatic temperature trim, WAGT responded with:[22]

Test Data for Automatic Temperature Trim for the J40-WE-1		
Test Engine	**ATT Unit Serial No.**	**Test Time - Hours**
J40-WE-22A	D-132	251 (included 150 hour endurance test)
J40-WE-22A	D-151	177.6 (included 150 hour official verification test)
J46-WE-8A	D-110	233 (included 150 hour qualification test)
J46-WE-8A	Unknown	68 (included 50 hour verification test)

On the same day the above data was sent, WAGT agreed to run an additional fail-safe wire connection for the Automatic Temperature Trim control per a BuAer request back in March. This situation was not mentioned in the response to the Air Force request for data nor was WADC copied on this second memo.[23]

Reinstallation of automatic squeal detection instrumentation was approved by the Air Force on August 17, 1953, demonstrating the constant flow of documentation on changes back and forth.[24]

The projected use of titanium discs in the last 14 engines to accomplish a 124.8 pound weight reduction ran into problems that would cause a serious delay in the delivery of these last 14 engines. WAGT requested the requirement for titanium discs be dropped and that a weight of 3,666 pounds be approved for acceptance of the batch of 14 engines.[25] The WADC approved the removal of the titanium components requirement, but disapproved the weight of 3,666 lbs for the entire batch of 30 engines. They again noted the additional weight impacted their supersonic soak time and stated they would consider the weight of each engine individually during its acceptance period.[26]

One (1) engine was accepted in August and two (2) more in September. All of these had the 726 in² nozzle and compressor sealing applied during production.

The ball bearing failures on the A/B fuel pump grounded the -1 until the bearings were replaced with the new bronze race type. The four engines delivered to NAA were modified in the field and returned to service by September 25.[27]

A conference on engine settings used during acceptance testing was held with the Air Force, and WAGT responded in part with this explanation: *"It is the Contractor's standard practice to set the fuel-air ratio in a range which will result in the maximum potential performance. The difference between the minimum and maximum values is so small that it is not practical to adjust all engines to fall within the optimum portion of the range. Likewise, to go beyond this range to obtain absolute maximum thrust is also not practical due to the insignificant amount of thrust gained as compared to the sharp increase in specific fuel consumption and approach to the stability limit of the afterburner. The optimum range of fuel-air ratio corresponds to a maximum thrust variation of approximately 200 pounds."*[28] It is believed the Air Force was seeking to obtain higher thrust ratings through adjustment to compensate for the additional weight beyond the specification of the engines they were receiving.

The issue of CID approvals flared up at this time and the many memos on this subject were previously discussed in a prior chapter on the -8.

Full specification compliance for radio interference was covered in a memo explaining the actions taken to date and the results of testing on WE030104 and WE030105. The changes made to date had brought the engine into compliance with AN-I-27A except for one frequency. Further investigation showed that the addition of another condenser in the junction box would bring the engine into full compliance. After testing of that change, a CID would be submitted to bring all engines into compliance by installing the changes. The memo noted that at NAA, Englewood, it was observed that NAA had installed a very long unshielded cable between the junction box and the

Auxiliary Electric Control. The interference from this wire would be considerable and would probably cause radio interference above the limits of AN-I-27A.[29]

A survey to change the A/B fuel line scupper from ¼" to 1/2" and incorporate it in production with engine WE030111 was circulated in mid-December. The reason for the change was to drain the A/B fuel lines quicker on shutdown. It was approved.[30]

Proof that higher thrusts from the basic -1 engine were being sought through operating at higher turbine inlet temperatures is obtained from a lengthy WAGT memo exploring "the maximum Safe Operating Temperatures Over Present Guarantee." The points and analysis conclusion of this memo are:[31]

1. Report A-1614 covered the J40-WE-8 Verification Test. The turbine inlet temperature was 1,465°F using a turbine unique to the first 16 -1 engines. Engine XJ40-WE-8-7 completed abbreviated over temperature testing in October. Running was accomplished in A/B for short periods (15 or less minutes over 5 runs) at temperatures ranging from 1,500 to 1,540°F. The turbine components were satisfactory on examination. The differences between the -8 and -1 configurations were considered so slight as to not require retesting using a -1 engine.

2. The current guarantee allowed unlimited operation at temperatures of 1,440°F or less consistent with maximum hot spot temperatures on each engine. Testing experience did not allow temperatures over 1,500°F to be used regardless of local hot spot temperatures.

3. A description of how a temperature traverse was taken was included.

4. A clarification of "one hour of operation under these high temperature conditions" stated it meant "one hour of operation at any average turbine inlet temperature between 1,440°F and the specified maximum average turbine inlet temperature." The hour could be spasmodic or continuous. After the hour was reached, the engine had to have a complete hot end inspection (including Zyglo inspection) and then was cleared for normal use not exceeding 1,440°F for any period within the overhaul period for that engine.

5. The engine temperature spreads were only affected in minor ways between different inlets and such differences could be ignored in setting the maximum limits.

6. The turbines were designed to a 1,500°F specification, but the engine met the performance guarantees at 1,440°F. The turbines were not changed, but the turbine inlet nozzles (guide vanes) were made hollow to lighten the engine, although they were still designed to handle the maximum hot spot temperatures. A cushion of at least 75°F existed in the current design.

7. The maximum inlet hot spot temperature was 2,060°F. Over 7,500 hours of running of -6 and -8 engines, a substantial amount at the hot spot temperature limits, indicated substantiation of the limits.

8. The Automatic Temperature Control operated as a "safety valve" by preventing abnormal peak temperatures from existing in any inlet duct configuration.

9. The maximum compressor inlet temperature was the same with and without the over-temperature assuming the compressor disc air leak fix was installed.

10. The thrust gain by operating at 1,550°F vs. 1,440°F would be approximately 300 pounds (2.7%) in non-A/B operation and 350 pounds (3.2%) in A/B at sea level static conditions.

11. The maximum allowable turbine operating temperature was determined by the maximum hot spot temperature.

The target price for the J40-WE-1 engine in 1953 was set at $292,458.[32]

In January, the overhaul periods on the -1 were lengthened to 150 hours, approved by BuAer, and inspection periods added within the 150 period on the turbine section.[33]

In late January, a conference was held to review the outstanding list of CID's (known to the AF as "Engine Notices or EN's) to be sure the status of all were known and to discuss whether back-fitting the early -1's would be appropriate. The information contained in the summary letter showed just how far behind the curve the paperwork was for the changes to the engines as field and testing changes continued at a rapid pace. Decisions summarized were:[34]

1. Due to the large number of EN's involved, back-fitting all the early -1 engines would be uneconomical.

2. The 58J474-4 assembly drawing (the latest build standard) would be complete as soon as possible so that the number of CID's on the Drawing Supplementary List would be minimized and the last block of 14 -1's could be defined ASAP.

3. Three methods of handling Engine Notices were identified:

a. Service Adjustment Fund changes that included EN's incorporated by WAGT at no expense to the government.

b. Cost Redetermination EN's on Contract NOa(s)-10385.

c. Air Force Call Contract.

4. After discussing additional CID's that required EN's to be written, they decided that the -1 had a total of 50 EN's identified to date. Of the 50, only 9 had been received as of that date.

5. Review of engine block numbers against the Drawing Supplemental Lists revealed that WE030110 had been built to the 58J474-2 drawing list, not the later 58J474-4 as had been previously believed. It was agreed after this review that WE030103-WE030110 were built to the 58J474-2 Supplemental Drawing list standard. The block demarcation point letter was shown to be in error.

6. A copy of all field letters regarding the -1 would be sent to WADC.

7. It was recommended that the delays in the Engine Notice generation process could be shortened by several months if the BAR approved CID (the basis for the EN) could be forwarded to the WAGT Service Department as soon as it was signed and reproduced. The new process would be applied to the backlog of EN's identified above. A revised Westinghouse Engine Notice chart entitled "XJ40-WE-1 Engine Notice and CID Info", including an expansion to include later engines was requested every three months and the distribution list was included.

8. To expedite future parts procurement of CID's involving the Air Force call contract, it was requested a specific notation be made on the CID submitted to WADC, Equipment Branch so that required parts could be ordered.

An updated model specification WAGT-X40E3A-1D dated April 2, 1954 in draft form was sent to BuAer on April 30, 1954 for review and approval.[35] The Air Force approved this specification for all 30 engines procured on June 17, 1954. In approving the specification they stated that it was "the best compromise obtainable" at that time. Of interest in this specification update was a listing of the net weights of critical materials used in the -1. As a result of the Korean War and rapid build-up of military aircraft due to the nuclear weapons arms race, the demand for these elements had become a carefully considered factor in procurement activities.

Critical Materials Used in the J40-WE-1 Engine (Net Weight-lbs)		
Material	Gross Weight	Net Weight
Chromium	1527.87	360.89
Cobalt	167.47	94.86
Columbium	3.46	1.50
Molybdenum	72.35	18.16
Nickel	944.58	272.74
Tungsten	33.73	11.80
Natural Rubber	No appreciable amount used.	

RPM control was now a standard requirement, with between 1 and 2% wander allowed, but no appreciable hunting. Cold/hot durability and functional testing requirements were included. Dry weight was 3,580 pounds. The SFC's were unchanged. The absolute maximum T_5 gas temperature was stated as 1,350°F, but each engine would have its own temperature listed on a decal and in the engine log book.

The BuAer grounding order issued April 5, 1954 for -22 and -22A engines had no effect on the -1 program regarding production or service use.

To immediately deal with continuing compressor stall reports, WAGT recommended opening up the first stage turbine nozzles by 16% as had been done on the -6 for the same problem, along with the pilot following certain minimum air speed requirements if the outside air temperature was below- 32°F and manually restricting the gas turbine temperature while in the stall region.[36] The engine would suffer from the same loss of performance and increase in SFC's as the -6 engine when such a solution was applied.

On May 14, WAGT dusted off a solution to the compressor stalling situation they had been working on for some time. This solution was called the "C-23" compressor, composed of 12 stages within the same external installation envelope and engine bearing points. The 12th stage to the compressor was Siamesed onto the back of the 11th stage and the diffusor modified accordingly. (See drawing in Chapter 9.) It would add 110.3 pounds to the weight of the engine and move the CG on the -1 to 72.12 inches aft of the front mount centerline. The idea was sent out as Engine Change No. 40 for the J40-WE-1.[37]

As of May 1954, 16 -1 engines had been produced. Of the balance of 14 on order, 2 were to be produced by Essington and 12 by Kansas City. Delays were being experienced waiting for parts to change to the 16% turbine nozzle opening to deal with compressor stalls.[38]

BuAer notified the Air Force at the end of June that additional funds in the amount of $307,622.00 were required to cover the final prices of the 14 -1 engines delivered during 1953 under Lot II, Item 1 of NOa(s) 10385.[39]

The June Monthly Production Progress Report production plan for -1's was unchanged.[40] The July report showed that one (1) additional -1 had shipped on the final lot of 14.[41] August's reports showed the engine reported shipped in July actually went out in August.[42]

A survey of the removal of the compressor seals on stages 9, 10 and 11 to facilitate production and maintenance service on the -1 was sent out in August to WADC.[43]

The survey on removal of the float valve assembly from the oil reservoir went out in September. (See the -8 production chapter.)[44]

September forecast three engines to be delivered but only one was actually delivered. The reason was that apparently there was a problem with the way the alloy of the turbine blades was heat treated and they were rejected after green and/or acceptance tests. The engines required penalty runs. The malfunction of controls during testing required their replacement and delayed further deliveries of engines.[45]

A survey was sent out in November covering the adding of a second A/B Igniter Relay Nozzle on the -1. This was being done to improve A/B light-off reliability. The CID was 43411A.[46]

The November Monthly Progress Report showed that all 14 -1 engines on the last lot had been delivered and it was hard to reconcile this report with the prior month's status. However, with the -22A and -22 engine program halted, it is likely they turned all their attentions to getting this backlog of -1 engines cleared.[47]

In late November, a survey was sent out to review the addition of a "trigger anticipator circuit" on the -1. The function would be to open the exhaust nozzle gates prior to an A/B light-off. The purpose was to improve the stall margin by eliminating excessive back pressure during an A/B light-off.[48] This survey only applied to the -1, indicating that development work on the -22/-22A engines had stopped.

The December Monthly Progress Report corrected the last month's report and showed 2 -1's still to be delivered. That would bring production up to 28 of the 30 engines ordered.[49] One additional -1 should have been delivered, but a front air seal failure during engine test late in the month resulted in a penalty run that delayed delivery into January. Parts shortages were reported as a result of failures during test on A/B and Dual Fuel Pumps. Dual Fuel Pumps were now being supplied by Pesco.

The January Monthly Progress Report showed one more -1 delivered and one more to go to complete the lot. Problems encountered in the month were a parts shortage for the A/B and lack of supplies of Resistoflex hose. The A/B had to be repaired and there were labor difficulties resulting in a work stoppage.[50] The last J40-WE-1 was shipped to the Air Force in February, 1955. The final 1954 price for the last 14 engines was $244,998.00 and the Air Force had to supply $160,720 dollars to cover the additional costs.

In October 1955, as the -22/-22A program was liquidated, the Air Force received 12 J40-WE-22 engines on BASO 68655 and 6856 at no apparent cost to the Air Force. Apart from using these for spare parts, what use the AF made of these engines is unknown. It is not believed any -22 engines ever flew in an X-10 since the configuration envelope did not match the X-10 installation requirements. The -22 performance was slightly higher than the published figures for the -1.[51] After the initial allocation, further memos followed up to see if additional engines were needed, but no record of further engine transfers are in the files.

The last known flight of an X-10 flight vehicle was on February 4, 1957. During flight number 10 the vehicle reached Mach 1.84 and on flight No. 19 it reached Mach 2.05. Only ten of the 13 flight vehicles were flown. North American's experience with the engines was very good. The engines were noted as being difficult to install because the interior engine space had various intrusions that required the engine be rocked both up and down while being inserted to allow it to go past these intrusions. The roller bearings on the installation rail were too small and slid rather than rolling when inserting the engine, adding to the installation effort. Performance of the missile was above specification in terms of acceleration and climb speed. NAA calculations indicated that the combined engines in A/B were producing almost 1,000 lbs of excess thrust over their rated thrust and above Mach 1.2 were producing in excess of 3,000 lbs of excess thrust. These numbers were approximate as there was a lack of certain airframe drag data that would have enabled determination if the increased performance of the airframe was due in part to lower than predicted airframe drag. Operation of the engines only caused a problem on flight No. 13 when both A/B's shut down a minute apart. Evaluation showed this was not due to compressor stall, but due to the airframe flying beyond the operational limit (altitude) of the A/B when the flameouts occurred. Restarts failed due to vaporization of fuel in the afterburner fuel rings and manifold. Actual run time of the -1 engines was low in general due to the short flight times of the tests, rarely exceeding 30 minutes. More run time was accomplished on the ground than in the air in preflight checks.

Two J40-WE-1 engines are extant along with their engine logs. The logs for serials WE030102 and WE030107 along with the X-10 airframe 19308 are in the Air Force Museum in Dayton, OH. The installed engines have never been removed from the airframe since the museum acquired the exhibit, so the serials of the engines have not been visually verified on the manufacturer's data plates, but are currently assumed to match the serials of the log books which were forwarded with the exhibit.

Chapter 8 – J40-WE-1 Procurement, Development and Production Citations

[1] MIPR 33-038 R-51-80N, January 6, 1951 (cited in contract NOa(s) 10385) **B305VE14RG72.3.2NACP**

[2] WAGT Memo, June 13, 1951, "NACA – J40 Afterburning Data; Transmittal of" **B937VE14RG72.3.2NACP**

[3] Wright Air Development Center Memo, August 4, 1952, "(Restr) Contract NOa(s) 10385, 52-403c, XJ40-WE-1 Engine Weights" **B937VE11RG72.3.2NACP**

[4] WAGT Memo, November 12, 1952, "Contracts NOa(s) 10385 and NOa(s) 52-043-C Installation Surveys Applicable to J40-WE-1 Engines" **B938VE12RG72.3.2NACP**

[5] WAGT Memo, December 30, 1952, "Contract NOa(s) 10385, J40-WE-1 First Stage Turbine Disc Assembly Our Reference WG-63300" **B938VE12RG72.3.2NACP**

[6] WAGT Memo, October 8, 1952, "Contract NOa(s) 10385, Change in Design 19243 Supplement A, New fuel control for the J40-WE-1,-8 and-22 Engines, Verification of" **B938VE12RG72.3.2NACP**

[7] BuAer Memo, October 27, 1952, "Contract NOa(s) 10385; J40-WE-1 engine acceptance" **B938VE12RG72.3.2NACP**

[8] WAGT Memo, November 7, 1952, 'Contract NOa(s) 10385 Westinghouse conf model specification WAGT-X40E3A-2B dtd October 1952 which describes the requirements for the XJ40-WE-1 turbojet engine." **B938VE12RG72.3.2NACP**

[9] AF Memo, November 28, 1952, "(Uncl) Contract NOa(s)-10385; XJ40-WE-1 Engine Acceptance" **B938VE12RG72.3.2NACP**

[10] WAGT Memo, October 30, 1952, "Contract NOa(s) 10385, J40-WE-8/-1 Engines Performance Guarantees" **B938VE12RG72.3.2NACP**

[11] BuAer Memo, July 25, 1952, "Contract NOa(s) 10385; acceptance of J40-WE-8/-1 engines." **B938VE12RG72.3.2NACP**

[12] BuAer Memo, June 7, 1951, "Westinghouse J40 Series Turbojet Engines for USAF Pilotless Aircraft Project MX-770." **B937VE8RG72.3.2NACP**

[13] WAGT Memo, October 31, 1952, "Contract NOa(s)-10385 Delivery Destinations for J40-WE-1 Engines Our Reference WG-6330" **B937VE12RG72.3.2NACP**

[14] AF Memo, December 15, 1952, "(Restr) Contract NOa(s)-10385, 52-403c, XJ40-WE-1 Engine Weight" **B938VE12RG72.3.2NACP**

[15] WAGT Memo, December 30, 1952, "J40 and J46 Engine Oil Temperature, Maximum Operating Limits, Verification of" **B938VE12RG72.3.2NACP**

[16] BAR telegram (replacement copy), August 4, 1953, concerning engine acceptances and shipments December 1952 **B939VE18RG72.3.2NACP**

[17] Air Force Memo, February 13, 1953, "(Restr) Contract NOa(s)-10385, Operating Limits and Acceptance of First Two (2) XJ40-WE-1 Engines." **B938VE12RG72.3.2NACP**

[18] Engine log for engine J40-WE-1 Serial WE030102.

[19] BuAer Memo, April 17, 1953, "Contract NOa(s) 10385, J40-WE-8/-1 engines, delivery of" **B938VE13RG72.3.2NACP**

[20] WAGT Memo, July 7, 1953, "Retractable Inlet Air Screens and Compressor Inlet Temperature Limiter Applicable to the J40-WE-1 Engines, Contract NOa(s)-10385" **B939VE16RG72.3.2NACP**

[21] BuAer Memo, August 7, 1953, "BuAer Contracts NOa(s) 10385 and NOa(s) 52-403-c; Westinghouse Electric Corporation" **B939VE16RG72.3.2NACP**

[22] WAGT Memo, August 13, 1953, "J40-WE-1 Automatic Temperature Trim – Substantiating Data" **B939VE18RG72.3.2NACP**

[23] WAGT Memo, August 13, 1953, "Modification of the Automatic Temperature Control." **B939VE18RG72.3.2NACP**

[24] Air Force Memo, August 17, 1953, "Contract NOa(s) 52-403c, 10385, 52-048. WEC Turbojet Engine Models J40-WE-1, 8, 22A, and 22. Survey Ltr W356J Reinstallation of Afterburner Squeal Detection Instrumentation Per CID 19421." **B939VE16RG72.3.2NACP**

[25] WAGT Memo, August 11, 1953, "Contract NOa(s) 10385 – Westinghouse Turbo-Jet Engine Model J40-WE-1. Incorporation of Weight Saving Titanium Engine Components. Information on" **B939VE16RG72.3.2NACP**

[26] WAGT Memo, October 15, 1953, "(Restricted) Contract NOa(s)-10385, Information on Incorporation of Weight Saving Titanium Engine Components in XJ40-WE-1 Engines" **B939VE16RG72.3.2NACP**

[27] WAGT Memo, September 25, 1953, "Contract NOa(s) 10385 – Failure of Ball Bearing On Afterburner Pump Drive Shaft, Information on" **B939VE18RG72.3.2NACP**

[28] WAGT Memo, September 21, 1953, "Contract NOa(s) 10385 Westinghouse Turbo-Jet Engine Model XJ40-WE-1. Performance Acceptance Test Procedure – Information on" **B939VE19RG72.3.2NACP**

[29] WAGT Memo, November 17, 1953, "Contract NOa(s) 10385 – Westinghouse Turbo-Jet Engine Model XJ40-WE-1 Radio Interference, Information on" **B943VE15RG72.3.2NACP**

[30] WAGT Memo, December 10, 1953, "Contract NOa(s) 10385. WEC Turbojet Engine Model J40-WE-1. Afterburner Fuel Lines Scupper Fitting, Survey of. Engine Change No. 37." **B943VE15RG72.3.2NACP**

[31] WAGT Memo, December 7, 1953, "Contract NOa(s) 10385 – Westinghouse Turbo-Jet Engine Model XJ40-WE-

1. Contractor's Agreement for Establishing Safe Operating Temperatures Over the Present Guarantee." **B939VE19RG72.3.2NACP**

[32] BuAer Memo, January 4, 1954, "Contracts NOa(s), 10653, and 10825; CY 1953 Target Prices, Information on" **B939VE18RG72.3.2NACP**

[33] WAGT Memo, January 13, 1954, "Contract NOa(s) 10385 F.P., WEC Turbo-Jet Engine Models J40-WE-1, J40-WE-6, and J40-WE-8 Engines. Operating Time Between Overhauls, Extension of." **B944VE20RG72.3.2NACP**

[34] WADC, Equipment Center Memo, February 15, 1954, "XJ40-WE-1 Engine Notice Conference, Decisions and Additional Desires of the Wright Air Development Center" **B943VE16RG72.3.2NACP**

[35] WAGT Specification WAGT-X40E3-1D dated April 2, 1954. **B944VE20RG72.3.2NACP**

[36] WAGT Memo, May 7, 1954, "Contract NOa(s) 10385: Incorporation of 16% Open First Stage Turbine Nozzle per CID 43386 on J40-WE-8, -22, -22A Engines, and per CID 43386A on J40-WE-1 Engines; Substantiating Data for and Effect on Performance of" **B939VE20RG72.3.2NACP**

[37] WAGT Memo, May 19, 1954, "Contracts NOa(s) 10385 and 52-048 – WECO Turbojet Engine Models J40-WE-1, -22A, -22 Proposed Compressor Surge Fix, Survey of Engine Change No. 40 for J40-WE-1 and 57 for J40-WE-22 Series" **B944VE2-RG72.3.2NACP**

[38] Monthly Production Progress Report for May, 1954, WAGT Produced. **B939VE20RG72.3.2NACP**

[39] BuAer Speed Letter, June 29, 1954, "Contract NOa(s), Westinghouse Electric Corporation: additional USAF funds required for" **B939VE20RG72.3.2NACP**

[40] Monthly Production Progress Report for June, 1954, WAGT Produced **B939VE20RG72.3.2NACP**

[41] Monthly Production Progress Report for July, 1954, WAGT Produced **B939VE20RG72.3.2NACP**

[42] Monthly Production Progress Report for August, 1954, WAGT Produced **B939VE20RG72.3.2NACP**

[43] WAGT Memo, August 24, 1954, "Contract NOa(s) 10385 – Removal of Mechanical Seals in Compressor Stages 9, 10, and 11 for J40-WE-1, -8, -22A, and -22 Engines" **B939VE20RG72.3.2NACP**

[44] WAGT Memo, September 15, 1954, "Contract NOa(s) 10385, 52-048. WECO Turbojet Engine Models J40-WE-1, -8, -22A, -22. Removal of Oil Reservoir Float Valve Assembly – Survey of." **B939VE20RG72.3.2NACP**

[45] Monthly Production Progress Report for September, 1954, WAGT Produced **B939VE20RG72.3.2NACP**

[46] WAGT Memo, November 15, 1954, "Contract NOa(s) 10385; WEC Turbojet Engine Model J40-WE-1; Additional Afterburner Igniter Relay Nozzle, Survey of." **B943VE14RG72.3.2NACP**

[47] Monthly Production Progress Report for November, 1954, WAGT Produced **B939VE20RG72.3.2NACP**

[48] WAGT Memo, November 18, 1954, "Contract NOa(s) 10385, WEC Turbojet Engine Model J40-WE-1; Trigger Anticipator Circuit, Survey of" **B944VE20RG72.3.2NACP**

[49] Monthly Production Progress Report for December, 1954, WAGT Produced **B944VE20RG72.3.2NACP**

[50] Monthly Production Progress Report for January, 1955, WAGT Produced **B944VE20RG72.3.2NACP**

[51] Written note on routing sheet attached to BuAer Memo, March 1, 1956, "J40 Type Engines, Possible Inter-Service Usage" **B934VFRG72.3.2NACP**

Chapter 9

J40-WE-22/-22A/-22B/-22A-G Procurement, Development and Production

--- ◆ ---

Procurement – J40-WE-22/-22A

These engines were the production versions of the J40-WE-8, procured under development contract NOa(s) 10385 to specification WAGT-X40E2-4C. The initial production contract was Amendment 5 to contract NOa(s)-10385 for 42 engines. Contract 52-048 was to be negotiated for an additional 158 engines to be produced at the Kansas City plant. Performance and weight were nearly identical to the -8 model. Originally, the -22 was procured, but an interim engine, the -22A, using the Simple Hydraulic Control of the -8 with the automatic trim feature of the -1 had to be substituted until the final Hydro-Mechanical Exhaust Nozzle Control system of the -22 was qualified. The main difference between the -8 and these engines was the control system. Also, a new gearbox based on the -10 gearbox design, was specified. This contained a constant speed generator transmission drive being developed by Sundstrand. The J40-WE-22B was provisionally assigned by WAGT in a proposal to build the -22 with an alternative afterburner design if the current design proved deficient. This designation was never assigned by BuAer, since the proposal was not acted upon. The -22A-G was ordered by the Power Plant Division as a gas generator for unspecified experiments. Those engines had the afterburner and all controls and components required for either the A/B or operation above sea level altitudes removed.

By October 23, 1951, the initial order under Amendment 5 was increased to 42 from Kansas City and 8 from Essington. Letter of Intent 52-048 now showed 158 engines produced at Kansas City and 68 at Essington.[1] Deliveries were to start in June, 1952 and build to a production rate of 33 engines a month by January 1953. The serial number block for these engines started at WE302501 regardless of the subtype.

Development

Development encompassed the new Hydro-Mechanical Control system, the Sundstrand constant speed transmission drive and other changes to control weight or address defects found in the -8/-1 engines.

It was recommended to change the 5th through 11th stage rotor discs for the block two engines (serials WE032511 and up) to redistribute disc mass, strengthen them and make them lighter.[2] The polar moment of inertia for the -22 would change from 717 lb/ft^2 to 758 lb/ft^2 as a result of this change. BuAer rejected the suggestion but would review it again after acceptance testing of the -22 was completed.[3]

EC 11 was proposed to modify the exhaust nozzle control to have a switch in the cockpit allowing the pilot to select airframe emergency battery power in the event of an electrical power failure. With the switch in the ON position, the pilot could elect to position the nozzle fully open or fully closed. It did not provide for any intermediate trimming functions.[4] This provided a manual override to offset the lack of a fail-safe failure mode under all circumstances, one of the issues with the Simple Hydraulic Control. This modified control was to be used on no more than the first 60 engines produced.[5]

Figure 1 (Left) J40-WE-22A Cutaway exhibit engine at the National Naval Aviation Museum, *Author*

At BuAer's request, WAGT submitted a new alternate afterburner design for the -22 to be used in the event the current compressor bleed cooling design failed in testing. The modified model was proposed as the -22B (interim). If the new afterburner design was incorporated, the intent was to produce the engine with its own model dash number. The new afterburner employed ejector cooling instead of compressor bleed air cooling and employed a new iris type exhaust nozzle with four hydraulic jacks driving a movable shroud that changed the ejector exit area in conjunction with the exhaust nozzle opening. The yoke, underslung actuator and connecting linkage were deleted. Fuel lines and afterburner mounting were unaffected. Removal clearance increased from 1.188 inches to 1.68 inches. The aft lifting pad was deleted and only the fore lifting pad was retained. Removal and handling would require the development of a sling or other type of lifting apparatus. The new design was 71 pounds heavier than the existing design. This change was surveyed as EC 101 to the -22.[6]

The constant speed drive was developed by Sundstrand under contract to Westinghouse. It consisted of a variable input speed drive that used a governor to control the angle of a wobble plate such that the output speed of the transmission was constant regardless of engine speed. A high pressure oil pump driven from the engine pumped oil to one side of the transmission and the wobble plate's angle was altered to produce the necessary constant speed regardless of the relative oil pressure on the drive side. The oil had its own fluid to fluid heat exchanger.

The gearbox was altered to mount the constant speed drive near the top of the rear face, driving a gear in the box that had a splined shaft access from the other side, the generator being mounted higher than previously. The oil reservoir mounting had to be modified to allow the mounting of the constant speed drive higher up.

The gearbox, drive, governor, oil pump and fluid-to-fluid heat exchanger had to undergo certification testing. Westinghouse proposed a standard 150 engine test with these components installed, with no pads loaded except for those required for engine operation. This test was to be complete by November 1952. A separate 150 hour laboratory test would be done and would be completed in December 1952.[7]

The new fuel control unit 58J851 qualification test approach approved for the -8 was also conditionally accepted for the -22 pending successful completion of the test. The new unit eliminated the possible condition of a clog in the acceleration valve chamber bleed orifice that could cause all fuel to be shut off to the engine. The new design bypassed fuel through the control unit allowing the pilot to manually control the engine using caution not to over-temp the engine. The throttle valve was relocated to the shutdown end of the throttle shaft for greater strength and for reduction in the angular freedom between them.[8]

The reports of aft engine compartment fires from McDonnell while using the -6 showed that many leakage conditions existed on the engine. These conditions affected all models either in use or development and WAGT responded with a long memo dealing with the actions being taken. These actions would be applicable to the -22 as well, but a formal survey on an EC for the -22 was not issued at that time due to several of the corrective actions needing to be tested or further developed.

J40-WE-6 Service Issues/Corrective Actions for -8/-22
Issue(s): Engine fuel leakage from anti-icing valve bolted flange, engine inlet and outlet temperature probe bosses, turbine outlet pressure rake bosses, and vertical flange connections to both the inlet and outlet ends of the rear bearing support assembly.
Corrective Action(s): -Improved gaskets would be used. Two types of high temperature gaskets were in test. If the tests were successful, all engines would be equipped with the new gaskets. -The vertical flange would not use gaskets due to dimensional issues if gaskets were inserted between the four flanges that were sandwiched together at this point. A scupper would be used that had an external overboard drain and if successful would be used on all engines.
Issue(s): Warning light chamber of the dual fuel pump leaking fuel past shaft.
Corrective Action(s): An O-ring seal was added to the warning light actuator shaft to prevent fuel from escaping if the diaphragm ruptured. All – 6 engines would be retro-fitted and -1, -8 and -22 engines were scheduled to include the modified part during production.
Issue(s): Oil leakage from around the packing seal of the exhaust nozzle actuator.
Corrective Action(s): Development of a good packing seal had been underway for some time but had not produced success as yet. A small scupper equipped with an overboard drain would be tested that would mount on the actuator. All engine models could accept this scupper if proven acceptable.

It was noticed that the design of the oil cooler had no provisions for mounting an integrally mounted fitting for measuring discharge oil pressure and temperature. EC 13 was surveyed to provide for such a fitting in the oil discharge line and reroute some piping for that change and to resolve some interference issues recently discovered.[9]

EC 14 was proposed to alter the drain fitting for both the overspeed relay and the afterburner fuel pump check valve from a "T" to a "cross" in order to add a connection for the new booster pump seal drain on the engine.[10]

Revised operating limits (see Appendix D) were published in October. The various curves consolidated into this chart were approved by BuAer in February 1953.[11]

The only changes on the chart were to remove the designation XJ40-WE-1 (a separate chart for that model was produced later) and the "Compressor Mechanical Limit" line was changed to say "Engine Mechanical Limit". (See the discussion in the -8 development chapter relative to the last change.)[12]

EC15 was surveyed on October 29. The EC proposed to change the duplex oil pump to be two pumps. The original duplex pump was mounted within the rear bearing support. It had a high pressure side to supply oil pressure to the exhaust nozzle actuator and the low pressure side aided the scavenging of the rear engine bearing. The oil flows through the two sides of the pump never interacted. The design of the high pressure side was difficult due to the increased pressure required by the Hydro-mechanical Exhaust Nozzle Control. WAGT proposed replacing the duplex pump with two pumps, the high pressure pump to be located and driven from the rear of the engine gearbox and the scavenge pump to be mounted in approximately the same location as the present duplex pump. The high pressure pump would use the pad freed up when the elimination of the electronic control also eliminated the governor alternator. The new pump location would require long high pressure hose lines. Development of these was considered minor relative to those expected if the present high pressure pump had to be developed to meet the new requirements. (At the time this EC was surveyed, the final design of the Hydro-mechanical Exhaust Control was a work in progress.)[13]

McDonnell complained of inaccessibility of several engine components, the first being the sparkplug locations in the XF3H-1 on the -6/-8 engines. WAGT was not able to determine exactly where the plugs could be relocated on the engine and still be inside the access window of the airframe, but proposed several new locations that could be tested on complete engines and in their high altitude test chamber, feeling the starting performance of the engines had proven well in excess of the specification and relocation would not likely change such performance in a significant way. BuAer agreed with the recommended approach.

The exhaust nozzle actuator filter was inaccessible in its present location. The recommendation was to move the filter to a point no more than 33 inches and not less than 31.8 inches forward of the rear engine mount on the A/B centerline and in the same approximate circumferential location. This change would be surveyed for the -22 engine.

The alternate spark plug locations and filter relocation were tested and found to cause no difference in performance. BuAer agreed with these changes and asked WAGT to include the new locations on the qualification test -22A engine, which would qualify the -22 as well.[14] WAGT argued that incorporating the change on the test engine would involve delaying the test and not prove the change was acceptable. They pointed out that the test results already submitted proved the change would be successful. A hand written note on this memo indicates that perhaps the first 10 or so -22A's would not incorporate the spark plug relocation change.[15]

On the airframe, the -6 engine exhaust nozzle actuator travel adjustment point was inaccessible. It was pointed out that for the -22, this adjustment was electrical, not mechanical and the adjustment was on the auxiliary electric control in nearly the same location as the -6 speed adjustment on the power regulator. It was recommended the planned -22 location be deemed satisfactory. BuAer noted that a lack of adequate installation drawings for the -22 and -22A engines prevented a review of the problem.[16] Provisional approval was given provided the control caused no accessibility problem on the F3H-1 aircraft.

Figure 2 NNAM exhibit engine, aft-end of the afterburner to the flame holder and spray ring assembly. *Author*

The designation J40-WE-22A begins to appear in the records in this time frame. The -22A did not incorporate the -22's planned Hydro-mechanical Exhaust Nozzle Control system, but instead utilized the -8's Simple Hydraulic Control modified with an Automatic Trim Control (as opposed to the -8's manual trim control).

WAGT responded to the need for installation drawings in a letter of January 20, 1953. They noted the -22A drawings were in final revision based on informal survey conversations with the airframe companies and would be complete by February 5. It would show the incorporated "Interim Exhaust Nozzle" control system. This drawing would be the first formal drawing of the J40-WE-22A engine.

Drawings for the J40-WE-22 engine required resolution of *"numerous detail problems"* associated with the Hydro-Mechanical control system, which had to be resolved through use of mock-up studies. Until that was done, it was difficult to predict an availability date for this drawing. By February 15 they would be in a better position to give a definite date.[17]

Damage to high pressure oil hoses on the -6 due to heat forced WAGT to move to stainless steel lines that could withstand the severe operating environment. They tested an Aeroquip hose O0303-02 (MIL-M-5511) for 32.5 hours at a surface temperature of 320°F +/-20°F. No leaking or damage was observed. The pressure in the hose was cycled from 0 to 1,700 psig at 6 cycles per minute throughout the test. At the end of the test the hose burst at a pressure in excess of its design value of 7,000 psi. Experience accumulated on J46 engines with scavenge oil temperatures of 300°F indicated that no detrimental effects to the hose would occur. It was recommended (and accepted) that this new hose be used as a replacement for the original hose. Also, the scavenge and oil temperature limits should be raised to 300°F.[18] (This also helped deal with oil cooling issues experienced at high altitudes under some circumstances.)

Complaints from Grumman over the design of the emergency exhaust nozzle control system planned for the -22A engines (only 60 of which were planned to be delivered) resulted in a redesign to eliminate the system continuously drawing power from the aircraft battery when the nozzle was fully closed or fully open. The provisions of the control design at that point were listed, stating that when the pilot selected the emergency control through the cockpit switch:

1. *Afterburner function continued.*
2. *Shut-off of afterburner fuel through the fuel control.*
3. *Re-initiation of afterburning was not possible under any conditions and re-ignition of the main combustion chamber was only possible if the battery supply was connected to the ignition circuit (airframe company option).*

4. *Neither manual nor automatic temperature trim was provided in emergency conditions (the pilot must monitor the instruments while moving the throttle).*
5. *Modulation of fuel flow in idle or afterburning without change in the nozzle opening.*

This memo noted that a report on the Hydro-Mechanical Control had been completed but not yet released as it was still under review internally.[19]

February brought a memo from BuAer approving the specification for the adapter gearbox for the -22 and reminding WAGT that the gearbox had to pass a 150 hour full load certification test to receive its qualification.[20] This gearbox would also be used on the -22A.

The installation drawing for the Model J40-WE-22A was released and sent out February 11, 1953. This drawing incorporated all EC's through number 18 (listed below) and in addition listed 11 other EC's (19-29) out for survey with the release of the drawing. This memo clearly states that the difference between the -22 and -22A was that the former used the Simple Hydraulic Control with the Hydro-Mechanical Exhaust Nozzle Control and the latter used the Simple Hydraulic Control with the Interim Emergency Exhaust Nozzle Control. (Additional unnumbered EC's were still in preparation.)[21]

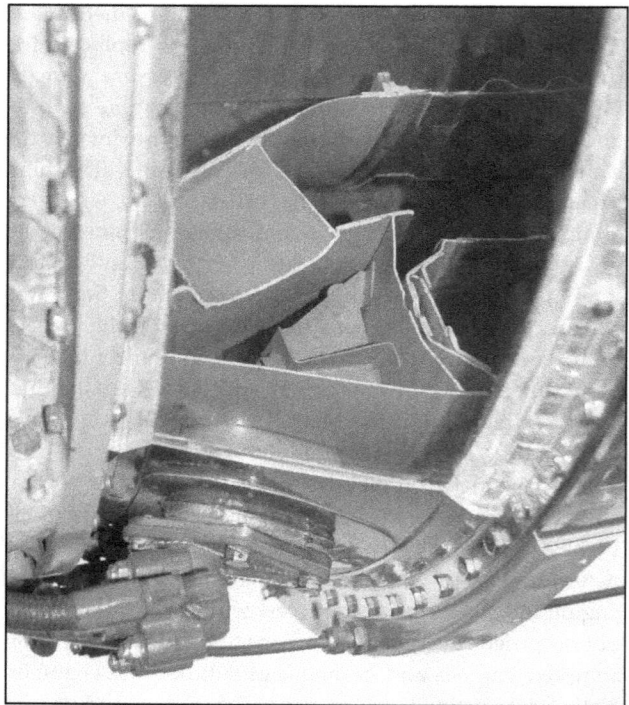

Figure 3 J40-WE-22A Cross section of diffuser exit into annular combustion chamber. *Author*

J40-WE-22A Engine Changes	
EC	Brief Description
1	Added T/C Circuit
2	Spark Plug Removal Clearance
3	Engine Mount Clearance
4	Clearance Envelope
5	T/C Cover and Junction Box Connection
6	Increase in Exhaust Nozzle Area
7	Auto Trim
8	Removal of Gearbox Lugs and Intake Duct Bosses
9	A/B Fuel Line Fittings
10	A/B Bleed Orifice (Cancelled)
11	Interim Control (Cancelled)
12	Alternative A/B Design (Cancelled – if used was to be -22B until new dash number assigned)
13	Oil Instrumentation
14	High Pressure Oil Pump Relocation
15	Boost Pump Seal Drain
16	Trim Box Mounts
17	Scuppers
18	Interim Control
19	Thread Alteration on Compressor Bleed Bosses
20	Oil Pump Inlet Filter Connection
-	Revised Piping Per 58J850) Rev. 0 (Except for piping associated with Interim Control and Aerotec Installation) (Comments requested – not surveyed)
-	Dimension on Drawing Revision 13 added (Not surveyed)
21	Gearbox Vent (Customer Connection)
22	A/B Cooling air S.O. Valve (Customer Connection)
-	Dimension Added on Drawing Revision 16
23	Junction Box Centerline Dimension
24	A/B Fuel Manifold Pressure Connection
25	Altered Starter Pad
-	Drawing Title Change – Drawing Revision 22
26	Revised Dimension and Tolerance of Rear Mounts
27	Alternative Spark Plug Locations (engine serial WE302511 and up)
28	Common Drains (revised back pressure +/- 2 psig.
29	Table of Adjustments

BuAer requested a backup verification test engine and WAGT responded that it was not possible to provide such an engine, since it would put the development plan at risk. The first J40-WE-22 was serial WE302501 and was to be used for the certification test. The first production -22(A) was E302502 and was to be used as a replacement for the 150 hour certification test if necessary.[22]

WAGT responded to the airframe constructor's questions relative to oil pressure and temperature instrumentation for the cockpits. The response noted the -22/-22A oil cooler discharge temperature range was -65°F to +300°F (not yet approved but approval expected) and the Installation Manual would contain this range.[23] WAGT report A-1546 from February 20, 1953 reported that tests of MIL-H-5511 hoses and Aeroquip 636 type hoses showed the Aeroquip 636 hose operated satisfactorily at the oil cooler discharge temperature of 300°F and a No. 2 rear bearing discharge temperature of 360°F. The -22A engine being readied for certification testing was using the Aeroquip 636 hose.[24]

Reports A-1456 and A-1510 relating to the tested actual behavior to prediction and the failure mode analysis of the Automatic Temperature Control system of the -22/-22A were accepted with one recommendation. It was noted that the failure of the circuits through pins "F" and "G" would cause issues. If "F" experienced an open condition, the exhaust nozzle would open with a possible reduction in thrust of 85% of the military value. If the pin "G" experienced an open condition it would have a similar effect and if shorted to ground, would have no effect. BuAer recommended that a second backup circuit to pin "G" be provided. A backup circuit to pin "F" was not desired, as it would increase the possibility of a short. Such a short was considered to more than offset the advantage of the reduced probability of an open condition. Special emphasis was recommended to ensure the integrity of the pin "F" circuit both in quality control and field service procedures.[25]

Figure 4 J40-WE-22A Afterburner, chromed on exhibit engine. Cutaway was painted and chromed to disguise materials used in construction. *Author*

EC 30 was for a pressure fitting for A/B squeal detection. It was to be included on all -22/-22A engines.

EC 31 on the booster pump seal drain was issued for re-survey on March 26. The cross drain fitting for the overspeed relay, afterburner fuel pump and booster pump seal drain showed the location of the cross for the -22A would cause interference with the airframe trolley systems. The drain was moved from 1 5/32 inches off the engine vertical centerline to 2 ¾ inches off the centerline. (The fitting would be a "T" on the -22, not a "cross" as on the -22A.)[26]

BuAer rejected proposed radiation interference limits for the -22/-22A from a WAGT report of February 25, noting the limits were unacceptably high compared to the limits in specification AN-I-27a. Given that the system was basically a hydro-mechanical device, the high proposed limits were found unacceptable. WAGT was requested to respond relative to what actions were being taken to reduce radio interference noise from the engine.[27]

In April, publications preparation related to the -8, -10, and -12 were redirected to the -22 only and BuAer asked what the costs were related to this change. WAGT responded that there were no additional costs for the change.[28]

WAGT had submitted specification WAGT-X40E-200 dated January 12, 1953 for afterburner insulation blankets and this specification was accepted with one minor change on May 8. Grumman, Douglas and McDonnell had assisted in developing the specification. The specification was general but described the requirements for a light weight, non-inflammable, sealed insulating blanket suitable for a specific turbojet engine at altitudes from sea level to 60,000 feet and temperatures of -67°F to +1,800°F.[29] BuAer determined that the blanket requirement might be moved to the -22 engine only. WAGT requested a mockup conference be held in Kansas City on June 11 to finalize the requirements and design.[30]

The relocation of the high pressure oil pump was included on the -22/-22A and the oil lines carefully routed in relation to the forward engine mount. This routing was surveyed and reviewed by all contractors and approved. This rerouting survey approval was needed before EC 14 could receive final approval.[31]

An American Bosch Corporation telegram in the files dated May 22, 1953 states that diversion of a Moorejig borer until October would negatively impact delivery of complex engine accessory (the engine control unit) by a minimum of 3 months. Contracts affected were NOa(s) 10385, NOa(s) 10825 (J46) and NOa(s) 52-048.[32]

In June, WAGT was able to report on the following 60 hour tests run on J40-WE-22A No. 3:

1. Bakelite sealing of the curvic clutch teeth.
2. Automatic Temperature Control 61E635-2 modified to incorporate improved shock mounting of the voltage and amplifier tube.
3. Exhaust Nozzle Fairing, EDSK228135, the same as 41E1-1 except made of .05 material.

The test program had been ten periods of: 5 minutes idle, 10 minutes afterburning, 5 minutes idle: Repeated 5 times with one 5 minute military (non-afterburning) inserted after the 4th cycle. In addition, there were five periods of the 10 hour Qualification Test Cycle specified Model Specification WAGTX40E2-4B.

The Automatic Temperature Control, serial 132, operated the engine satisfactorily and at no time did any malfunctions exist or were any adjustments made. The engine operated within a 15 degree band of turbine outlet temperature, which 90% of the time held at exactly 1,180°F, the original setting.

The exhaust nozzle fairing cracked at 39 hours along the aft flange of the channel along the weld. The crack was stop-drilled and the test continued satisfactorily. The cause of the crack was rubbing by the afterburner eyelids. A well-worn house afterburner had been used to preclude delays while waiting for a new one to become available. The aft circumferential eyelid stiffeners had separated from the basic part of the welds. It was considered that if a new afterburner had been used in the test, the rubbing would not have occurred. The fairings were continuing to be tested to ensure no further problems existed.[33]

Based on these tests, WAGT recommended that the Automatic Temperature Control and Fairing be considered qualified for both the J40-WE-1 and -8 production engines. The Bakelite tests were covered in a letter dated June 1 (covered in the -8 Chapter).

At the end of March, WAGT had reported that the Sundstrand Constant Speed Transmission had failed in test twice due to thrust bearing failures. It was determined that the bearings could not withstand the 85 HP continuous loading required by the specification. They recommended reducing the specification to 75 HP continuous and retest the unit. They noted that the unit would still more than meet the requirements of the three airframe manufacturers at the 75 HP rating. The BAR endorsement was in concurrence with the recommendation. In June, BuAer approved the specification change. A handwritten note on the WAGT March 31 memo noted that the McDonnell project was the "only remaining application". This is a clear indication that Douglas had initiated their engine change(s) to the J57 for production. The Grumman F10F-1 production contract had been cancelled.[34]

In early July, WAGT reported on the status of the Hydro-Mechanical Exhaust Nozzle Control for the -22 and the plans for flight testing the -22 engine under a modified B-45 aircraft in a program referred to as the Flight Suitability Test. It was anticipated the tests would begin in late August, 1953. The in-flight testing was necessary since the altitude control system at the AEL or NACA facilities had too slow a response rate to properly demonstrate the Hydro-Mechanical Exhaust Nozzle Control could manage the engine under the transient conditions experienced in actual flight.

Ground testing of the control appeared to be promising. Fifty hours of engine and 200 hours of laboratory testing had been accumulated with no major troubles to date. It was expected that 300 hours of engine test time would be accumulated prior to the flight verification test beginning. Production of the engines would begin before the flight verification test was conducted and "any improvement to the control found desirable through flight testing will be incorporated as soon as possible". This statement is flagged on the memo by BuAer.[35]

Figure 5 J40-WE-22A Exhaust nozzle component hinge arrangement and construction. *Author*

In August, WAGT submitted a detailed qualification test plan for the Hydro-Mechanical Exhaust Nozzle Control to be used on the -22. This 500 hour test would be conducted in Westinghouse's altitude test chamber and include cold soak and hot soak portions. BuAer approved the test plan with three questions/comments:

1. The temperature trim motor had now become part of the exhaust nozzle control box and would not be interchangeable with the present Manning, Maxwell, and Moore temperature trim amplifier (used on the -1 & -22A). What was the plan for qualifying the new amplifier?

2. Clarification on wording relative to the two throttle positions indicating "Trim" and "Basic" was requested, as the conditions where "Trim" was indicated appeared to be such that no trim was actually available for that condition.

3. BuAer had found the drawings for the -8 showing the control wiring, auxiliary electric control, and temperature amplifier were very helpful in understanding the interaction of the control system parts and requested similar drawings be furnished to BuAer.[36]

WAGT answered in September with the statement that the Electrical Control Assembly was being tested for 500 hours and the Automatic Trim Amplifier and the sequencing relays were part of that Assembly. A separate report would be issued on this test. The wording of the proposed test was corrected and the drawings BuAer requested were provided.[37]

A transmittal of drawings and other materials to BuAer included a complete listing of EC's at that point for the -22/-22A engines. The extended list beyond EC 31 covered above were:

J40-WE-22A Engine Changes 32-41	
EC	Brief Description
32	Booster pump drain seal (resurvey)
33	Installation features of pressure sensitive A/B detection switch (Aerotec)
34	Addition of gearbox bosses
35	Revisions to auto trim box
36	Removal of solenoid operated afterburner bleed valve (resurvey)
37	Removal of fire shield
38	Relocation of oil filter
39	Envelope violation (-22)
40	Booster pump vent relocation (-22)
41	Check and surge-check System

First 150 Hour Verification Test

A 150 hour verification test was attempted beginning on August 11, 1953 using development engine J40-WE-22A #5. The test was stopped after 28 hours due to a cracked transmission cover.[38] This Kansas City built engine used a Kansas City built afterburner and a modified (improved) combustion chamber liner header. In the memo describing the engine to be used in the test, Westinghouse noted that some Essington built afterburners might have to be used for some J40-WE-22A engines on a limited basis, and, since that afterburner was only qualified on the -1 and -8 engine requested that it be approved as qualified on the -22A as well without a separate verification test being run.[39] This request was denied by BuAer on the grounds that the verification engine had to be equipped with the afterburner with which it would be produced and, due to delays in verification of the -22A, the need for Essington produced afterburners appeared to be eliminated.[40]

August's report on the progress of correcting the compressor disc overheating problems due to leakage past the curvic clutches and around the through bolts of the compressor spool indicated that the verification test engine had newly designed aluminum 1st and 5th stage compressor discs and a new steel 7th stage disc. The leakage paths were sealed for Stages 1-8 with a rolled edge, stainless steel seal ring fitted into a groove machined in the face of the curvic clutch.

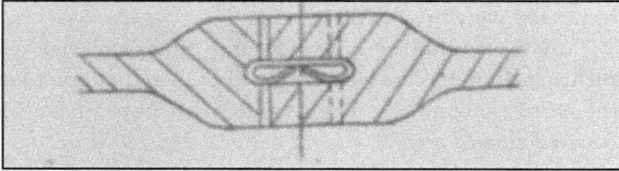

Figure 6 Compressor Sealing Stages 1-8, J40-WE-22A Verification Engine. *(Inset from WAGT EDSK 230146, 8/19/1953)*

Stages 8-11 were sealed (Figure 7) with a flat, stainless steel seal ring fitted into a groove on the ID of the curvic clutches. The corners of the groove and the seal ring were sealed with high temperature air curing Silicon rubber.

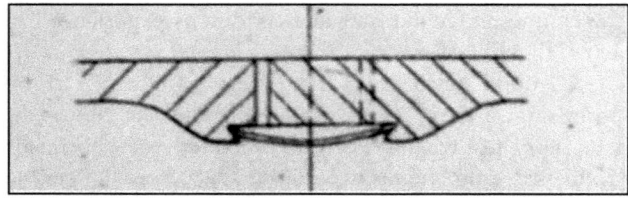

Figure 7 Compressor Sealing Stages 8-11, J40-WE-22A Verification Engine. *(Inset from WAGT EDSK 230146, 8/19/1953)*

Engines serial WE302511 and up would be built with these types of seals and prior builds were to use Bakelite sealing as tested and proved on the -1 and -8. All engines using the Bakelite sealing would be retro-fitted with the improved sealing at their first overhaul. As related in the -8 development story, at first Westinghouse was recommending that all stages be sealed in the belief that leakage throughout the compressor was lowering compressor efficiency, but this proved not to be the case.[41]

Specification WAGT-X40E2-4C Amendment 1: J40-WE-22A Performance Requirements				
Ratings	Thrust lbs	RPM	SFC lb/hr/lb	Fuel Flow lb/hr
Maximum (A/B)	10,900	7,260	2.70	-
Military	7,250	7,260	1.078	-
Normal	6,500	7,260	1.035	-
90% Normal	5,850	7,260	1.02	-
75% Normal	4,880	7,260	1.04	-
Idle	450	3,000	-	1,800 (Max.)

The BAR recommended WAGT's solution be accepted and approved.[42] BuAer responded that the seals appeared adequate and approved them for the -22A and -22.[43]

The release to production date for anti-icing valves was moved from November 1 to February 1 due to rescheduling of various components by the supplier.[44] BuAer later asked for the cost and weight savings that would accrue from deleting the anti-icing requirement. There were no cost savings but 12.5 pounds of weight would be saved. WAGT asked that the specification be changed if the anti-icing was to be deleted.[45]

The final specification for the -22 (WAGT-X40E2-4C) with Amendment 1 covering the -22A was submitted for approval on October 1, 1953. The dry engine weight was given as 3,615 pounds for the -22A and 3,451 for the -22.

The afterburner shutoff valve was eliminated and cooling air was now bled continuously from the compressor during engine operation regardless of whether the afterburner was in operation.[46]

BuAer approved the specification with the following changes. The Hot Soak had to be run before the Cold Soak

to properly simulate the aging of components. The temperature control electrical requirements had to be changed to say "not required" during startup. Production engines should follow the reduced final acceptance run specified in MIL-E-5010A and the specification changed to reflect that new requirement.[47]

Second 150 Hour Verification Test

Following the verification testing, BuAer issued a conditional acceptance of the J40-WE-22A manufactured at the Kansas City facility based on an inspection on October 14, 1953. The acceptance memo contained numerous conditions, included below in their entirety to give the reader a full understanding of the performance of the test engine and the overall program in completing the voluminous other tests and reports of those tests.

1. Reference (a) summarizes testing accomplished on a Kansas City manufactured Model J40-WE-22A engine to demonstrate compliance with the requirements of reference (b). References (c) and (d) describe the detailed parts and manufacturing source of such parts as used in the test engine. Bureau of Aeronautics inspection of the test engine subsequent to completion of the 150-hour test schedule was conducted at the Westinghouse Kansas City plant on 14 October 1953.

2. Based on a review of references (a) through (d) and on observations made during the engine inspection, the Chief of the Bureau of Aeronautics considers that the Kansas City manufactured Model J40-WE-22A engine, conforming to the Parts List of the engine used in the verification test reported in reference (a), satisfied the requirements of reference (b), except as set forth below:

a. Mechanical discrepancies:

i. Combustion chamber liner failure – The Chief of the Bureau of Aeronautics is willing to accept engines containing combustion chamber liners of the configuration tested in the verification engine until 1 January, 1954 only, at which time an improved durability liner must be available in production engines.

ii. High pressure oil supply line – The tested steel line is acceptable until no later than 1 January, 1954 at which time a silicone rubber line must be incorporated in production engines.

iii. Eleventh Stage Compressor Outlet Vane Assembly – Only a revised part, with spot welds between every vane rather than between every other vane, is acceptable.

b. Specification requirements for installation of 60 KVA alternators. Since the engine test was conducted with a 15 KVA General Electric Model 2CM-212-A2 alternator, operation of all J40-WE-22A engines accepted under the subject contract is authorized with this specific alternator only.

c. Certain below listed reports concerning control system component laboratory tests (equally applicable to J40-WE-8 and -22A engines) are currently unapproved by this bureau:

i. Automatic temperature control, P/N 61E635-4
ii. Power scheduler, P/N 61E902-1
iii. Constant speed drive, P/N 54J304-1
iv. Constant speed governor, P/N 61E681-3
v. Shuttle valve, P/N 68H52-1
vi. Shut-off valve, P/N 64F404-1
vii. Four-way valve, P/N 64F838-1
viii. Exhaust nozzle actuator, P/N 58J54-8

Although these reports should be available and approved by the chief of the Bureau of Aeronautics prior to acceptance of the Model J40-WE-22A engine as complying with reference (b), this bureau is willing to waive this requirement based on prior acceptance of the Model J40-WE-8 engines without availability of such reports and on the contractor's stated plans to expedite preparation and transmission of such reports as have not been submitted. It is the understanding of this bureau that all specification required tests have been completed satisfactorily except for the power scheduler, P/N 61E902-1, which is progressing satisfactorily.

d. Laboratory test report on the Accessory Gearbox, P/N 58J800-3 has not been received in this bureau.

Although this report should be available and approved by the Chief of the Bureau of Aeronautics prior to acceptance of the Model J40-WE-22A engine as complying with reference (b), this bureau is willing to waive this requirement based on the understanding that this test has been completed satisfactorily.

e. Specification requirements relative to radio interference: Design changes necessary to enable compliance with model specification requirements must be incorporated by 1 January 1954.

f. Certain J40-WE-22A control system components listed below bear part numbers which are different from corresponding J40-WE-8 control system component part numbers and for which the contractor has not presented laboratory qualification reports:

i. Power scheduler, P/N 61E902-1
ii. Afterburner fuel control, P/N 61E208-5
iii. Fuel regulator, P/N 58J851-5
iv. Overspeed relay, P/N 63F376-3
v. Fuel Booster pump, P/N 58J147-7
vi. Shut-off valve, P/N 64F404-3
vii. Exhaust Nozzle actuator, P/N 58J54-27

The Chief of the Bureau of Aeronautics is willing to accept the above control system components based on the minor changes made from the configuration of approved J40-WE-8 components.

g. Weight – The test engine exceeded specification weight of 3,451 pounds. Reference (e) proposed a specification weight change to 3,790.8 pounds. This change is hereby approved.

Upon Receipt of the contractor's agreement with the actions indicated in items (a) through (g) above, the Model J22-WE-22A will be considered verified.

3. While the following items are not causes for rejecting the verification test, they are nevertheless considered unsatisfactory and should be corrected as soon as practicable:

a. Items A.1.(d), A.1.(e), A.2.(a), A.2.(e), A.2.(c), A.2.(e) of Section XI of reference (a).
b. Inlet duct and inlet cuff cracks.
c. Overspeed relay spline wear.
d. Inadequate protective treatment both inside and outside of the compressor casing.
e. Afterburner linkage rod end rollers worn.
f. Galling on thrust side of outer ring of No. 1 main bearing.
g. Carbonizing of afterburner fuel lines.
h. Erosion of main fuel pump.
i. Galling of compressor case by inlet guide vanes.
j. Cracked turbine inlet guide vanes.
k. Shear of compressor through bolts upon disassembly.
l. Means should be provided to permit replacement of broken or cracked afterburner sealing strips.
m. Poor quality hardware parts used in engine components.

n. Early replacement of oil filters due to the use of MIL-L-7806 oil.

4. This bureau concurs with the contractor's recommendation (d) of reference (a) to install a transitory potentiometer in the exhaust nozzle actuator. Operation of J40-WE-22A engines accepted under the subject contract is not authorized until this change is made.

5. The Model J40-WE-22A engine discussed above will be authorized for 150 hours of operation prior to rejection to overhaul. It is planned to not consider either military time or afterburner time as regards rejecting the engine to overhaul. If not in agreement, the contractor's recommendation on this item is requested.

6. It is understood that an up-to-date version of the "Preliminary Operating Recommendations" will be forwarded to this bureau for review and approval in approximately two weeks. Also, it is requested that this publication clearly state the various operating restrictions, particularly in regards to the operating time limit before overhaul and the replacement of the high pressure oil supply line and oil filters at 50 and 15 hours, respectively.

7. The contractor's comments or concurrence in the condition of acceptance of the J40-WE-22A engine as outlined above are requested. Also, information is desired as to when the above listed component test reports and the required publications will be received in the Bureau of Aeronautics.
(End Memo)[48]

WAGT responded to this conditional acceptance letter on November 19, stating:

1. This contractor wishes to express agreement with the items set forth in the referenced letter, paragraph 2 items (a) through (g). It should be pointed out with reference to item (2a_2) that the J40-WE-22A engines exclusive of those two (the first two flight cleared engines – auth.) which are limited to 50 hours, will include the new silicone rubber line to replace the steel line in the high pressure oil system. Item (2a-3) has been complied with on all engines. Item (2b), the contractor agrees with authorizing the use of a General Electric model 2CM-212-82 15 KVA alternator on the subject engines at present. Recent development indicates that in the near future sufficient substantiating data will be available to allow this restriction to be removed. Item (2e) which requires compliance with model specification radio interference limits by 1 January 1954 is acceptable to this contractor, since Change in Designs are now being initiated which by the addition of filter elements in the Auxiliary Electric Control and the junction box will reduce the radio interference level to acceptable limits. Item (2a-1) requires that an improved combustion chamber liner be available for J40-WE-22A production engines by 1 January 1954. This contractor's development program on combustion chamber liners is progressing favorably and at the present time it looks like it will be possible to release to the Production Department an improved combustion chamber liner by 1 December 1953. Release of an improved configuration by this time will allow incorporation of the improved part by the specified date.

2. With respect to the preliminary operating recommendations requested in the referenced letter, paragraph 6, this publication was transmitted to the Bureau of Aeronautics, Power Plant Division, on 6 November 1953. It was clearly stated in this publication that the operating restrictions between overhauls were for such things as replacement of oil filters, etc.

3. With reference to paragraph 7 in the reference letter, the contractor wishes to inform the Bureau of Aeronautics that the latest information regarding the required publications listed in paragraph 2c of the reference letter will be supplied by separate correspondence. Purpose of this letter is to transmit the contractor's concurrence and comments as requested.
(End Memo)[49]

The exchange of memos regarding acceptance of the engine as verified ended with this memo from the Kansas City BAR:

1. Forwarded (Item 1 of prior memo)
2. This office considers that the basic letter satisfies the minimum requirements of reference (a) of the basic letter and thereby establishes the J40-WE-22A as a verified engine model.
3. Production engines built and tested in conformance with reference (a) and applicable specifications will be accepted by this office unless advised to the contrary.
(End memo.)[50]

Work continued on producing the necessary documentation in support of production and field use. The equipment list for the J40-WE-22A and -22 engines dated October 1, 1953 raised the question with BuAer as to why accessory equipment such as fuel pumps and the constant speed drive transmission were included in the Handbook of Overhaul Instructions for the 58J800 Gearbox Assembly. WAGT's reply noted that various seals, bearings and shafts for these accessories were actually part of the transmission and to test the accessory you would have to test the entire gearbox as well. A written note on the routing sheet about the inclusion of seals, bearings and shafts for accessories notes that this type of design could lead to trouble and WAGT should be made to make accessories that could be tested independently. It would appear this type of direction was very late in coming, since this type of design was used from the very beginning and BuAer raised no objection.[51]

(The details of EC's 42-51 were not found.)

EC 53 was surveyed in January 1954, regarding the elimination of 4 unused holes and connected bosses in the rear mounting ring to simplify manufacturing, the existing rings to be used until exhausted.[52]

EC 52 was surveyed in February. It proposed changing the three underslung afterburner fuel lines with a single overhead main fuel line and two smaller auxiliary fuel lines. The change would deal with the afterburner fuel leakage problem at shut down (removing the necessity for an afterburner fuel-lines scupper) and the afterburner ignition delay problem.

The internal afterburner fuel rings would be modified to sub-divide afterburner fuel within the afterburner shell. A fuel shut-off valve (still in development) would be located in that vicinity of the quick-disconnect clamp between the aft end of the proposed fuel line and the single external elbow. The two auxiliary fuel lines approximating ¼ inch in size would be located adjacent to the main fuel line. The first would act as a return line for fuel which would continuously circulate from the fuel control to the shut-off valve by way of the main fuel line (regardless of throttle position) and the second operating the actuating line for the shut-off valve when afterburner operation was selected. The shut-off valve would be fuel actuated. The fuel line changes would be incorporated as part of the hydro-mechanical control on the J40-WE-22 engines serial WE302561 and up. All J40-WE-22A engines would be back-fitted if the change was approved.[53]

Figure 8 Sundstrand constant speed transmission and governor (lower right), WAGT P49675, 6/29/1953. *Courtesy Hagley Museum and Library*

The verification 150 hour test on the Sundstrand constant speed transmission, transmission governor and shuttle valve was run in October 1953 and report A-1656Z dated 10/24/1953 was submitted by WAGT on November 11, 1953 reporting on this test. A short description of this transmission is warranted here, as the proper functioning of the constant speed transmission was critical to the supply of a constant electric to the systems of the aircraft using advanced radar and associated equipment.

The unit was a cartridge-type mechanical-hydraulic unit which mounted into the gearbox in such a manner that it received its input power by direct gearing from the engine rotor and fed its output to two generator pads of the gearbox. Its purpose was to drive an alternator at constant speed to supply the airframe with 400 cycles per second 120/208 volt alternating current power and to drive a direct current generator. The constant speed was maintained for engine rotor speeds from idle to military. The transmission was governed by a flyweight governor mounted on the gearbox and sensed the transmission output speed by direct gearing.

The transmission was integral with the gearbox and engine lubrication system, receiving its charge oil directly from the engine main oil pump and discharging the oil back to the oil lubrication system. Internal oil lines were provided in the housing which mated with internal oil lines in the gearbox supply to supply lubrication oil to the transmission and to transport the drain oil for scavenging by the engine's scavenge pump. The governor also received its lubrication oil by internal lines in the gearbox. The verification test report summary indicated that at the completion of the 150-hour full load test, all parts were disassembled, inspected and found to be in excellent condition except one motor pushrod, four pump push rods and the pump wobbler plate. The defects in these parts did not affect the transmission operation and were not evident until disassembly. Test on similar parts under more realistic operating conditions, i.e. frequent shutdowns, resulted in a more desirable wear pattern on other test units. The test had been run in conjunction with the 150-hour test of the transmission.[54] Post verification test, WAGT added a heavier counter weight in the accessory transmission to reduce resident vibration experience during the engine verification test. This had no effect in testing on the bearings of the constant speed transmission. On February 8, 1954, BuAer accepted the test results and released the gearbox and the three components for production on both the -22 and -22A.[55]

The qualification test on the -22 was next up and WAGT apparently wanted the requirement to produce a test plan be moved to BuAer. BuAer rejected this approach, stating: *"Section 3(b) of the General Provisions of the subject contract will show that the contractor is required to provide all reasonable facilities and assistance for the convenience of the local Navy Inspectors in the performance of their duties. As such it is considered that the contractor should, for any engine being tested in the subject contract, provide the Bureau of Aeronautics Representative with the details of how the test is to be conducted (test specification) and results therefrom. As such, the BAR has at all times the prerogative and responsibility for reviewing the contractor's test specification and disapproving any provisions therein which do not meet Model Specification requirements as approved by this bureau. Since the current procedure as outlined above provides for full inspection of engines to be delivered to the Navy, no benefit can be gained from making the test specification a Bureau of Aeronautics requirement."[56]*

In addition, BuAer noted that any "interim" parts included in the test would be the parts required in production until a change-in-design (CID) was approved by the Navy. It was WAGT's prerogative to use such interim parts as they chose.

Lastly, BuAer noted that many items needed approval before the start of the qualification test. *"First the engine must be presented for inspection as regards compliance with Model Specification requirements for weight, envelope, radio interference, and other factors not involving engine operation. If approved the engine may then be calibrated and the results of this calibration presented to the Navy for approval. If this calibration is approved, the endurance running part of the Model Specification requirements for qualification may be commenced."*

WAGT sent in report A-1753 in March, recommending removal of the hydraulic locks from the hydro-mechanical control system and replacing them with a simple single check valve. This would effectively lock the exhaust nozzle in the position it had at the time of a hydraulic failure without the complexity and weight of the multiple control locks. A slight loss in power would occur due to loss of temperature trim but would be small and not a hazard to safe flight. The modified system provided a safer overall flight control than the existing system with two hydraulic locks in the circuits. The BAR endorsed the recommendation.[57,58] The change was surveyed as EC 56.[59] The report was updated August 10 with additional information and WAGT requested BuAer's approval to make the change.[60] It was rejected for the -22A based on initial indications of the amount of possible thrust loss, but WAGT submitted updated information in the referenced updated report. In October WAGT sent a request asking when a reply might be received on this recommendation.[61]

EC 54 was surveyed in March to change the Junction Box dimensions to reflect the actual production part. The new dimensions exceeded the installation envelope and WAGT invited comments from the airframe manufacturers on any interference issues discovered.[62]

On May 7, WAGT recommended all -22A and -22 engines be fitted with a 16% opened first stage turbine nozzle compared to the standard nozzle. This change was identical to the one made to the -6 and -8 engines for compressor stall reduction/elimination. Testing demonstrated a thrust reduction of 75 lbs in Military power and 150 lbs in A/B, but the engine still met the guarantees. They felt a verification test was not necessary since the materials, etc. had not changed and the change had been verified on the -6 and -8 engines previously. The SFC's were slightly worse than normal production figures, but still met the specifications at all engine speeds. The included charts enable us to determine the standard actual production thrusts and SFC's of -22 and -22A engines at that point:[63]

Modified engines were flight tested under the B-45 on at least 4 test flights. The aircraft had a Mach limit of 0.71 at that time and the full compressor stall range had not been explored. It required flights in the F3H-1N to help extend the speed envelope to enable completion of the charts.

Even with the nozzle change, the engine was compressor limited. The air intake stagnation temperature could not be lower than -32°F and flight speeds had to be adjusted to ensure this temperature was met or exceeded. Also, the required decrease in turbine outlet temperature when a stall range was entered was included on an enclosure. The aircraft had to be placarded in the cockpit to make the pilot aware of the limits.

On May 12, a memo was submitted reporting on the results of a Special Engine Test to Prove Integrity of Piping. XJ40-WE-22A #4 was built up as a production -22A with the following modifications:

1. A new exhaust nozzle actuator scupper and exhaust nozzle actuator packing nut designed to provide positive locking.

Performance Impacts of 16% Open 1st Stage Nozzles on -22/-22A			
Test Exit Thrust (lbs)	SFC Specification	SFC Normal Test Exit	SFC w/16% Nozzle
4,880	1.040	0.983	1.040
5,850	1.020	0.955	0.995
6,500	1.035	0.961	0.991
7,680	1.078	1.000	1.015
11,400 A/B	2.700	2.300	
11,300	2.700	---	2.350

2. The rigid tubing which comprised the J40-WE-22A emergency system piping was re-bracketed to provide added rigidity.

3. The rigid lines of the J40-WE-22A emergency system had redesigned end fittings which gave added fatigue resistance as well as a positive leak-proof fitting even in the absence of torque on the jam nut.

4. The relief valve which formed a part of the emergency system for the J40-WE-22A had an additional support added to provide increased rigidity.

5. The four-way valve which was a component of the J40-WE-22A emergency system had its supporting brackets stiffened to provide additional rigidity.

6. One first stage turbine blade had 5.8 ounces of its mass removed to create an imbalance in the engine during starting, acceleration and high speed continuous running.

7. The piping under scrutiny had a water and Bon Ami solution chalk-like coating to reveal any leaks that developed.

The engine was run through five start-ups and then run up to full speed for a total of 30 minutes and no leaks were observed. The vibration was 200 mils at start-up and 25 mils at full speed. Normal vibration was about 5 mils. WAGT submitted the modifications proved the integrity of the design of the new emergency system piping.[64]

May 19 brought the survey of EC 57, which proposed the raising of the compressor ratio by using a twelfth stage added to the compressor as a solution to the compressor stalls. The survey does not mention a particular compressor modification design, but it is likely this was the same design used for the C-23 compressor proposed earlier for the -8 engine compressor stalls. This approach was recommended over opening up the turbine nozzles, as it did not lower performance either of SFC's or thrust. However, it would have added weight of 110.3 lbs to the engine.[65] A formal proposal for the C-23 compressor modification was finally made in February, 1955. This had significantly more information in it. To accommodate the longer compressor with its extra stage, the diffuser was shortened and an extra section was bolted to the rear of the compressor case. If volume production occurred, a new case incorporating the 12th stage as an integral part would reduce the weight of the engine to 3,959 lbs. The proposal indicated a development engine had been flown with this compressor and demonstrated surge free performance up to 60,000 feet.[1] Engine performance with this compressor based on actual

test results is in the table below. Altitude is assumed to be sea level but was not stated.

A written note on the circulation cover sheet indicated that a BuAer representative had suggested to WAGT as far back as January, 1954 to submit this compressor modification as a cure for compressor stalls.

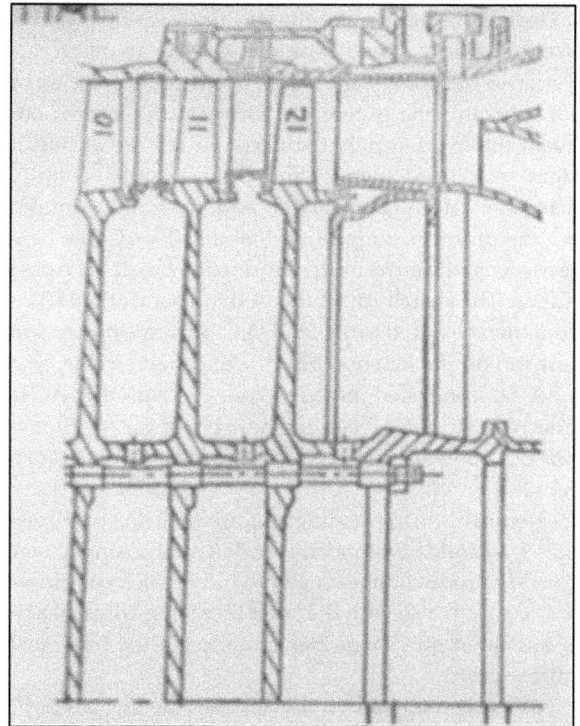

Figure 9 C-23 Compressor drawing showing attachment of 12th Stage, CR 6.0, also proposed for -8's and -1's. Not accepted for either the -8 or -22 engines. *WAGT Memo, 5/14/1954, Unnumbered Drawing*[65]

J40-WE-22 Engine Comparison with and without the C23 Compressor Modification				
	Thrust lbs.		SFC	
	J40/C23	J40-WE-22	J40/C23	J40-WE-22
Maximum (A/B)	11,650	10,900	2.410	2.570
Military	8,000	7,250	1.039	1.078
Normal	7,000	6,500	1.010	1.035
90% Normal	6,400	5,850	1.000	1.020
Weight	3,974	3,911		

The intent of the proposal was to produce modification kits to retroactively upgrade all -22 and -22A engines with the modified compressor, case extension and shortened diffuser.[66] The proposal was formally rejected by BuAer in

---

[1] Not supported by the flight test data available. See Chapter 14.

April, due to the fact the F3H-1N program was not to be continued and not enough data existed to predict the improved performance of the airframe with the modified engine.[67]

EC 58 was surveyed in July. It added an AC circuit to the filaments in the automatic temperature control to eliminate the delay while the auto control filaments warmed up. During this period, the engine nozzle would sometimes move to the fully closed position, causing an overtemp of the engine. Currently the pilots could reduce the throttle before turning the automatic temperature control on to prevent the overtemp, but suffered lower thrust until the system stabilized again and the throttle could be advanced.[68] In August WAGT added another option to solve the problem, suggesting that a 110 VAC line be run directly from the airframe to the junction box cable from the airframe. The switch currently used to operate the 110 VAC would then be wired to the 28 V.D.C. airframe power and to a new pin on the junction box. In the engine cable, a wire would be connected between pin "G" of the AGT at connection 3102-18-1P and the new pin in the junction box for 28 VDC power supply. Responses were requested in two weeks.[69]

Removal of the sealing requirement for compressor stages 9, 10 and 11 was recommended after testing of twelve engines in production testing with fully sealed compressors (all stages) and also with the latter three stages unsealed had demonstrated no change performance to the fully sealed compressors.[70]

In early September, BuAer finally got around to looking at a report regarding the use of hydraulic fuses in the hydraulic lines making up the exhaust control nozzles. This report, A-1776 dated May 17, 1954, was a follow-on to report A-1595 which investigated possible ways to prevent oil loss from the lubrication system of the engine if a hydraulic line in the exhaust nozzle system failed. In the latest report, WAGT recommended that fuses not be used as they offered no protection if a line failed and might in fact cause the nozzle to malfunction.[71]

In September, the elimination of the float valve on the oil reservoir top filler port was surveyed. No EC number was mentioned in this memo.[72]

In August a warning light was surveyed for the cockpit that would illuminate if the compressor outlet pressure differential to the engine compartment exceeded the model specification. The warning light had been requested by McDonnell but the circumstances behind why it was needed were not explained in the files.[73]

A specification weight change was requested in August based on the many changes to the engine since the qualification of the -22A was approved, that engine weighing 3,847 pounds. The additional weight of the changes was approximately 25 pounds and a 10 pound variation factor was added to accommodate minor variations in production from one engine to another. This brought the new specification weight to 3,882 pounds.[74] BuAer approved the new weight October 11 but noted no further weight increases on engines would be considered unless accompanied by statistics on recent production engines.[75] The changes affecting weight were listed as:

Weight Changes - Post J40-WE-22A Certification Test		
CID	Description	Weight Change (lbs)
19898*	Replaced all paper filters with metal filters	0.35
43148A	Added clamps to prevent fuel leakage at the A/B fuel ring inlet port	4.00
43149C	Added stiffeners to the exhaust collector	0.39
43230	Added shields to prevent the thermocouple leads shorting out against the insulation blanket	0.55
43290A	Added channels to cover the protruding bolt ends on the A/B in order to prevent damage to the insulation blanket	5.20
43295	Lengthened the A/B insulation blanket	2.70
43323	Increased the size of the rear bearing strut bosses	0.77
43333	Revised the spacers on the retaining screws for the compressor vane assembly	0.51
43364	Revised the tubing arrangement to prevent leakage	2.80
43383	Added a trigger anticipator circuit to provide better A/B light offs	5.97
43411	Added an additional A/B igniter relay nozzle	0.51
43438	Strengthened the oil supply and scavenge tubes at the number two bearing	0.21
43444A*	Removed a portion of the compressor blade roots in the first four stages	-2.47
43460*	Added additional fixtures to provide the proper amount of fuel for acceleration at different altitudes	3.70
*CID's not approved at time of memo	TOTAL	25.19

EC 45 was surveyed in September. This would eliminate the drain box on the gearbox, provided the pad was used with a hydraulic pump instead of a generator. The airframe manufacturers were using hydraulic pumps with self-contained lubricating systems and the boss was not needed. Removal would eliminate abrupt casting sections in that region and reduce possible stress concentrations.[76]

The incorporation of a two-step acceleration fuel flow was surveyed in September as well. This was requested by McDonnell to limit the acceleration fuel available during acceleration at the higher altitudes. The design incorporated a pressure valve, switch and tubing to integrate it into the fuel supply system.[77]

Engineering changes to correct operating deficiencies in production engines were developed and applied in Kansas City and not all of the EC's for these were preserved in the contract files. These changes are visible in the production story below as they occurred.

A requirement to produce the last five -22 engines using titanium discs for the last five compressor stages was dropped in 1955 due to cost. The long delays in developing manufacturing processes capable of producing these discs reliably out of titanium stock directly delayed the discovery that the discs in test, made from available titanium alloy stock already in Westinghouse's possession, demonstrated excessive disc growth. It was not possible to determine if the rate of growth (presumably due to a lack of sufficient material to make discs) could be statistically determined so as to set a safe life limit. A new alloy was available but procurement would take 6 to 12 months and the delay would not make these discs available in time to be used in production, given that all J40 engines would have been completed by the end of 1955. The finally pricing for production J40 engines was done without inclusion of the costs for titanium discs.[78]

Production

Production schedules continued to slip, with production to start in early 1953, pending delivery and installation of five machines from the Ex-Cell-O Corporation of Detroit at least four months prior to the start of production. The nature of these machines was not specified.[79]

The work force at the Kansas City plant had to be recruited and trained on the production and testing process, a major effort that relied on the Essington team for leadership and support. This effort and its progress were at times visible, causing delays and even impacting available service use hours of individual engines due to excessive testing time needed to pass engines out to service once production began.

In November, the decision was taken to curtail the -8 program and switch to the -22 for production. Handbooks and accessories handbooks for the former model would not be required. However, engine service instructions would be required comparable to the -6 manuals already furnished.[80]

The BAR in Kansas City requested the allocation and shipping requirements for the first engines due to be produced in January 1953.[81] Due to the delays in finalizing the design and testing of the Hydro-Mechanical Control for the -22, a short 60 engine production run of the interim -22A model using the Simple Hydraulic Control of the -8 but incorporating an automatic trim control was planned. As of February 4, 1953, the engine allocation was determined for these first 60 engines:

Initial -22A Engine Allocation Plan	
Allocation	# Engines
Grumman F10F-1, installations and spares	4
Douglas F4D-1, installations and spares	25
McDonnell F3H-1, installations and spares	21
BuAer Maintenance Training Requirements	4
Ford Motor Company Factory Training Requirements	1
BuShips Marine Turbine Project	5

The 60 engines were to be shipped beginning in March and through the month of August. Forty seven (47) were to be charged against contract NOa(s) 10385 and thirteen (13) against contract NOa(s) 52-048.

On February 18, BuAer noted that difficulties were being experienced in getting production started and that WAGT had forwarded a new production schedule.[82] In May, with the signing of Amendment 5, the projected schedule changed again.[83]

Revised Projected Production Schedules for the J40-WE-22A										
	Jan	Feb	Mar	Apr	May	Jun	Jul	Aug	Sep	Oct
Initial	0	0	6	6	9	13	-	-	-	-
Feb Mod	0	0	3	7	11	13	-	-	-	-
May Mod	0	0	0	0	1	4	10	13	19	3

Amendment 5 (Fixed Price) replaced the Letter of Intent for these engines included in the above new schedule. Interestingly, this amendment notes the 50 engines involved were for use in the F10F and F3H program. The F4D program is not mentioned, likely having shifted to the J57 for production airframes.[84] The other 10 engines mentioned earlier are not referenced in these memos or amendments.

A production schedule produced in June by WAGT reflected the current total 477 engines still on order for both the -22A and -22 as follows:

June 25, 1953 -22A and -22 Production Schedule 1953/54																		
	J	J	A	S	O	N	D	J	F	M	A	M	J	J	A	S	O	N	D
-22A	3	8	15	19	11														
-22					3	21	40	50	50	50	49	47	45	29	20	17			

With the reduction in programs requiring the J40 engines, BuAer issued a new production requirement for 1953, requiring a total of only 33 J40-WE-22A engines and 3 J40-WE-22 engines during that year. The memo states that the new schedule stretched out the -22A production one month longer and the -22 production by about 9 months. Peak production rate dropped from 19 to 11 and from 50 to 35 for J40—22A and -22 respectively. (The new schedule attachment is not in the file.) A handwritten note on the circulation cover sheet notes that the order for J40-WE-24 engines was in the process of being terminated.[85]

Actual production was delayed while verification testing was delayed over and over for various reasons (see development above). The Kansas City BAR wrote a memo in September stating that a pre-verification J40-WE-22A serial WE302506 was estimated to be shipped to McDonnell the week of September 28 for ground running, and, subject to final clearance, for limited flight operations. The verification test was waived for shipment of a J40-WE-22A power section (WE302507) to be used for the BuShips Marine Turbine Project. The estimated shipment date for this engine was October 9. It appeared that one or more additional engines could be prepared for shipment during the week of October 19.[86] That schedule then slipped to October 30 for the first engine and November 2 for the second.[87]

With the engine verification test considered passed and engines being prepared to ship, the first two flight cleared engines (WE302508, WE302509) had been limited to 50 flight hours before rejection to overhaul due to inclusion of marginal parts. WAGT recommended the rejection to overhaul be raised to 150 hours if the following inspection

and actions were taken at 50 hours: 1) Incorporate CID 43162 which replaced the rigid tube and flex section combination high pressure oil line with a new high pressure silicone rubber hose. 2) Incorporate CID 43216 which added a new heavier counterweight to the constant speed drive rotor to effect a sizable decrease in induced vibration; and 3) Inspection of the 58J793-2 exhaust collector and if found damaged, replace with the new 77J27-1 exhaust collector, which successfully passed the 150 hour verification test.[88] The BAR approved the recommendation.[89]

Figure 10 Power control unit with throttle attachment arm in front left, WAGT P48641, 1/30/53. *Courtesy Hagley Museum and Library*

A correction to the proper oil to be used in the engine had to be sent out as CID 19931, stating the proper oil for the engine was MIL-L-7808. A WAGT memo from October stated in error that the oil to be used was MIL-L-7808A.[90]

Field maintenance of the engines required a qualification from WAGT regarding the compressor elements, turbine discs and exhaust nozzle eyelid replacement as stated in a BuAer letter of December 31, 1953.

On compressor blades: It was physically impossible to remove the compressor blades from the first four stages of the compressor without unstacking the entire rotor. The compressor straightening vanes (mixer) required the almost complete disassembly of the engine. All other compressor elements could be easily removed. New first through fourth stage compressor blades were in verification at that time which would allow removal without unstacking the rotor. It was anticipated the new blades would be available in production by May 15, 1954.

On turbine discs: No procedure existed to effect replacement of either the first or second stage turbine wheel as a unit without completely rebalancing the rotor. Individual blades could be replaced. Work was in hand to hopefully present within three months a rebalancing procedure the field could utilize in replacing individual turbine discs.

Exhaust nozzle eyelids: Trouble was anticipated if the field replaced these, as they were ("must be") individually fitted at assembly. However, experience had shown the eyelids had a much greater life than the afterburner assembly.[91]

The McDonnell F3H-1N using the J40-WE-22A was cleared for flight in February.[92]

In March, engine serial WE302509 was returned to Kansas City for investigation. The condition of the returned engine was detailed in a memo from the BAR Quality Control Kansas City to the BAR St. Louis with the comment and reminder that engines should be in operating order when returned to the plant. The deficiencies noted were:

1. Five container bolts missing.
2. Three bellows on shipping containers deflated.
3. Three lines disconnected and left open in the vicinity of the oil pump.
4. One open port in oil pump.
5. Two open ports on fuel line cross fittings.
6. Cables connecting the two power scheduler units were twisted and badly crushed at forward end near Y-junction where they hooked up to main control cable.
7. Magnetic valve part number 64F404-3 had a loose cable connection. Part was safety-wired with the wire pulling in the wrong direction.

8. When engine was turned over in "roll over dolly" oil leaked out of the ATC cover.
9. Pressure rake cover missing from compressor outlet.
10. Screw missing from fuse cover plate of ATC.
11. Large insulation blanket covering combustion chamber housing was gouged and cut open for a length of approximately 9 inches at the bottom aft location.
12. Three filters and the A/B assembly missing.

Damaged or missing parts were replaced by stripping other engines and the engine was tested and returned without the filters or A/B assembly.[93] Parts required to return the engine to running condition that were not part of the normal overhaul were charged to the contract.

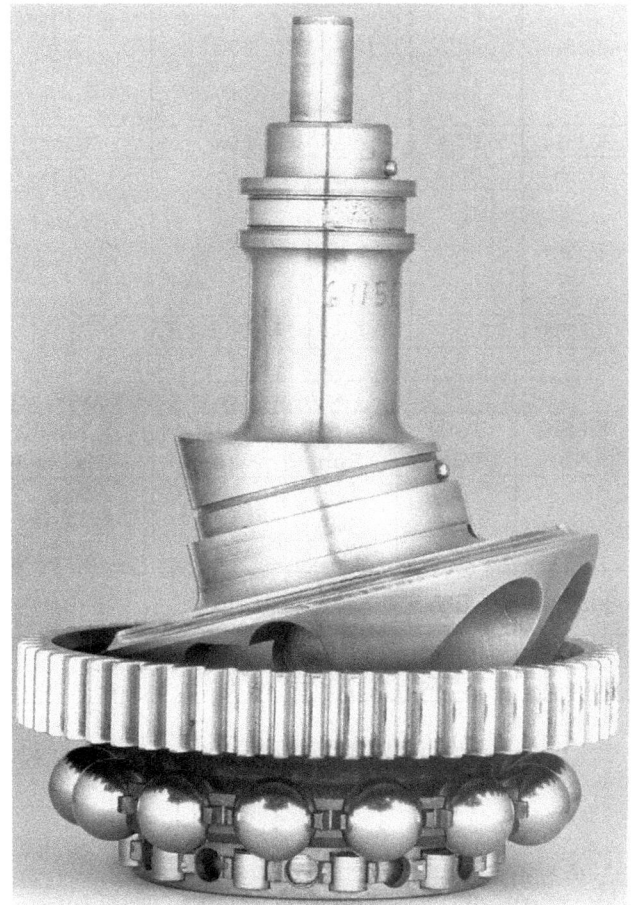

Figure 11 Wobble shaft from the constant speed transmission after the second acceptance test, WAGT P49160, 4/13/1953. *Courtesy Hagley Museum and Library*

BuAer raised the issue of acceptance testing time reporting and asked that WAGT supply reports showing the test times for each engine, quality control issues and actions taken to address such issues. On February 4, WAGT submitted reports covering December and January. In the original reports, WAGT called initial check out and first runs as "Green Runs" and BuAer corrected them saying the term should be "Initial Run". The chart as reported has been modified by the author to conform with later charts on the same subject below.

Kansas City Engine Acceptance Test Run Times (Hrs) – December 1953								
Engine No.	Total Time	Initial Run	Special Runs	Final Checkout	Final Runs	Leakage Check	Penalty Time	Penalty Reason(s)
302507	24.9	7.94	1.26	12.55	2.0	1.35	0.0	
302508	40.2	13.7	0.0	19.4	6.5	0.3	0.0	
302509	28.9	6.9	0.75	11.75	1.8	8.5	0.0	
302510	48.55	11.5	0.0	30.5	2.6	5.3	0.0	
302511	22.5	9.2	0.0	7.47	2.37	3.48	0.0	
302512	30.5	6.2	1.05	19.9	1.8	2.0	0.0	
302513	41.5	10.9	1.32	13.45	3.0	3.8	9.0	Oil leak gearbox and reservoir. Teleflex cable.
302514	27.4	6.45	4.06	12.4	2.17	2.25	0.0	
302556	11.3	4.8	0.0	2.3	2.0	2.2	0.0	
302557	15.3	6.3	0.0	2.2	2.0	0.4	4.4	Oil cooler outlet line, oil consumption
302558	10.5	5.6	0.0	2.9	1.5	0.5	0.0	

Kansas City Engine Acceptance Test Run Times (Hrs) – January 1954								
Engine No.	Total Time	Initial Run	Special Runs	Final Checkout	Final Runs	Leakage Check	Penalty Time	Penalty Reason(s)
302506	23.7	0.0	0.7	5.9	2.7	5.2	3.0	Fuel leak, carbon seal on boost pump shaft
302510	19.6	0.0	0.0	10.0	1.8	3.0	4.8	Overspeed relay replaced
320513	0.0	0.0	0.0	0.88	13.17	0.63	0.25	Throttle torque
302515	25.54	7.5	0.0	0.833	16.153	1.05	0.0	

WAGT reported in detail some of the root causes for the excessive test times, summarized here:

1. Learning curve of people operating on new production line(s).
2. The testing of house engines did not provide the experiences needed for new production.
3. Accessory testing stand facilities not totally in place due to late deliveries from vendors.
4. Testing specifications (definition of "normal") not sufficient in all cases.
5. Controls malfunctioning, all vendors.
6. Lack of specialized test equipment for troubleshooting.

Corrective actions had been taken or were being taken in every case to eliminate or control the root problems. In spite of the actions WAGT was reporting, the testing preparation time required for acceptance and shipment was still being strongly criticized by BuAer after WAGT had put their improvement plan in place to lower the run time of the engines. Results for engines up to that point were included in a chart, here reproduced. Westinghouse argued that the trend shown in the data demonstrated (scatter charts curves were provided) that the improvement plan to reduce running time was taking effect.[94]

Kansas City Engine Acceptance Test Run Times (hrs) – February 1954								
Engine No.	Total Time	Initial Run	Special Runs	Final Checkout	Final Runs	Leakage Check	Penalty Time	Penalty Reason(s)
302506	28.0	6.2	0.7	5.9	2.7	5.2	7.3	Cracked exhaust collector
302515	29.04	7.5		16.153	0.833	1.05	3.5	Cracked exhaust collector
302517	25.4	5.1		10.33	0.85	0.92	8.2	Cracked exhaust collector
302518	24.15	5.3		11.3	0.85	3.0	3.7	Cracked exhaust collector
302519	36.2	5.3		18.6	2.4	4.9	5.0	Cracked exhaust collector
302520	14.0	4.3		8.1	1.4	0.2	0.0	Cracked exhaust collector, dual fuel pump

On March 16, engine WE302512 suffered a first stage turbine blade failure in flight causing an engine seizure. The aircraft, an F3H-1N, landed successfully, damaging the wheels and tires, but the balance of the airframe was undamaged. Westinghouse submitted an initial teardown report dated March 24, 1954 that lacked a definite cause for the blade failure. The engine had approximately 74 hours of run time since acceptance at the time of failure. The engine was part of the compressor stall investigation and had an opened first stage turbine nozzle.[95']

On April 7, WAGT sent a memo explaining the late delivery of support handbooks for the engine had been mainly due to the delays in finalizing the design. They now predicted the final support and overhaul manuals would be delivered to BuAer for reproduction to begin by July 1.[96]

In early April, WAGT requested permission to retain WE302509 as part of the compressor stall elimination study. The engine was to be replaced in overhauled condition or a new engine substituted at WAGT's expense at the end of the study. The BAR endorsed the request.[97] The engine was bailed to WAGT under contract 54-402b for development testing.[98]

On April 13, engine serial WE302507 was returned to the plant because of oil leakage from the No. 1 bearing. It was to be repaired and returned to the BAR St. Louis.[99]

In April, acceptance and delivery of J40-WE-22A or -22 engines was stopped due to turbine blade breakage, fluid leaks and compressor surge. Westinghouse was asked to present recommendations to be forwarded to BuAer as to the extent of fabrication and assembly of parts involved in the deficiencies. The BAR was asked to provide a list by April 12 of the time and location of all reported engine fluid leaks in the F3H-1 airplane.[100]

A written note on the transmittal sheet for the above telegram notes that the 120 J40 engines scheduled for delivery in 1954 represented 19% of Westinghouse's total engine delivery program across all engine types in production by Westinghouse. The sizable overhead of Westinghouse meant that the unit costs on these engines would go up even higher. The reasons for the J40 defects were unknown and the time to eliminate them could not be estimated at that time.

In spite of the stop acceptance order, the BAR in St. Louis requested 5 engines be delivered in a wire dated April 22, 1954. This was in lieu of the 6 engines previously requested. Serial WE302511 was reported as struck off charge.[101] (Reason not stated. – Auth.)

A production status wire dated April 7 showed: 1 Engine Accepted, None shipped, 2 prepared for shipping, 1 completed final test, 1 in final build, 3 undergoing penalty runs, 6 in initial build. Numerous parts had been built for possible compressor stall elimination and turbine durability improvement. They were: 1st stage turbine nozzle, 11th stage compressor disc, compressor through bolts, 1st stage turbine blades.[102]

WAGT completed report A-1686, 61E500-2 Adapter Gearbox Qualification Test Report on March 22, 1954. The gearbox was used to drive the Vickers variable delivery pump used on J40-WE-22 production engines. The 150-hour test was run in the WAGT test cell and the gearbox was in good shape at the end of the test, operating without issues.[103] The test results were accepted and production of the adapter gearbox approved by BuAer May 13, 1954.[104]

May's production report showed 44 (of 50 on order) J40-WE-22A's had been manufactured in Kansas City. This report is confusing, as the Essington portion of the report shows 29 J40-WE-22's on order to be shipped from Kansas

City, but the Kansas City report stated these engines were J40-WE-22A's.[105] The June Production Progress Status Report shows no further production, manufacturing being delayed while waiting for "Comstal" fix components. The discrepancy between the Kansas City and Essington reports regarding numbers of engines on order was now resolved, all being called J40-WE-22A's.[106] In July's report, the discrepancy (-22 vs. -22A) re-appears regarding these 29 engines. Ninety five (95) J40-WE-22's now show as being on order, with 3 completed to date.[107]

WAGT recommended to BuAer that in order to lengthen the overhaul period from the current 150 hours to 180 hours, the incorporation of CID 43605 on an engine be required on or before the 150 hours was achieved. This CID covered the replacement of the No. 2 oil pump, 60E197-9 with a more rugged design 60E197-13 which eliminated the cracking of the earlier pumps then being discovered during overhaul.[108]

The 150-hour verification test report A-1802 on the J40-WE-22 gearbox was completed on June 10, 1954 and was submitted for approval. The tests included two 150-hour runs of the basic gearbox and another to qualify alternate parts and alternate suppliers of parts. Both tests were completely successful, with the exception of two alternate bearings from the latter test which were considered not qualified.[109] BuAer accepted the reports and approved the gearbox for production of the -22 on August 18, 1954.[110] The same gearbox manufactured by Spencer-Lycoming passed its 150 hour test and was reported in report A-1806 dated August 3, 1955 sent to BuAer September 13, 1955.[111]

On June 30, BuAer was informed by wire that engine WE302519 was shipped to McDonnell on June 26 by truck.[112]

July 1954 found WAGT requesting that the first stage nozzle vanes, part no. 60E13-6, shipped to the field as spares, be returned to be modified to the 60E13-7 level so that they were compatible with the 16% open nozzle assembly configuration. This was approved.[113]

August brought a memo from WAGT stating that -22 production would begin in September and it requested allocations for the engines.[114] The August Production Progress Status Report showed all 50 -22A engines on order were completed and 33 of the 95 -22 engines on order were completed.[115] (Lack of properly manufactured power controls held up shipments.) In September, BuAer responded to the allocation request: The 95 -22 engines on order were partially allocated to St. Louis as follows:[116]

Month	Number
October 1954	2
November 1954	5
December 1954	8

A report on production issues discussed the accessory problems encountered and the actions and plans in place to resolve them.[117]

1. Porosity of castings, such as the adapter pads for the boost pump and afterburner pump. An impregnation specification was added but was voluntary. Leakages of parts on hand in test slowed availability of qualified parts. A flagging procedure was now in place to identify specification changes needed for immediate incorporation in order to avoid wasted motion and delays.

2. Improper bench calibration of fuel controls. Statistical correlation between the test stands and engine performance of the controls was established. While establishing control over the situation, limited test bench capacity needed to be addressed and additional test stands were being activated.

3. High pressure oil pump. Leakage on test in a large number of cases was encountered. A new gasket design proved effective in a large percentage of cases, but the issue was not totally resolved.

4. Aneroid acceleration switch bellows failures. A new, more rugged design was in place with the supplier, but the new parts were not available yet in production.

5. Trigger anticipator solenoid. Failure of the electrical connector due to handling and excessive specification requirements. The specification was revised and a new design giving added strength of assembly was completed. New parts incorporated the change.

6. Damage of electrical cable assemblies prior to and during installation on the engine. Greater care in assembly and a redesign sub-dividing the assembly into smaller components were undertaken to reduce vulnerability to handling damage.

7. Afterburner fuel control. Had been in short supply due to supplier production difficulties. Some problems had been addressed through minor redesign.

8. Engine assembly deficiencies causing malfunction in test. Inspection practices had been improved.

9. Limitation of Test Cell capacity. Additional cells were made available for testing and several were being sound-proofed to allow 16 hour operation. The backlog of engines needing testing contributed to the bottleneck in test cell availability.

The September Monthly Production Progress Report shows 44 J40-WE-22 engines now complete, but only 5 delivered in September. The problems were noted by WAGT in the above report in part with the added comment that foreign materials passed through two engines during test. The BAR added that inexperienced operators and a large number of design changes were affecting production in addition to the issues listed by the contractor.[118]

In October, WAGT informed the BAR Kansas City that they were working on a three shift basis to provide engines and additional engines to the current production outlook would be available if the necessary accessories became available.[119] This shows that many engines were assembled waiting for the modified parts to be able to get through test for acceptance. The determination of exactly how many engines had been shipped to BuAer at any one point is difficult because the production reports always show more engines complete than actually got shipped at any one point.

December 29, 1954 Engine Allocations*	
Priority 1, Item 2, Lot XIX, Quantity 4	January, 1955, McDonnell St. Louis, NOa(s) 51-640
Priority 2, Item 1B, Lot XX, Quantity 5	January, 1955, Same
Priority 3, Item 1A, Lot XX, Quantity 10	February, 1955, Same
Priority 4, Item 2, Lot XX, Quantity 14	March, 1955, Same

*Under contract – 95, Previously allocated – 15, This allocation – 33: Total 48 allocated.[127]

Production Outlook as of October 1954					
	Oct.	Nov.	Dec.	Jan.('55)	Feb.('55)
J40-WE-22A	7	7	7	3	0
J40-WE-22	1	2	4	9	12
Total New Production	8	9	11	12	12
Reworked Engines	6	4	2	0	0
Total	14	13	13	12	12

Wooden afterburner shipping containers were coming back from McDonnell upon receipt as part of engine deliveries. They were to be reused at the factory and had to be in good condition. A wire to the BAR St. Louis asked for 15 containers to allow them to make their shipping dates.[120] In October the BAR asked that of the 40 GFE shipping containers to be supplied, they begin arriving December 1 at a rate of 10-15 per month and that if none were to be supplied, WAGT would be notified in time for WAGT to procure containers by other means.[121] In late October instructions were given to the BAR St. Louis to begin shipping wooden A/B shipping containers to WAGT Kansas City at a rate of 15 per month, the first shipment to arrive not later than December 1.[122]

Quick change kits were requested from WAGT to allow interchangeability between -22A and -22 engines in the field.[123] In the same time period, the BAR St. Louis requested an additional calibrated engine for education purposes be allocated to the BAR Kansas City. It was noted that at that time 48 engines had been allocated and this would allocate one of those to Kansas City.[124]

Engine WE302535 suffered an unspecified problem and originally was to be returned to Kansas City for investigation and repair but WAGT determined they would repair it on site in St. Louis. The telegram authorizing its return was cancelled.[125] Later it was determined that WE302535 had suffered an internal fire due to oil starvation. It and WE302531 (excessive oil consumption) were returned to WAGT Kansas City for repair under the warranty.[126]

The November Production Progress Report shows that 69 J40-WE-22 engines had been constructed at that point. The BAR comments sheet on this report was not located.[128]

The Automatic Temperature Control (ATC) on engine WE302525 had been reported in September by McDonnell as malfunctioning, the engine being unable to achieve its T_5 redline during flight line run-up. The immediate corrective action taken was to place another ATC in the wing of the aircraft and connect it to the AEC and junction box with jumper cables. The malfunctioning unit was not immediately removed due to inaccessibility of the unit. Later the engine was changed due to foreign material damage and the unit was returned to WAGT for inspection. The unit, part no. 61E635-4 (rebuilt to a Westinghouse Engineering sketch EDSK238616A) was not a new ATC, but modified experimentally to be similar to the 61E635-6 current production unit. It had operated on J40-WE-22A WE302517 at McDonnell for 20 hours and on the subject engine for 17 hours with no difficulties on either engine reported. On investigation, the unit operated perfectly. It was suspected that a poor contact in one of the ATC connectors was responsible for the malfunction.[129]

The December Production Progress Report showed 81 J40-WE-22 engines completed. Production problems reported were replacement of soft No. 1 bearing housing liners (root issue addressed and not expected to recur) and A/B cracking during test causing a shortage of afterburners. This shortage was anticipated to continue into January. Leakage was improved but not fully resolved. Corrective actions already noted were being applied.[130]

The January Production Progress Report showed 93 J40-WE-22 engines completed. Problems were severe and numerous, as noted by the BAR in the comments section. *"There were no J40-WE-22 engines shipped in January although five (5) were forecast. Continued malfunctioning of camboxes, afterburner regulators, and microjets resulted in excessive penalty runs. The leak problem also continued to be a time consuming factor although improvements in assembly operations and incorporation of design changes have improved this condition. A vibration problem developed late in the month as of this date is under* engineering study. Although some parts shortages existed the main reasons for delay of engines were the continued failure and subsequent replacement of control items and work stoppages in shops. The five engines forecast for January shipment were lost in the following manner:

Figure 12 J40-WE-22 Accessories gearbox exploded view, with front on the top and the top of the gearbox to the left, WAGT P49763, 7/16/1953. *Courtesy Hagley Museum and Library*

1. Rejected on overtime test running due to cam box and afterburner regulator failures.

2. Leaks and electric control assembly failures delayed the engine in final test. A crack in the rear bearing support and an oil leak in the rear bearing on January 28 prevented January shipment.

3. Leaks and control failures held engine from completing final test. The cam box had to be changed five times. On January 26, an engine was rejected for overspeed caused by installation of a checkout regulator which did not function properly. When cause was discovered it was too late to re-assemble and re-test for January shipment.

4. Cam box and afterburner shortage delayed the engine to final test. While on test, the engine developed numerous leaks. These and the failure of a bracket weld on

the combustion chamber sent the engine to re-work area on January 26, 1955.

5. Manufacturing problems in the form of blade clearances, a warped mixer assembly, and failure of mating parts to fit delayed the final buildup. Shortage of afterburner and cambox delayed movement to final test. While on test, the engine encountered afterburner light-off trouble, collapse of exhaust collector and a leak at the oil reservoir, preventing January shipment.[131]

The February Production Progress Report showed all 50 -22A and 95 -22 parts for engines had been manufactured, but actual shipments lagged for many quality issues. The monthly production progress reports and actual accepted engine quantities never agreed, the production report apparently meaning the major/all components for the engines were in hand. The author has tried to reconcile the picture but discrepancies remain.

The J40-WE-22A shipment issues were listed as:

1. Only two engines were shipped against the forecast of three.

2. One of the above was delayed by mandatory engineering changes that had to be made before any further engines of that type could be shipped. The changes included:

   a. Replacement of second stage nozzle seals – nozzles had to be removed from assembled engines and sent to shop for rework.

   b. Installation of a new high pressure oil filter.

   c. Rear bearing support oil tubes checked, replacing earlier light wall tubes with heavier tube types.

   d. Installation of new heavier type fittings on rear bearing support.

   e. Installation of modified high pressure relief valve.

3. Shortage of afterburners and yokes.

4. Rework of afterburners prior to final run.

5. Work stoppage in shops interfering with movement of parts.

6. Rejection of yokes for cracks and unsatisfactory repairs being rejected by Quality Control.

7. Recently identified problem with exhaust collectors would impact deliveries in March and most likely April as well.

The J40-WE-22 shipment issues listed were:

1. No engines shipped against a forecast of six.

2. Excessive vibration was encountered resulting in the failure of rigid tube assemblies and brackets which required engineering design changes. The new designs incorporated new flex hose lines and a new spring to

modify the variable delivery pump. Late procurement of these parts delayed engines entering test.

3. Several second stage nozzle seals failed necessitating disassembly of all J40 engines to replace the subject seals. While in work, the rear bearing supports had to be re-inspected to make certain that oil tubes of heavier wall thickness had been installed.

4. Engineering changes originally scheduled to begin being incorporated beginning on the twenty first production engine instead had to be incorporated on the first engines prior to shipment. Field requirements had to be met before engines in process could be supplied with the new parts.

5. Employee walk-outs and disciplinary furloughs in engine assembly, accessory test and engine testing prevented any engines from shipping in February.

6. Recently discovered problems with the exhaust collector were threatening engine shipments in March.[132]

No engines were shipped in March while the problems identified in February were resolved. No engines were predicted to ship in April either, awaiting receipt of the new exhaust collectors, the first of which was due to be received April 4. Only the exhaust collector needed completion of field requirements, all other parts having completed this process.[133]

April's Production Progress Report showed a trickle of engines being shipped. Two -22A's were shipped against none forecast. Three -22's were shipped against none forecast. The new exhaust collectors were being incorporated; the engine builds were completed and were passed to test. Some penalty runs were experienced which were due to a mandatory change to correct high VD Pump Case Pressure. Also, a delay in testing was due to a question on the Thrust Cradle Calibration in the test cells. These issues restricted the completion of more engines. A new method of testing gearbox and accessories assemblies prior to engine installation was expected to reduce penalty runs due to leakage.[134]

On May 9, WE302545 was returned to Kansas City from St. Louis for repair due to turbine damage from a cracked blade and investigation into blade cracking problems.[135]

The May Status Report shows the last -22A due to be shipped was not shipped due to high oil consumption and accessory trouble in testing. All -22 engines were overdue on the contractual schedule. Eleven engines out of 14 forecast were shipped. The problems in May were:

1. A shortage of gearbox and accessories assemblies due to testing prior to installation. However, the pretesting had already demonstrated its effectiveness in the reduction of engines failing their acceptance testing.

2. Contamination discovered in several accessories required re-inspection and a security investigation.

3. Cracked turbine discs in final test resulted in three engines being rejected and triggered an engineering program to identify the problem and find a solution.[136]

In June, a wire showed one -22A left in penalty test and 78 -22's left to deliver under the contract. This gives some idea of how many actual -22's ordered actually shipped to the customer by this date.[137] The June Production Status Report showed no further engines forecast for delivery, indicating that production was halted in terms of delivery to the customer.[138] The BAR comments sheet was not attached to the June report in the files.

Flame holder welds failing in the afterburners had been reported back in April and investigation showed that the repair manual was not detailed enough for the repair to be consistently accomplished. The repair manual was updated. The frequency of failure was low and the original design was considered adequate.[139] A memo a week later explained that thermal expansion was causing the weld cracks and that only by changing from intermittent welds to continuous welds could this type of cracking be completely eliminated.[140]

On July 25, 1955, the termination in part of Lot XX Item 2 (11 engines) and Item 3 (27 engines) eliminated 38 J40-WE-22's from the program. Based on the (confusing) engine production status as of the end of June, a total of 20 additional engines were left to be completed through acceptance test and then delivered to complete the J40 production program. In total, there were 50 J40-WE-22A's (all delivered except 1, the last possibly delivered later but not included in the termination negotiations) and 57 J40-WE-22's under contract, (37 accepted) at the time of this termination wire.[141]

In September, 1955, with the contract cancelled, BuAer agreed with WAGT's proposal to deliver the 40 engines in current production status as follows:

Ten (10) engines – Category I (after satisfactorily completing the initial and final acceptance test), to be delivered in accordance with the specifications at the contract price.

Eighteen (18) engines - Category II (after satisfactorily completing the initial acceptance test, disassembly and reassembly), to be delivered prior to final testing at a reduction in contract price of $5,600 per engine.

Twelve (12) engines - Category III (after satisfactorily completing the initial acceptance test), to be delivered upon completion of initial test, omitting the operations associated with disassembly, reassembly and final testing at a reduction in the contract price of $7,991 per engine.[142]

The above brought the total complete and assembled engines J40-WE-22 delivered to BuAer to seventy seven (77).

The total -22A and -22 delivered production was 127 engines. The balance of the materials was consigned to scrap. Twelve (12) of the completed -22 engines were transferred to the Air Force. What use the Air Force made of these engines is unknown. Their most likely use was as a spare parts source for the -1's in use on the MX770 project.

In January 1956, WAGT wrote a memo citing the various performance records the J40 had achieved in various airframes and taking issue with the Navy's determination that the engines were not *"suitable for flight in aircraft."* WAGT believed it had met all expressed and implied warranties and guarantees for the engines.[143]

On January 9, 1956 WAGT requested that J40-WE-22A serial WE032516 then currently at the Kansas City plant be assigned to Westinghouse (with proper security maintained) to be used for displays and educational purposes.[144] In spite of much argument noted on the circulation cover sheet of this request, it appears this engine was in fact bailed to Westinghouse and suitably modified as a cutaway exhibit. It is likely this is the engine now in the National Naval Aviation Museum collection and on display in Pensacola, Florida. This cutaway exhibit is definitely a J40-WE-22A, but lacks its manufacturing plate, the insulation blanket around the A/B and the exhaust gas temperature decal.

J40-WE-22/-22A/-22B/-22A-G Mechanical Descriptions

J40-WE-22/-22A

These engines were closely similar to the J40-WE-8 except for modifications to improve reliability, compressor stall performance, airframe thermal protection, afterburner light-off performance and the use of two different control systems. The -22A used the Simple Hydraulic Control with automatic temperature trim. The -22 used the Simple Hydraulic Control with a Hydro-Mechanical Temperature Control System. All of the control system changes were made to lower the need for airframe electrical power to control any part of the engine either on primary or emergency control and to provide for improved fail-safe operation.

The first 16 -22A engines had slightly different first stage turbine discs due to a manufacturing error. These discs were reused rather than discarding them. A continuous stream of EC's were applied to both the -22 and -22A during and after production.

J40-WE-22B

The -22B (WAGT provisional designation) was a rejected proposal to substitute a new afterburner design using four hydraulic rams to operate an iris type multi-flap exhaust nozzle. None were ever built. It is possible and even likely that this nozzle design came from the -10 design work, but the remaining -22 files are silent on the subject. *(See Chapter 10)* The -22B model designation was not approved by BuAer.

J40-WE-22A-G

The -22A-G was ordered by the Power Plant Division to WAGT specification WAGT-240F as a gas generator for unspecified experiments. The engines were J40-WE-22A's with the afterburner and all controls and components required for either the A/B or operation above sea level altitude being removed. The -22A gearbox was replaced with a J40-WE-8 type since the constant speed generator capability of the -22A transmission was not needed.[145] Five of these engines were delivered to the Power Plant Division. Their use after delivery is unknown. The serial numbers are not preserved in the records, but their manufacturing totals were included in production volumes for the -22 series of engine as quoted at the House of Representatives sub-committee hearing.

Figure 13 XJ40-WE-22 Mockup, bottom view of piping and hose runs, WAGT 12/3/1952.
Courtesy Hagley Museum and Library

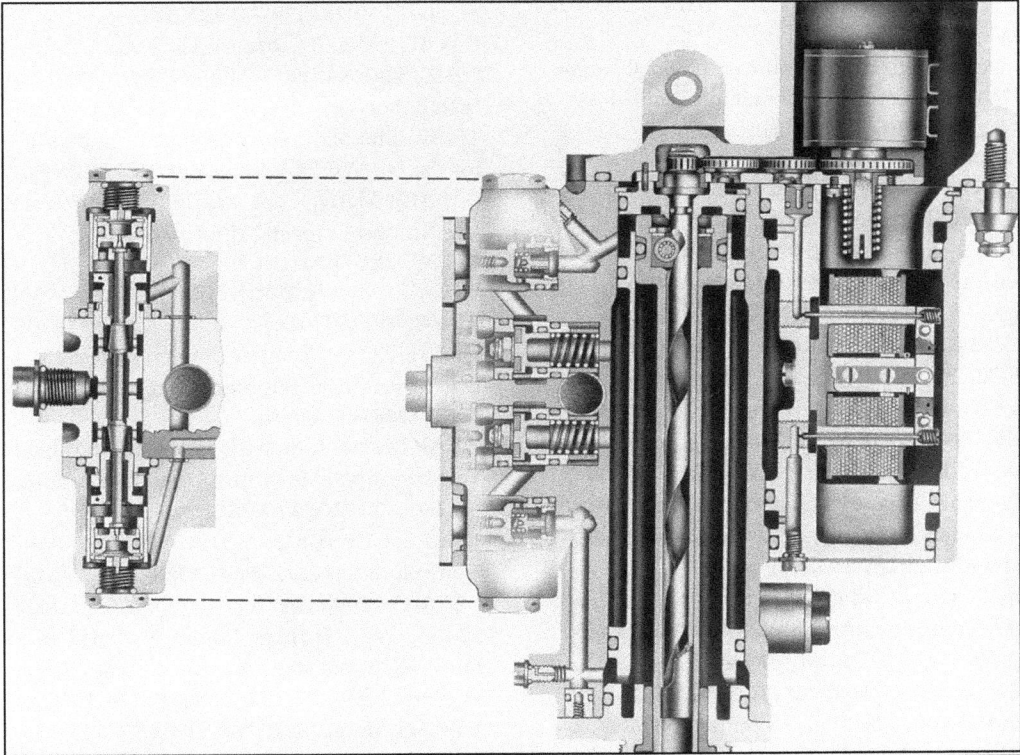

Figure 14 XJ40-WE-22 Exhaust nozzle actuator internal component design, WAGT P48298, 12/12/1952.
Courtesy Hagley Museum and Library

Chapter 9 - J40-WE-22/-22A/-22B/-22A-G Citations

[1] WAGT Memo, October 23, 1951, "Westinghouse Engine Delivery Schedules" **B937VE8RG72.3.2NACP**

[2] WAGT Memo, August 13, 1952, "Contract NOa(s) 10385, Starting Torque requirements for the J40-WE-22 engines" **B937VE11RG72.3.2NACP**

[3] BAR Memo, August 27, 1952, "Contract NOa(s) 10385, Starting Torque Requirements for the J40WE-22 Engines" **B937VE11RG72.3.2NACP**

[4] WAGT Memo, September 2, 1952, "Contract NOa(s) 10385; 52-048, WEC Turbojet Engine Model J40-We-22, Interim Emergency Exhaust Nozzle Control, Survey of. Engine Change No. 11." **B938VE12RG72.3.2NACP**

[5] BuAer Memo, May 12, 1952, "Contracts NOa(s) 10385 and NOa(s) 52-048, J40-WE-22 Engines" **B937VE11RG72.3.2NACP**

[6] BuAer Memo, September 16, 1952, "Contract NOa(s) 10385; 52-048. WEC Turbojet Engine Model J40-WE-22. Alternative Afterburner Design, Survey of Engine Change No. 101." **B937VE11RG72.3.2NACP**

[7] WAGT Memo, October 8, 1952, "Contract NOa(s), Qualification of Gearbox and New Parts Associated with Application of the Constant Speed Drive to the J40-We-22 Engine." **B942VE9RG72.3.2NACP**

[8] BAR Memo, October 10, 1952, "Contract 10385, Change in Design #19243, Supplement A, New Fuel Control for the J40WE-1, -8 and -22 Engines; Verification of" **B942VE9RG72.3.2NACP**

[9] WAGT Memo, October 22, 1952, "Contract NOa(s) 10385; 52-048. WEC Turbojet Engine Model J40-WE-22. Oil Temperature & Pressure Instrumentation, Survey of. Engine Change No. 13." **B937VE11RG72.3.2NACP**

[10] WAGT Memo, October 29, 1952, "Contract NOa(s) 10385; 52-403c, 52-048. WEC Turbojet Engine Models J40-WE-8, J40-WE-22. Booster Pump Seal Drain, Survey of. Engine Change Numbers 101 and 14 for the J40-WE-8 and -22 respectively." **B937VE11RG72.3.2NACP**

[11] BuAer Memo, February 26, 1953, "Contracts NOa(s) 10385 and NOa(s) 52-403c: revised performance curves for Westinghouse Models J40-WE-8, -22 and -22A engines; approval of" **B938VE12RG72.3.2NACP**

[12] WAGT Memo, October 7, 1952, "Contract NOa(s) 10385 Westinghouse Model Specification WAGT-X40E2-4 which describes the XJ40-WE-8, -22 turbojet engine: revised operating limits curve for" **B937VE11RG72.3.2NACP**

[13] WAGT Memo, October 29, 1952, "Contract NOa(s) 10385; 52-048. WEC Turbojet Engine Model J40-WE-22. Replacement and Relocation of High Pressure Oil Pump, Survey of. Engine Change No. 15." **B937VE11RG72.3.2NACP**

[14] BuAer Memo, February 26, 1953, "Accessibility of J40-WE-22 engine components in the Model F3H-1 airplane" **B938VE12RG72.3.2NACP**

[15] WAGT Memo, April 23, 1953, "Contract NOa(s) 10385. WEC Turbojet Engine Models J40-WE-22; Accessibility of Engine Components in Model F3H-1 Airplane." **B938VE12RG72.3.2NACP**

[16] WAGT Memo, November 7, 1952, "Contract NOa(s) 10385 – Accessibility of J40-WE-22 Engine Components in the Model F3H-1 Airplane." **B938VE12RG72.3.2NACP**

[17] WAGT Memo, January 20, 1953, "Contract NOa(s) 10385. WEC Turbojet Engine Model J40-WE-22A and J40-WE-22; Installation Drawings for" **B942VE11RG72.3.2NACP**

[18] WAGT Memo, October 15, 1952, "Contract NOa(s) 10385, Temperature Limits for Oil Hoses in Turbine Area of J40-WE-1, -6, -8 and -22 Engines." **B942VE11RG72.3.2NACP**

[19] WAGT Memo, January 20, 1953, "NOa(s) 10385; 52-048. WEC Turbojet Engine Model J40-WE-22A. Interim Emergency Exhaust Nozzle Control, Information On." **B938VE12RG72.3.2NACP**

[20] BuAer Memo, February 6, 1953, "Westinghouse Model Specification WAGT-X40E2-50 dtd 31 Oct 1952 which describes the adapter gearbox for XJ40-WE-22 turbojet engines" **B942VE12RG72.3.2NACP**

[21] WAGT Memo February 11, 1953, "Contract NOa(s) 10385; 52-048 WEC Turbojet Engine Model J40-WE-22A. Installation Drawing for J40-WE-22A, Survey and Transmittal of. Engine Change Numbers 19-29 inclusive for the J40-WE-22A." **B938VE12RG72.3.2NACP**

[22] WAGT Memo, February 26, 1953, "J40WE-22A Verification Program; Information on" **B942VE12RG72.3.2NACP**

[23] WAGT Memo, March 4, 1953, "Contract NOa(s) 10385, 52-408. WEC Turbojet Engine Models J40-WE-22A and J40-WE-22. Oil Temperature and Pressure Instrumentation, Supplementary Information On." **B938VE12RG72.3.2NACP**

[24] BAR Memo, March 23, 1953, "J40 and J46 Engine Oil Temperatures, Maximum Operating Limits, Verification of" **B938VE12RG72.3.2NACP**

[25] BuAer Memo, March 31, 1953, "Automatic Trim Control, reports on" **B938VE12RG72.3.2NACP**

[26] WAGT Memo, March 26, 1953, "Contract NOa(s) 10385, 52-048. WEC Turbojet Engine Models J40-WE-22A and J40-WE-22. Booster Pump Seal Drain, Resurvey of. Engine Change No. 31." **B938VE12RG72.3.2NACP**

[27] BuAer Memo, March 31, 1953, "Contract NOa(s) 10385, tentative radio interference limits for J40-WE-22 engines" **B938VE12RG72.3.2NACP**

[28] WAGT Memo, April 17, 1953, "Publication Requirements for Contract NOa(s)-10385 Our Reference WG-63300" **B942VE12RG72.3.2NACP**

[29] WAGT Specification WAGT-X40E-200, Turbo-Jet Insulation Blanket Specification **B942VE12RG72.3.2NACP**

[30] BAR Telegram, June 5, 1953, Insulation Blanket J40-WE-22A Mockup Conference Request **B942VE12RG72.3.2NACP**

[31] WAGT Memo, May 14, 1953, "Contract NOa(s) 10385, 42-408 WEC Turbojet Engine Model J40-WE-22 Replacement and Relocation of High Pressure Oil Pump, Supplementary Information on." **B938VE12RG72.3.2NACP**

[32] American Bosch Corp telegram, May 22, 1953, diversion of Moorejig borer **B942VE12RG72.3.2NACP**

[33] WAGT Memo, June 3, 1953, "J40 Component Qualification Test, Information on" **B942VE13RG72.3.2NACP**

[34] BuAer Memo, June 10, 1953, "Contracts NOa(s) 10385, WG-63200 and NOa(s) 52-048 WG-57780: constant speed transmission unit for Westinghouse Models J40-WE-22, -22A, -24 and -26 engines" **B938VE12RG72.3.2NACP**

[35] WAGT Memo, July 2, 1953, "J40-WE-22 and J46-WE-8 Flight Suitability Tests" **B942VE13RG72.3.2NACP**

[36] BuAer Memo, August 18, 1953, "Contract NOa(s) 10385; J40-WE-22, Qualification test of the hydro-mechanical exhaust nozzle control 64J1-2" **B949VE16RG72.3.2NACP**

[37] WAGT Memo, September 18, 1953, Contract NOa(s) 10385; J40-WE-22 Qualification Test of the Hydro-Mechanical Exhaust Nozzle Control 64J1-2" **B949VE16RG72.3.2NACP**

[38] BAR Memo, August 20, 1953, "Contract NOa(s) 10385 – J40-WE-22A Verification Test; Information on" **B949VE16RG72.3.2NACP**

[39] WAGT Memo, July 20, 1953, "Contract NOa(s) 10385. J40-WE-22A Verification Test. Information on." **B949VE16RG72.3.2NACP**

[40] BuAer Memo, September 21, 1953, "Contract NOa(s) 10385 – J40-WE-22A Verification Test" **B942VE13RG72.3.2NACP**

[41] WAGT Memo, August 19, 1953, "Contract NOa(s) 10385, Acceptance of Mechanically Sealed Compressor Rotor for J40-WE-1, -8, -22A and -22 Engines; Request for" **B949VE17RG72.3.2NACP**

[42] BAR Memo, September 1, 1953, "Contract NOa(s) 10385, Acceptance of Mechanically Sealed Compressor Rotor for J40-WE-1, -8, -22A and -22 Engines; Request for" **B949VE17RG72.3.2NACP**

[43] BuAer Memo, October 12, 1953, "Mechanical Seals for J40-WE-8, -22A and -22 engines; approval of" **B949VE18RG72.3.2NACP**

[44] WAGT Memo, September 16, 1953, "Contract NOa(s) 10385 – Westinghouse Turbo-Jet Engine Models J40-WE-1, -8, -22A and -22. Contractor's Schedule for Incorporation of Anti-Icing Provisions." **B943VE14RG72.3.2NACP**

[45] WAGT Memo, October 26, 1953, "Contract NOa(s)-10385 and 52-048 Removal of Anti-Icing Provisions from J40-WE-6, -8, -22 and -22A Engines Our Reference WG-63300 and WG-67750" **B943VE14RG72.3.2NACP**

[46] Specification WAGT-X40E2-4C for J40-WE-22 and J40-WE-22A (Amendment 1) Dated October 1, 1953. **B948VE19RG72.3.2NACP**

[47] BuAer Memo, October 27, 1953, "Contracts NOa(s)-10385 and 52-403c, Westinghouse Specification WAGT-X40E2-4C of 20 May 53 and Amendment -1 thereto covering the Models J40-We-8, -22 and -22A engines; approval of" **B948VE19RG72.3.2NACP**

[48] BuAer Memo, November 10, 1953, "Contract NOa(s) 10385; Model J40-We-22A engine, acceptance of" **B949VE18RG72.3.2NACP**

[49] WAGT Memo, November 19, 1953, "Contractor NOa(s) 10385 – Model J40-WE-22A Engine." **B949VE19RG72.3.2NACP**

[50] BAR Memo, November 25, 1953, "Verification of J40-WE-22A Engine Model." **B943VE14RG72.3.2NACP**

[51] WAGT Memo, November 23, 1953, "Contract NOa(s) 10385. Contractor-Furnished Equipment List for J40-WE-22A and -22 Engines of 1 October 1953." **B943VE15ARG72.3.2NACP**

[52] WAGT Survey Memo, January 26, 1954, "Contract NOa(s) 10385, 52-048, WEC Turbojet Engine Models J40-WE-8, -22A, -22. Alteration to Engine Mount Ring, Survey of. Engine Change No. 117, 53 for the J40-WE-8 and the J40-WE-22 Series Respectively." **B943VE16RG72.3.2NACP**

[53] WAGT Memo, February 10, 1954, "Contract NOa(s) 10385, 52-048. WEC Turbojet Engine Models J40-WE-22A, -22. Revised Afterburner Fuel System, Survey of. Engine Change No. 52." **B943VE16RG72.3.2NACP**

[54] WAGT Report A-1656Z, October 24, 1953. "150-Hour Load Test of the Constant Speed Transmission, Transmission Governor, and Transmission Shuttle Valve" **B943VE16RG72.3.2NACP**

[55] BuAer Memo, February 8, 1954, "Contract NOa(s) 10385 – Qualification test of Constant Speed Transmission used on J40-WE-22A and -22 Engines; Approval of" **B949VE20RG72.3.2NACP**

[56] BuAer Memo, February 23, 1954, "Contract NOa(s) 10385 – J40-WE-22 Engine Qualification Test" **B943VE16RG72.3.2NACP**

[57] BAR Memo, March 23, 1954, "Contracts NOa(s) 10385, 10825; Hydraulic Lock, (Exhaust Nozzle Control System) Removal from J40-WE-22, J46-WE-8 and -18 engines" **B938VE12RG72.3.2NACP**

[58] WAGT Report A-1753, March 3, 1954, "A Study to Determine the Feasibility of Removing the Hydraulic Locks from the Hydro-Mechanical Exhaust Nozzle Actuator Control System" **B938VE12RG72.3.2NACP**

[59] WAGT Memo, March 26, 1954, "Contract NOa(s) 10385, 52-048. WEC Turbojet Engine Model J40-WE-22. Removal of Hydraulic Locks, Survey of. Engine Change No. 56." **B949VE20RG72.3.2NACP**

[60] WAGT Memo, August 20, 1954, "Contracts NOa(s) 10385, 10825; Hydraulic Lock, (Exhaust Nozzle Control System) Removal from J40-WE-22, J46-WE-8 and -18 Engines."WE-22, J46-WE-8 and -18 Engines." **B949VE20RG72.3.2NACP**

[61] WAGT Memo, October 5, 1954, "Contract NOa(s) 10385, 10825; Hydraulic Lock, (Exhaust Nozzle Control System) Removal from J40-WE-22, J46-WE-8 and -18 Engines." **B944VE20RG72.3.2NACP**

[62] WAGT Memo, March 26, 1954, "Contract NOa(s) 10385, 52-048. WEC Turbojet Engine Model J40-WE-22. Revised Junction Box Dimensions, Survey of. Engine Change No. 54." **B943VE17RG72.3.2NACP**

[63] WAGT Memo, May 7, 1954, "Contract NOa(s) 10385: Incorporation of 16% Open First Stage Turbine Nozzle per CID 43386 on J40-WE-8, -22, and -22a Engines, and per CID 43386A on J40-WE-1 Engines; Substantiating Data for and Effect on Performance of" **B949VE20RG72.3.2NACP**

[64] WAGT Memo, May 12, 1954, "Contract NOa(s) 10385 – Results of Special Engine Test to Prove Integrity of Piping." **B949VE20RG72.3.2NACP**

[65] WAGT Memo, May 14, 1954, "Contract NOa(s) 10385, 52-048. WEC Turbojet Engine Models J40-WE-1, -22A, -22. Proposed Compressor Fix, Survey of. Engine Change No. 40 for J40-WE-1 and 57 for J40-WE-22 series." **B944VE20RG72.3.2NACP**

[66] WAGT Memo, February 29, 1955, "Production of J40/C23 Engine; Proposal AGT 123" **B944VE20RG72.3.2NACP**

[67] BuAer Memo, April 14, 1955, "Westinghouse Proposal AGT 123, J40 Turbojet Engine with Twelve Stage Compressor" **B944VE20RG72.3.2NACP**

[68] WAGT Memo, July 12, 1954, "Contract NOa(s) 10385. WEC Turbojet Engine Model J40-WE-22A. Revision of Automatic Temperature Control Electrical System, Survey of. Engine Change No. 58." **B944VE20RG72.3.2NACP**

[69] WAGT Memo, August 3, 1954, "Contract NOa(s) 10385. WEC Turbo-jet Engine Model J40-WE-22A. Revision of Automatic Temperature Control Electrical System, Additional Information On." **B944VE20RG72.3.2NACP**

[70] WAGT Memo, July 21, 1954, "Contract NOa(s) 10385 – Removal of Mechanical Seals in Compressor Stages 9, 10 and 11 for J40-WE-1, -8, -22A and -22 Engines." **B949VE20RG72.3.2NACP**

[71] WAGT Report, A-1776, May 17, 1954, "Evaluation of Hydraulic Fuses as Protection Against Loss of Lubricating Oil in Exhaust Nozzle Actuator Systems." **B944VE20RG72.3.2NACP**

[72] WAGT Memo, September 15, 1954, "Contract NOa(s) 10385, 52-048. WECO Turbojet Engine Models J40-WE-1, -8, -22A, -22. Removal of Oil Reservoir Float Valve Assembly – Survey of" **B944VE20RG72.3.2NACP**

[73] WAGT Memo, August 27, 1954, "Contract NOa(s) – WEC Turbojet Engine Model J40-WE-22A. External Configuration of Compressor Pressure Limiting Warning Switch – Survey of" **B944VE20RG72.3.2NACP**

[74] WAGT Memo, August 31, 1954, "Contract NOa(s) 10385 – WEC Turbojet Engine Model J40-WE-22A Model Specification Weight Increase – Request for" **B949VE20RG72.3.2NACP**

[75] BuAer Memo, October 11, 1954, "Contract NOa(s) 10385, Westinghouse Specification WAGT-40E2-4D covering the Model J40-WE-22A engine" **B949VE20RG72.3.2NACP**

[76] WAGT Memo, September 16, 1954, "Contract NOa(s) 10385, 52-048. WEC Turbojet Engine Model J40-WE-22, -22. Modification and Removal of Hydraulic Pump Drain from Gearbox, Survey of Engine Change No. 45." **B944VE20RG72.3.2NACP**

[77] WAGT Memo, September 9, 1954, "Contract NOa(s) 10385; WEC Turbojet Engine Model J40-WE-22A, Incorporation of Two Step Acceleration Fuel Flow – Survey of" **B944VE20RG72.3.2NACP**

[78] WAGT Memo, May 5, 1955, "NOa(s) 10385 Deletion of Requirement for Five J40-WE-22 Turbojet Engines with Titanium Compressor Discs Our Reference WG63300" **B949VE21RG72.3.2NACP**

[79] WAGT Memo, September 1952, "Westinghouse Electric Corporation, Kansas City Works, Contracts NOa(s) 10385 and NOa(s) 52-048, J-40 Engines, Production of" **B942VE9RG72.3.2NACP**

[80] BuAer Memo, November 6, 1952, "Supplemental Contractor-Furnished Equipment List for J40-WE-8 and -22 Engines dated 1 October 1952" **B942VE11RG72.3.2NACP**

[81] BAR Memo, January 13, 1953, Contract NOa(s) 10385 – Westinghouse Electric Corporation, Kansas City, J40-WE-22 Turbo-Jet Engines; allocation of, request for" **B942VE11RG72.3.2NACP**

[82] BuAer Memo, February 18, 1953, "Westinghouse J-40 Engine Procurement to 13 February 1953" **B942VE11RG72.3.2NACP**

[83] WAGT Memo, May 25, 1953, "Contract NOa(s)-10385 Amendment No. 5 Our Reference WG-63300" **B938VE12RG72.3.2NACP**

[84] BuAer Contract NOa(s) 10385 Amendment 5, February 25, 1953. **B944VFRG72.3.2NACP**

[85] BuAer Memo, August 10, 1953, "J40-WE-22/22A Engine Production Schedule" **B949VE17RG72.3.2NACP**

[86] BAR Memo, September 22, 1953, "Contact NOa(s) 10385 and 52-048, J40-WE-22A Engines; Delivery of" **B949VE17RG72.3.2NACP**

87 Bar Memo, October 22, 1953, "Contract NOa(s) 10385 – Delivery of Pre Verification J40-WE-22A Engine Our Reference WG-63300" **B949VE18RG72.3.2NACP**

88 WAGT Memo, December 1, 1953, "Contract NOa(s) 10385 Initial two J40-WE-22A Engines, Operating Limits for" **B943VE15RG72.3.2NACP**

89 BAR Endorsement Memo, December 8, 1953, "Contract NOa(s) 10385 Initial two J40-WE-22A Engines, Operating Limits for" **B943VE15RG72.3.2NACP**

90 BuAer telegram, December 19, 1953, corrects WAGT error on proper oil to use in J40-WE-22A engine. **B943VE15RG72.3.2NACP**

91 WAGT Memo, February 24, 1954, "Class "C" Line Maintenance for J40-WE-22A and -22 Engines; Scope of"

92 BuAer Telegram, February 15, 1954 (approx.), Clearance of F3H-1N and J40 for Test Flying **B943VE17RG72.3.2NACP**

93 BuAer Memo, March 5, 1954, "Contract NOa(s) 10385 Westinghouse Electric Corporation, Kansas City, J40-WE-22A engine serial 302509; condition of, report on" **B943VE17RG72.3.2NACP**

94 WAGT Memo, March 3, 1954, "Excessive Engine Time on J40 & J46 Engines During Acceptance Testing" **B943VE17RG72.3.2NACP**

95 WAGT Memo, March 24, 1954, "Contract NOa(s) 10385, WEC Turbo-jet Engine Model J40-WE-22A, WE302512, 1st stage turbine blade, failure on." **B943VE17RG72.3.2NACP**

96 WAGT Memo, April 7, 1954, "Publications for Westinghouse Gas Turbine Engines – Contracts NOa(s)-10385, NOa(s-52-048, and NOa(s)-10825." **B943VE18RG72.3.2NACP**

97 WAGT Memo, April 7, 1954, "NOa(s) 10385; J40-WE-22A Serial 302509" **B943VE17RG72.3.2NACP**

98 BuAer Telegram, April 14, 1954, Loan of WE302509 to WAGT for Develop Testing **B943VE18RG72.3.2NACP**

99 BuAer Telegram, April 13, 1954, Return of WE302507 to Kansas City for Oil Leak in No. 1 bearing. **B943VE18RG72.3.2NACP**

100 BuAer PP Telegram, April 5, 1954, Acceptance and delivery of J40-WE-22A and -22 Engines Halted. **B943VE17RG72.3.2NACP**

101 BAR Telegram, April 22, 1954, Request for 5 J40-WE-22A engines. **B949VE20RG72.3.2NACP**

102 BAR Kansas City Telegram, April 7, 1954, Status of Engine and Parts Manufacture for the J40 program. **B9449VE20RG72.3.2NACP**

103 WAGT Report A-1686, "61E500-2 Adapter Gearbox Qualification Test Report" **B943VE16RG72.3.2NACP**

104 BuAer Memo, May 13, 1954, "Contract NOa(s) 10385, 150-hour qualification test of the 61E500-2 adapter gearbox for use on J40-WE-2 engines" **B944VE20RG72.3.2NACP**

105 WAGT Monthly Production Progress Report for May, 1954. **B949VE20RG72.3.2NACP**

106 WAGT Monthly Production Progress Report, June, 1954. **B949VE20RG72.3.2NACP**

107 WAGT Monthly Production Progress Report, July, 1954. **B949VE20RG72.3.2NACP**

108 WAGT Memo, June 20, 1954, "Contract NOa(s) 10385 – Westinghouse Turbojet Engine Model J40-WE-22A. Extension of Overhaul time from 150 to 180 hours of Engine Operation – Contractor's Comments on" **B949VE21RG72.3.2NACP**

109 WAGT Report A-1802, June 10, 1954, "150 Hour Full Load Parts Verification Test of the J40-WE-22 Gearbox" **B944VE20RG72.3.2NACP**

110 BAR Memo, August 18, 1954, Contract NOa(s) 10385 – Report A-1802 – "150 Hour Full Load Parts Verification Test of the J40-WE-22 Gearbox, Submittal of" **B944VE20RG72.3.2NACP**

111 WAGT Report A-1806, August 3, 1955, "150 Hour Verification Test of a Lycoming-Spencer Manufactured J40-WE-22, Accessory Gearbox Part Number 58J800-15" **B949VE21RG72.3.2NACP**

112 BAR Telegram, June 30, 1954, Shipment of WE302519 by truck to McDonnell in St. Louis **B944VE20RG72.3.2NACP**

113 WAGT Memo, July 22, 1954, "Contract NOa(s) 10385 Westinghouse Turbo-Jet Engine Model J40-WE-22A – Modification of First Stage Nozzle Vanes P/N 60E13-5 to P/N 60E13-7." **B944VE20RG72.3.2NACP**

114 WAGT Memo, August 26, 1954, "Contract NOa(s) 10385; Westinghouse Electric Corporation, Kansas City, J40-WE-22 Turbo-Jet Engines, allocation of, request for" **B944VE20RG72.3.2NACP**

115 WAGT Monthly Production Progress Report, August, 1954. **B949VE20RG72.3.2NACP**

116 BuAer Memo, September 22, 1954, "Contract NOa(s) 10385, Lot XIX, Items 1 and 2, J40-WE-22 Engines; partial allocation of" **B944VE20RG72.3.2NACP**

117 WAGT Memo, September 28, 1954, "NOa(s) 10385; J40-WE-22A Engines, Control Problems Affecting Delivery of" **B944VE20RG72.3.2NACP**

118 WAGT Monthly Production Progress Report, September 1954 **B949VE20RG72.3.2NACP**

119 WAGT Memo, October 15, 1954, "Delivery of J40-WE-22A/-22 Turbojet Engines." **B949VE18RG72.3.2NACP**

120 BuAer telegram, October 22, 1954, "Wooden J40 A/B Shipping Containers, 40 ea." **B944VE20RG72.3.2NACP**

121 WAGT Memo, October 4, 1954, "Contract NOa(s) 10385 Reuse of Wooden J40 A/B Shipping Containers" **B944VE20RG72.3.2NACP**

122 BuAer Telegram, October 22, 1954, "Wooden J40 A/B Shipping Containers" **B944VE20RG72.3.2NACP**

123 BAR St. Louis Telegram, October 26, 1954, "Quick change kits, -22A/-22" **B944VE20RG72.3.2NACP**

124 WAGT Memo, October 26, 1954, "Contract NOa(s) 10385, J40-WE-22A engines; additional allocation, request for" **B944VE20RG72.3.2NACP**

125 BuAer Telegram, October 27, 1954, Shipment of WE302535 Cancelled **B944VE20RG72.3.2NACP**

126 BuAer Memo, November 18, 1954, J40-WE-22A: Turbo Jet Engine Servos 302531 and 302535 **B944VE20RG72.3.2NACP**

127 BuAer Memo, December 29, 1954, "Contract NOa(s) 10385, Lot XIX, Item 2: Lot XX Items 1A, 1B and 2, J40-WE-22 Engines; partial allocation of" **B944VE20RG72.3.2NACP**

128 WAGT Monthly Production Progress Report, November, 1954. **B949VE20RG72.3.2NACP**

129 WAGT Memo, December 1, 1954, "Contract NOa(s) 10385, WEC Turbo-Jet Engine Model J40-WE-22A, WE302525 – Automatic Temperature Control – Malfunction Of." **B944VE20RG72.3.2NACP**

130 WAFT Monthly Production Progress Report, December, 1954 **B949VE21RG72.3.2NACP**

131 WAGT Monthly Production Progress Report, January, 1955 **B944VE20RG72.3.2NACP**

132 WAGT Monthly Production Progress Report, February, 1955 **B944VE20RG72.3.2NACP**

133 WAGT Monthly Production Progress Report, March, 1955 **B944VE20RG72.3.2NACP**

134 WAGT Monthly Production Progress Report, April, 1955 **B944VE20RG72.3.2NACP**

135 BuAer Telegram, May 9, 1955, Shipment of J40-WE-22A Serial WE302545 to Kansas City

136 WAGT Monthly Production Progress Report, May 1955 **B9494VE21RG72.3.2NACP**

137 BuAer Telegram, June 16, 1955, Quantity of -22A and -22 Engines left under contract. **B944VFRG72.3.2NACP**

138 WAGT Monthly Production Progress Report, June 1955 **B949VE21RG72.3.2NACP**

139 WAGT Memo, July 8, 1955, "Contract NOa(s) 10385 WEC Turbo-Jet Engine Model J40-WE-22A – Afterburner Flameholder Failures, Contractor's Comments On." **B9494VE21RG72.3.2NACP**

140 WAGT Memo, July 15, 1955, "Contract NOa(s) 10385 WEC Turbo-Jet Engine Model J40-WE-22A. Board of Inspection and Survey Yellow Sheet – Contractor's Comments On." **B949VE21RG72.3.2NACP**

141 BuAer Memo, July 25, 1955, Contract NOa(s) Termination in Part Lot XX Items 2 and 3 **B949VE21RG72.3.2NACP**

142 WAGT Memo, September 13, 1955, "Price Revision for Forty J40-WE-22 Engines, Contract NOa(s) 10385" **B949VE21RG72.3.2NACP**

143 WAGT Memo, January 26, 1956, "Contract NOa(s) 10385, J40-WE-22/22A – Reliability thereof" **B949VE21RG72.3.2NACP**

144 BuAer Memo, January 9, 1956, "J40-WE-22A engine serial 302516, Disposition of, Request for" **B949VE21RG72.3.2NACP**

145 BuAer Memo, August 31, 1953, "Contract NOa(s) 52-048; Model Specification WAGT-240F which describes the J40-WE-22A-G Hot Gas Generator, Transmittal of." **B3VE9RG72.4.3NACP**

Chapter 10

XJ40-WE-10 and XJ40-WE-12 (XJ40-WE-3) Procurement and Development

Procurement

Improved higher thrust J40's were proposed and accepted for development in 1948 as the WAGT model X40E8A-2. These engines became the prime focus for airframe manufacturers. The earlier models (-6/-8) were now relegated to the role of experimental airframe testing until the later higher power models became available. The basic non-afterburning engine was designated J40-WE-12 by BuAer and the afterburning version the J40-WE-10. The specification assigned was WAGT-X40E8A-2, submitted November 26, 1949. Initially the original engines as bid offered only modest power increases over the -6 and -8 versions along with some improvement in specific fuel consumption. A study continued after the initial proposal and in later communications WAGT offered even larger thrust growth, albeit with a longer development schedule. In December, 1948 BuAer added a requirement for a 50-hour flight substantiation test for the -12 engine to expedite the engine's availability for flight test.[1]

The development history of the higher power -10/-12 engines is complex and somewhat obscure. The increased airflow and compression ratio moved the compressor design challenge beyond the simpler -6/-8 and into unknown territory. As severe problems in compressor development occurred, WAGT proposed several de-rated engines in several production "Block" designations, each block moving the production engine closer to the final -10/-12 objectives. These de-rated engine blocks were given their own model designations as -24 and -26 engines. The de-rated engines are covered in Chapter 11.

It is likely that the original -12 and/or the revised -12a proposed engines were also discussed with the Air Force as a possible "-3" version of the J40, but no data was found referencing a -3 model in the extant files. In support of the original -12 engines being the -3, the later de-rated Block II and Block III versions of the -12 were referred to in several early WAGT documents as the J40-WE-5. No official BuAer documentation approving the J40-WE-5 designation was found.

The -10/-12 engines as initially proposed would have had a fourteen stage single spool axial compressor, annular combustion chamber and two stage turbine, the first stage turbine guide vanes being variable. The variable guide vanes on the first stage turbine (first prototype -12 only) were included to enable the engine to avoid off-design stalls.[2] The exhaust nozzle remained an eyelid type design. In May of 1949, the specification was formally upgraded to provide more power. BuAer also increased the ram pressure ratio to 1.89 (from 1.7) and the operating altitude to 60,000 ft (from 50,000 ft). The last two changes came out of a study for the J46 program.[3] Both models were to be installationally interchangeable with their lower power -6 and -8 equivalents. Those earlier engine designs were later changed accordingly to achieve interchangeability, particularly the -8/-22 length.[4] The development delay to achieve the newly offered increased performance was estimated at two months and BuAer accepted the delay. The memo clearly states that the intent was to upgrade the production F3H, F10F and F4D to the -10 version. The long range bomber designs being developed by Douglas and Curtis Wright were to use the -12. Thus it can be stated definitively that these two engine models were the ones sought to power the coming next generation of fighters and bombers for the Navy. The airframe manufacturers began to design to the higher levels of thrust and improved fuel consumption numbers.

The actual specification written in November 1949 states the engine would be developed to MIL-E-5007 General Specification for Aircraft Turbo-jet (dated 19 July, 1949) to the revised (above) higher performance specifications and the Qualification Test conducted under MIL-E-5009 (modified as described in the specification). The temporary designation of J40-WE-10a and J40-WE-12a for the increased power versions became just J40-WE-10 and -12 when the specification was accepted.[5] The internal WAGT model designation of X40E7A for the -12 was not used in later communications, becoming Appendix A of the specification for the X40E8A.

J40-WE-10, -12 Initial Performance Targets vs. Growth Targets				
	-10 Model X40E8-2		-12 Model X40E7-2	
	Original (-10)	Revised (-10a)	Original (-12)	Revised (-12a)
Thrust (A/B)	11,750 lbs	13,700 lbs	-	-
Thrust Military	7,920 lbs	9,275 lbs	8,100 lbs	9,500 lbs
Engine Length	266 inches	288 inches	181 inches	197 inches
Engine Weight	3,575 lbs	3,968 lbs	3,210 lbs	3,396 lbs

Model X40E8-2 Specification Performance[6]			
Ratings	Thrust (lbs)	RPM	SFC (lbs/hr/lb)
Maximum (5 min.) RPR 1.89[#,*]	13,700*	7,260	2.500
Military (30 min.) RPR 1.89	9,275	7,260	0.950
Normal (Cont.) RPR 1.79	8,330	7,260	0.893
90% Normal	7,500		0.850
75% Normal	6,250		0.805
Idle	555	3,000	-

* In afterburner
# Ram Pressure Ratio

The upgraded engine configuration remained a split intake, 14 stage single spool axial, annular combustion chamber with 2 stage reaction turbine with the exhaust nozzle a clamshell type driven by a yoke. Except for the first -12 development engine, all would have fixed first stage turbine nozzles.

The basic engine would operate up to (at least) 60,000 feet with air starts up to 45,000 feet. At sea level, the engine could function up to a peak flight ram air pressure ratio of 1.89 down to -65°F for intervals of 2 minutes each for a sum of 30 minutes (down from the original 5 minutes) during its total service life before being rejected to overhaul. Normal continuous operation up to 1.79 ram pressure ratio at sea level between -65°F and 100°F (assuming a 95% inlet duct pressure recovery) was guaranteed. It can be seen that the higher thrusts were achieved by increasing the airflow to 160 lbs/sec and allowing the engines to operate at increased working temperatures for short periods. Such operation effectively traded thermal fatigue life of the engine for

higher power settings. The net result in service use would likely be shorter periods before rejection to overhaul.

The afterburner would be capable of air starts up to 40,000 feet, with absolute operating altitude of not less than 45,000 feet.

The specification included stringent cold and hot soak conditions for satisfactory operation. Engine operation for acceleration from 40 percent military to 100% maximum thrust available would not exceed 6 seconds up to 6,000 feet. From 6,000 feet to operational altitude, from idle to 100% maximum thrust, not to exceed 20 seconds.

Maximum gas temperature for take-off would be 710°C (1,310°F) and the same for starting. Normal rating and below would be 649°C (1,200°F). Maximum allowable gas temperature never to exceed would be 816°C (1,500°F) for not more than 5 seconds and 760°C (1,400°F) the balance of the time. Other key installation metrics were:

- Weight (-10 A/B version): 3,968 lbs
- Length: Room Temp - 287.0 inches, Max Operating Temp – 289.0 inches
- Basic Diameter: Room Temp – 42.5 inches, Max Operating Temp – 42.7 inches

Simulated Flight Maneuver Loads: Able to withstand 1.5 times the static loads equivalent to the values specified in Gyroscopic Moments and Flight Maneuver Forces. The engine would be able to operate in "0" G flight maneuvers for up to 10 seconds.

Appendix A described model X40E7 for what would become the J40-WE-12, the basic engine. Its performance specifications were:

Specification X40E7-2 Specification Performance[7]			
Ratings	Thrust (lbs)	RPM	SFC (lbs/hr/lb Thrust)
Maximum	9,500	7,260	0.930
Military (30 min.)	9,500	7,260	0.930
Normal (Cont.)	8,550	7,260	0.870
90% Normal	7,700		0.820
75% Normal	6,400		0.767
Idle	560	3,000	-

- Weight (-12 non-A/B version): 3,396 lbs
- Length: Room Temp - 197.0 inches, Max Operating Temp – 198.0 inches
- Basic Diameter: Room Temp – 42.25 inches, Max Operating Temp – 42.5 inches

The contract to develop both of these engine models was NOa(s) 10067.

Development

From the beginning of development of the compressor for the -10/-12, BuAer was notified that WAGT required an upgrade to their test tunnel to provide the higher power needed to test both single and dual-spool compressors (the latter being developed under Contract NOa(s) 10123 covering the -14/-16). WAGT asked for funds to provide for the added power supplies for the tunnel.[8]

The Monthly Progress Reports (A-910, A-918, A-925) for October through December state that work was progressing on layouts and investigations with aluminum, Discaloy, titanium alloys and various types of bearings. No hardware had been ordered.[9,10,11]

A modification in the engine starting approach was requested by BuAer in a major requirement change.[12] In a memo dated January 30, 1950, BuAer noted that they had been sponsoring development of a solid propellant starter capable of cranking a J34 engine to 3,200 RPM in 2.5 seconds. The development was very promising and they had decided to develop a similar starter for the XJ40-WE-10 engines because the engines were scheduled for use in interceptor aircraft. The starter characteristics were expected to be:

1. The engine would be cranked to 3,000 RPM (idle) in 3.5 seconds.
2. The starter weight would be approximately 50 pounds.
3. The starter was estimated to fit within an envelope approximately 7.25 inches in diameter and 13 inches long.
4. The maximum torque would be close to the beginning of the starting cycle and was estimated to be on the order of 2,200 lb.-ft. at the engine rotor.
5. A 5-inch diameter bolt circle starter pad would be acceptable. The use of a 12 tooth starter jaw would be required.
6. The combustion chamber for the propellant would be remote from the starter and connect via a 1.25 inch inside diameter duct. The pressure within the duct would be on the order of 1,500 psi.
7. The exhaust gases from the starter were to be ducted overboard from the aircraft in one 4-inch inside diameter or two equivalent area ducts.
8. The present gear ratio of 1.5:1 between the starter drive and the engine rotor was preferred, but a 1:1 gear ratio would be acceptable if necessary. A 1:1 ratio would likely increase the weight and size of the starter above the figures above.

9. The starter pad would not be retained in the location that it had on the J40-WE-8 gearbox. The position of the starter pad originally on the XJ40-WE-2 engine appeared to provide the best location, provided there was access to the starter and space for the starter intake and exhaust ducts.

WAGT was requested to provide:

1. The estimated increase in engine weight, if any, required for use of a solid propellant starter.
2. The estimated time to develop an accessory gear box suitable for use with the starter.
3. A cost proposal including a cost breakdown, covering the gearbox development.

WAGT responded to the requirements for a "Fast Starter" with a memo outlining the approach they would take if agreement could be reached on the specifications by June 1:[13]

1. Redesign the -10 gearbox to include a starter pad on the front using a gear ratio of 1:1. This would be a direct drive to the engine shaft as the existing gearbox could not transmit the high torque load. The design was not expected to affect the installation envelope currently planned.
2. The electronic control of the primary system took 15 seconds to warm up prior to engine start commencing. It was planned to start the engine on the backup control and switch over to the primary once the warm up period was accomplished. The sequence would be automatic and not require pilot involvement.
3. Engine weight would go up about 10 pounds.
4. Development to certification of parts would require 18 months from authority to proceed being received; the starter could be included in the -10 type test engine if the authority to proceed on both the qualifications of the -10 and -12 engines and the starter were received by June 1, 1950. (The -12 would not have the starter available for its 50 hour flight certification test, which would precede the -10 test.)
5. The price of the cost-plus-fixed-fee contract would be $246,227.00 under the current contract and $190,343 if developed under MIL-E-5007.
6. If the mock-up of the gearbox with solid propellant starter was to be included in the -10/-12 mockup, a delay of the mock-up review would be required to allow for the expected two months of mock-up preparation time required.

Consolidated Vultee responded to a survey concerning the removal of a forward combustion chamber drain system dump valve by recommending it be retained. They were planning to use the -10 in the "Skate" aircraft (Contract NOa(s) 10507) and noted that in the beaching gear position, the engine was in a nose down attitude of approximately 1.5 degrees. Also, in an air restart, it was likely the aircraft would be in a nose down attitude and a false start might load the engine with fuel, causing fire and explosive hazards. This response was endorsed by the BAR.[14] WAGT determined in May that a dump valve was necessary to prevent fuel from oozing from nozzles after shut down.[15]

The February and March progress reports noted that work continued on layout and design of various elements. Test work was being accomplished under various engineering projects which had been set up and were under way.[16,17] The April report showed that layout and design work continued and was progressing satisfactorily. Some purchasing and manufacturing on a minor scale had started, along with development testing on a small scale.[18]

Supporting justification for the change from the AN to MIL specifications increased the total cost of the -10/-12 development at that point to $3,329,440. This included the increase of $731,341 for the change in specification to MIL-E-5007.[19] The increase in contract price and delivery schedules was handled in Amendment 1 to the contract.[20]

The first installation drawings were sent out March 13, noting the major changes were:

1. A revised, larger diameter compressor
2. A revised, shorter combustion chamber
3. A revised, larger diameter turbine
4. A revised afterburner outer shell
5. Relocated engine mounts.

Two possible engine mount locations were provided with comments requested. One set was interchangeable with the current -8 mounts and the other set was designed for 40G crash loadings as well as maneuver loads. The 40G mounts had been moved forward to provide the additional material and to accommodate heavier aircraft mating structures. Only one set would be provided on the engine.[21]

The engine performance curves for specification WAGT-X408A-2 for models XJ40-WE-10 and -12 were approved and distributed in June.[22]

The May 1950 progress report showed that a good majority of the progress was still in the design and layout stages. The -10 mock-up under item 2 had been started with assembly underway.[23]

In June, BuAer wrote a long memo to WAGT on the subject of control systems for variable area exhaust nozzle turbo-jet engines. They acknowledged that such controls were of an order of magnitude more difficult to develop than engines of the past. The testing of the experimental electronic control on the J34-WE-38 at BuAer's AEL was proving the point. BuAer did not want to be committed to only one type of control for all variable area J34, J46 and J40 engines. Immediate action was to be initiated. The contract for the -10 and -12 was under consideration to be modified to allow the development of an alternate control system. The system would be basically interchangeable with the current WAGT design. In order to take advantage of all possible solutions, a separate contractor or WAGT subcontractor was desired to handle the design, as this would minimize additional work load on the engine contractor. Alternate electronic designs were not ruled out, but any design should not include components inherently sensitive to wide variation in ambient temperature, pressure, and power supply voltage or extreme conditions of vibration and shock loading. The memo supplied the names of companies that had done direct contract work with BuAer which should have been applicable to the proposed alternative control program:

1. Chandler Evans Division of Niles-Bement, Pond Company
2. Holley Carburetor Company
3. Manning, Maxwell and Moore, Inc.
4. Simmonds Aerocessories, Inc.

The alternative controls system should include all fuel handling components from, but not including, the pump discharge to the fuel manifolds and should include the exhaust nozzle actuator but not the power source for the actuator. The system should be designed to meet the requirements of Specifications MIL-E-5007, 8, 9, and 10 and complete the required tests. The development would be in five phases with the next phase not to be started until the prior phase was successfully completed and BuAer approval received. Monthly progress reports were to be provided. The two final designs were to be delivered complete to AEL for testing. An informal proposal was requested to be submitted to BuAer by November 1, 1950 for extension of the current contract to include the alternate control program described.[24]

Report A-974 showing the calculations and results for the bearings required for the -10 and -12 shows that the bearing diameter could be the same as the -6 engine for both the No. 1 and No. 2 bearings. This report shows for the first time that the compressor now had only 13 stages and had a design compression ratio of about 6.00.[25]

In July, BuAer reminded WAGT that they had earlier been advised that the generator drive requirement for the -10/-12 engines was the same and requested that WAGT revise the specification WAGT-X40E8A-2A dated June 7, 1950 to reflect that the generator drive type should be type XII-E with a continuous torque rating of 600 pound-inches.[26]

The June Monthly Progress Report showed that progress on design, component testing and materials was continuing. The mock up was expected to be available for inspection the week of September 11. Overall progress was not considered good, with the mock-up being approximately 9 months late. The possibility of having an actual engine for initial evaluation by the end of the year was considered remote. This was considered to impact the fifty hour (-12) engine scheduled for completion April 30, 1951.[27]

WAGT published new delivery dates for the various contract items at the beginning of September.[28]

Adjusted Contract Deliverable Dates as of September 1, 1950		
Contract Item	Description	New Delivery Date
1	Engineering design study	10/31/1950
2	-10 Mockup review	09/30/1950
3	-10 Mockup delivery	11/30/1950
4	-10 150-hour qualification test and report	03/31/1952
5	-10 150-hour engine	03/31/1952
6a	-10 final design drawings	04/30/1952
6b	-10 engine operating instructions	04/30/1952
6c	-10 material and process specifications	04/30/1952
7c	-10 final summary report	04/30/1952
8	-12 conversion mockup parts review	09/30/1950
8	-12 conversion mockup parts delivery	11/30/1950
9	-12 prototype engine evaluation test report	04/30/1951
10	(Item cancelled)	N/A
11	-12 50-hour test and report	09/30/1951
12	-12 50-hour engine	09/30/1951
13a	-12 final design drawings	10/31/1951
13b	-12 engine operating instructions	10/31/1951
13c	-12 material and process specifications	10/31/1951
14c	-12 final summary report	04/30/1952

The August activity report showed steady improvement on the contract progress to the satisfaction of the BAR. Many accessories (not named) common to the -8 and -10 engines had been ordered from the suppliers. The main engine components such as the compressor, turbine, combustor, diffuser, blades, discs, gearbox and bearing supports were in manufacturing or placed on order.[29]

Dissatisfaction with the WAGT monthly status reports content was expressed in late September. BuAer noted that the reports through July only covered the -12 engine. Parts unique to the -10 were absent. No status reporting had been received on the afterburner cooling system design, afterburner exhaust nozzle design, and XJ40-WE-10 engine weight.

They requested detailed information on:

1. The high pressure oil pump installation – pump to be used, space available for the unit at the location intended and method of cooling.
2. The variable turbine nozzle system planned for the first XJ40-WE-12 – mechanical design, linkage and description of the control system. (Only the first XJ40-WE-12 was approved to have the variable turbine nozzles until a test proved such a system was necessary and that other means, such as variable compressor stators, could not achieve the same component matching.)

Beginning in future reports, a separate section covering the -10 components was requested.[30]

Mock Up Review

The mock up review board for both the J40-WE-10 and J40-WE-12 was held at the Essington, PA location September 11 and 12, 1950. The physical description of the engine now reflected the change to the 13 stage single spool axial compressor. Fuel was to be JP-4 or 100/130 avgas as an alternate. The overall impression of the mock-up was considered very good. The summary comments were:

A. Various changes on the accessory gearbox arrangement were requested and it was determined that further study of the accessory drive questions by BuAer was required.
B. The routing and support of piping was unsatisfactory and six "rules" for piping were drawn out.
C. A tentative decision was made to provide a GFE AND 20006 Type XVI D pad mounted adapter gearbox to the airframe contractors.
D. Another mock up review board of the accessory gearbox configuration was planned once coordination was completed.
E. Mock-up changes 2, 11, 27, 33 and 57 were considered to have possible effects on engine weight.

F. Additional action on some items would be necessary as noted in the item.

G. The mock-up(s) were approved subject to ultimate completion of the action items and approval of the accessory gearbox.

Performance of the engines was the same as the later first specification, but the maximum performance was listed as two minutes to a total of 30 minutes cumulative total before rejection to overhaul. A later specification change raised this to 5 minutes with a 30 minute cumulative total before overhaul.

Mock-up Board Daily Summary Items were:

Item 1 – (not listed in report)

Item 2 - Investigate various methods of actuating afterburning cooling air flow valves in ambient temperature of 250°F including hydraulic, pneumatic, electrical

A/B operation should be inhibited if cooling air is not present and cockpit indicators if A/B cooling is not adequate.
Action: A WAGT study to be submitted.

Item 3 – (not listed in report)

Item 4 - Provide cockpit indicators of A/B cooling motor valves position
Action: Include in study for Item 2

Item 5 - Actuate A/B cooling air valve with afterburner fuel pressure instead of current electrical actuation.
Action: Include in study for Item 2

Item 6 - Provide more adequate and positive support for fuel and oil lines.

Reroute or support ignition cable to prevent chafing against flexible hoses.

Following rules to apply:

1. There must be no interference between control rods or moving parts and tubing or flex hose.
2. Where lines come together they must be bundled together or separated by clamps.
3. Lines must not come in contact with sharp corners of any parts of the engine.
4. Electrical conduits shall not be bundled together with fuel and oil lines.

---

5. Metal and hose lines shall not be bundled together.
6. Support all lines and hoses to prevent chafing and possible failure.

Action: Change mock-up to show rerouting and supports in compliance with the above "rules".

Item 7 - Redesign piping and electrical routing to eliminate running a group of each through a common clamp and if possible route electrical cabling well clear of fuel and oil hoses.
Action: Redesign mock-up

Item 8 - Eliminate interference between fluid conduits and throttle control linkage to afterburner regulator

Make the linkage itself more substantial and provide protection against inadvertent damage or misalignment by people working on the engine.
Action: Redesign mock-up

Item 9 - Support the electrical conduit to exhaust nozzle actuator adequately.
Action: See Item 6

Item 10 - Improve support of starting fuel solenoid.
Action: Redesign mock-up.

Item 11 - Relocate or redesign all brackets, clips, rods, lines, aft of the firewall to clear engine shell by ¾" minimum and alter the design to accommodate installation of a 5/8" insulating blanket.

Design and fabricate a prototype blanket.
Action: Change the mock-up, blanket under lines at #2 bearing support to be provided. Anti-icing lines and valves not to be redesigned to clear engine by ¾".

Coordinate with blanket vendor

Item 12 - Provide "customer connection" to give thrust indication
Action: BuAer to advise WAGT of type of connection required when info was available.

Item 13 - Request Contractor eliminate "false-start" detent requirement on throttle quadrant.
Action: Not a valid requirement from BuAer.

Item 14 - Thermocouple harness leads to control – support adequately where they join large conduit.
Action: Change mock-up

Item 15 – Eliminate Resistoflex hose elbow couplings from all applications on the engine
Action: To be re-reviewed after services coordination of standard couplings.

Item 16 - Provide shield over openings for Turbine Nozzle pushrod levers in A/B diffuser section.
Action: Change mock-up

Item 17 - Relocate spark plugs so as to be confined in an area 20° below and 60° above the horizontal centerline.
Action: Rejected by mock-up board.

Item 18 - On forward end mount recommend the two studs on centerline of engine be located on two aft corners of pad
Action: Study use of tapped hole or Helicoil type insert and cap screws in place of centerline studs.

Item 19 - Replace solid propellant starter with a type offering fewer ducting problems. Present design requires exhaust duct 6" in diameter.
Action: Rejected. Not a mock-up board cognizance item

Item 20 - Provide clearance envelope in accordance with AND10305 for Type XII-E D.C. generator drive on aft side of gearbox.
Action: Rejected. To be reviewed in conjunction with Item 23.

Item 21 - The DC generator pad on the accessory gearbox is not exactly per AND 10305 specification (clearance envelope).
Action: Rejected. Action to be taken contingent on BuAer study. See item 23.

Item 22 - Provide a heat shield beneath starter.
Action: Rejected

Item 23 - Conduct study to determine the changes that will be involved in changing torque rating of AC generator drive from 2,500 to 4,200 pound inches.
Action: A policy from BuAer to be forthcoming.

Item 24 - Provide additional accessory drive pads for Additional Generator and Additional low speed pads for pneumatic pump.
Action: See Item 23

Item 25 - Request BuAer review accessory drive generator speeds so as to provide generator output during taxi operation of aircraft.
Action: See Item 23

Item 26 - Recommend that engine gearbox be changed to include power take-off as follows:

  a. AND 20002 Type XII J 1 Required
  b. AND 20002 Type XII A 1 Required
  c. AND 20002 Type XII E 3 Required

Pads required by McDonnell, Grumman, Douglas.
Action: See Item 23

Item 27 - Provide bulkhead fittings at fire seal and change all AN-H-24 hose back of fire seal to stainless steel tubing or metal flexible hose.
Action: BuAer to furnish info on metal flex lines.
Action: Contractor to submit study.

Item 28 - Color code all lines and hose in accordance with AND10375.
Action: Approved. Specification to be changed.

Item 29 - Change oil tank to stainless steel for better fire resistance.
Action: Modify specification

Item 30 - Provide a better bracket to support the eyelid actuator oil filter.
Action: Change mock-up

Item 31 - Relocate exhaust nozzle actuator to reduce required clearance envelope
Action: Change mock-up

Item 32 - Insure no interference between exhaust nozzle actuator piston rod and aft air line fitting.
Action: Change mock-up

Item 33 - Request fuel system accessory drains be consolidated into a single drain connection with a straight thread.
Actions: Change mock-up

Item 34 - Clean up exhaust nozzle control and high pressure oil line tubing to decrease line length as much as possible.
Action: Change mock-up

Item 35 - Fuel nozzle inlet elbows – delete small diameter solid tubing elbow and change to banjo or other type of direct hose connection.
Action: Change mock-up

Item 36 - Redesign engine boost pump to permit air starts at a reasonable altitude, say 10-15,000 ft. with airplane boost pump inoperative. Engine boost pump should allow maximum power at sea level with airplane boost pump inoperative.
Action: Specification change needed

Item 37 - Extend operating range to reasonable values. A/B should be capable of operation at Mn.5 up to 55,000 ft. The -10 is worse than the -8 in that respect.
Action: Previously requested study on this issue underway by WAGT.

Item 38 - Provide cannon plug or similar type quick-disconnect fitting at end of thermocouple harness for replacement and inspection of thermocouple harness
Action: Change mock-up

Item 39 - Rotate anti-icing valve 180°.

Move electrical lead that is near anti-icing valve closer to engine.

Locate all lines in this region as close to the vertical centerline as practicable.
Action: First rejected, second and third change mock-up

Item 40 –

1. Spark plug ignition leads should be supported by a common bracket where the lines cross the compressor flange.
2. Fuel and oil lines should be supported at mid-compressor underside where now only a clamp is used.
3. Compressor out pressure limiter line starboard should be rerouted or supported.
4. Anti-icing line running from starboard side of engine to the "T" joint at bottom on engine should be supported.
5. Vent line from oil-filler to oil reservoir needs an additional clamp.
6. Afterburner fuel supply line and fuel lines need bracket in vicinity of reservoir drain valve.

Action: All included in Item 6

Item 41 - Provide adequate insulation around anti-icing air line forward of the fire seal where it passes fuel and oil lines and the fuel flow distributor.
Action: Change mock-up

Item 42 - Use of rubber covered clamps on anti-icing lines is objected to.
Action: Change mock-up and use high temperature material instead of rubber

Item 43 - Provide moisture proof electrical conduit connectors at all locations.
Action: Change mock-up

Item 44 - Provide splash cover over electrical junction box starboard side. Relocate junction box if a splash shield is not possible.
Action: Change mock-up

Item 45 - Provide shield or other protective arrangement to prevent oil spillage from reaching electrical junction box.
Action: See Item 44

Item 46 - Interferences on fluid lines and electrical conduits must be prevented by means other than selective orientation of end fittings.
Action: Rule #7 added to Item 6

Item 47 - Comply with Section 3.26 14.5 of specification MIL-E-5007, applying to fill fitting sizes on tanks.
Action: WAGT to comply

Item 48 - Provide for mounting of flow meter transmitter that will meter total engine flow rather than main flow only.
Action: BuAer to provide data on envelope and mounting provision when information is available.

Item 49 - Move fuel flow meter forward "X" inches to permit a flow meter on the -10 engine large enough to measure A/B fuel flow as well as engine fuel flow.
Action: See Item 48

Item 50 - Request engine to be furnished with flow meter and wiring to electrical junction box.
Action: Rejected

Item 51 - Request BuAer coordinate design of flow meter transmitter so that transmitter stays within engine fire wall diameter.
Action: BuAer to attempt to coordinate
Item 52 - Provide drain for any H_2O in low point of air inlet duct.
Action: Rejected

Item 53 - Request electrical and fuel lines be rerouted on right side of engine to provide room for fuel inlet fitting.
Action: WAGT to submit study

Item 54 - Provide for removable sump in bottom of oil tank.
Action: Rejected

Item 55 - Study the possibility of relocating the following items:

a. Ignition Unit
b. Starting fuel solenoid
c. Compressor pressure limiter

Action: Rejected

Item 56 - Request provide governor alternator and power regulator cooling air supply and required ducting from inlet air ducts to those units plus return ducts if required.
Action: Rejected

Item 57 - Increase allowable pressure differential across walls of governor alternator to permit operation with a positive pressure differential of engine max. ram pressure differential of -2 psi.
Action: WAGT to submit study

Item 58 - Install indicator graduated in degrees to provide position of power lever.
Action: Change mock-up

Item 59 - Change the following adjustments so as to make changes heretofore only possible at time of control overhaul:

a. Emergency speed governor compensation
b. Emergency acceleration fuel flow
c. Emergency minimum fuel flow
d. Power lever travel

Mark the title of the adjustments on each of the following:

a. Primary idle
b. Primary temp
c. Primary rpm
d. Emergency idle
e. Emergency max and min nozzle area
f. Emergency rpm

Action: Change mock-up

Item 60 - Reduce height of proposed turbine nozzle actuator control to stay within firewall diameter.
Action: Change mock-up

Item 61 - Recommend the overspeed relay adjustment be made more easily adjustable and more accessible if possible.
Action: Change mock-up

Item 62 - Increase allowable back pressure on anti-icing exhaust to +2 psig.
Action: Approved. Mock-up change not needed.

Figure 1 Mock-up of XJ40-WE-10 engine right side. Exhaust nozzle detail seems to reflect the change from an eyelid design to an iris type with multiple hydraulic rams, WAGT P43138, 9/19/1952. *Courtesy Hagley Museum and Library*

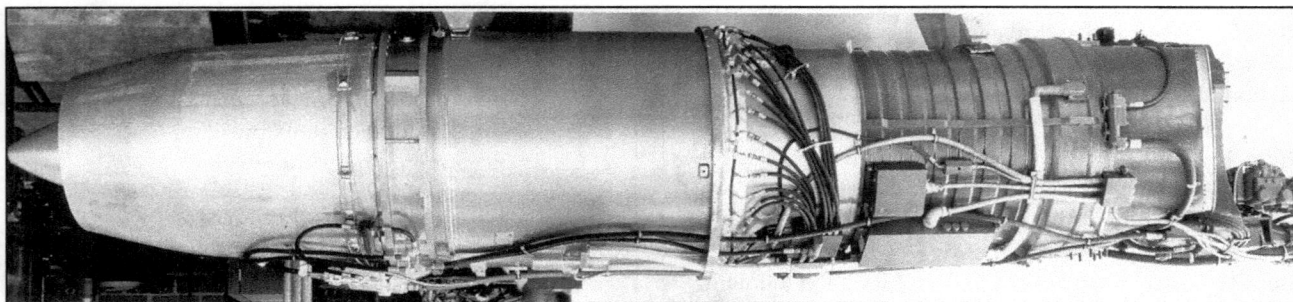

Figure 2 Mock-up of XJ40-WE-12 right side as of 9/4/1950 – shows the exhaust nozzle type to be movable center cone much like the J34 at this stage. *Courtesy Hagley Museum and Library*

The Power Plant Division of BuAer noted that the mock-up review board report Item 48 indicated that the fuel flow meter would measure the main engine fuel flow only. They commented that since the -10 version of the engine in the planned interceptors would be operating in A/B a considerable amount of the time, the fuel flow measurements should be modified to include both the main engine and A/B so as to provide the pilot with more accurate information on how much fuel was left available.[31]

Mock up board Item 27 requested that the contractor change all AN-H-24 hose back of the fireseal to stainless steel tubing or metal flexible hose. The Airborne Equipment Division was requested to furnish WAGT with various data on metal flexible hose.[32] The BAR reported in November on this issue, stating that no metal flexible hose was currently in use in airframe applications because AN-H-24 hose had been satisfactory. The Air Force had tested Chicago Metal Hose type RF-51 with unsatisfactory results under vibration and pressure test conditions. Any engine installation of metal hose had been the responsibility of the engine manufacturer.[33] This information was reported to WAGT on November 14.[34]

The September activity report summary indicated the program had improved steadily over the last three months. Overtime authorization had improved with more detail parts being released for manufacture. The program for the XJ40-WE-12 looked better than it had in the past. The accessories outlook also looked encouraging since the design was in parallel with only slight deviations to that used on the XJ40-WE-8 engine, some of which had already been evaluated. The outlook for having a -12 engine ready for initial testing during February 1951 looked promising.[35]

In late October WAGT notified that the alternate control specification had been written and distributed to the enclosed list of possible subcontractors. BuAer's response to this was that they never got a copy of the specifications for review. The specifications had been sent out to: Chandler Evans; Holley; Simmonds Aerocessories; Manning, Maxwell and Moore; Bendix, Teterboro; Bendix, South Bend;

Hamilton Standard; Marquardt Aircraft and Sperry Gyroscope Co.[36]

WAGT notified BuAer in November of their design changes to the gearbox to accommodate the constant speed drive. The drive was the same Sundstrand unit as that later used in the -22 gearbox, including the layout and governor setup. The gearbox would provide the same number of constant speed and variable speed generator pads.[37] If immediate approval was received, it would be incorporated into the -10 gearbox of the qualification engine. Engine weight increased by 75 pounds for a single constant speed transmission on each gearbox.

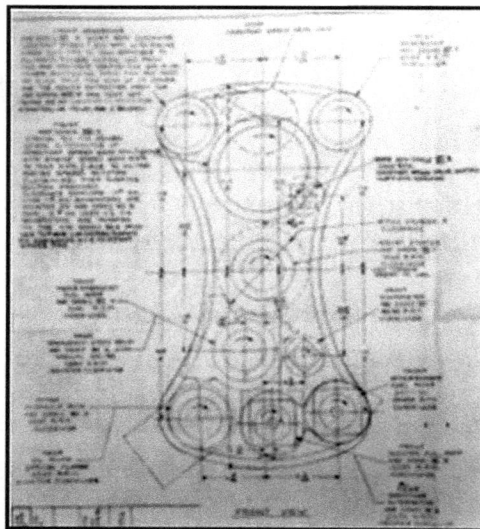

Figure 3 XJ40-WE-10 Constant speed transmission gearbox proposal outline - WAGT Drawing 62F732 [37]

The development of the modified gearbox and control system for the -10/-12 to accommodate the "fast" starter was added to the contract as Item 15 on November as Amendment #2, for a total estimated cost of $185,331.[38]

Amendment 2 was immediately followed by Amendment 3, which modified Amendment 2, Item 15 to add the requirement that the gearbox also accommodate a constant speed drive. The gearbox would be identical for both the -10 and -12 but on the -12 the constant speed drive pads would be covered with blank plates. This amendment added another $156,881 to the total cost of the contract.[39]

An earlier BuAer request to modify the gearbox to allow a different DC generator envelope clearance was made before the addition of the constant speed drive requirement had resulted in WAGT modifying the gearbox layout for the DC generator. BuAer rejected the changes due to the fact that the entire gearbox had to be modified to accommodate the constant speed generator requirement.[40]

The use of titanium in engines by several other manufacturers raised the concern that insufficient supplies of the material might arise. The -8 engine was projected to use titanium for some compressor discs to save 105 lbs and that non-availability of titanium for the prototype -10 and -12 engines was already impacting that program. WAGT was requested to "advise the Bureau as to whether the contractor had determined that titanium will be available in the quantity and at the times required for schedule delivery of the various engines currently on order, and whether a satisfactory source of supply has been established to meet possible future requirements for the material."[41]

In response to a BuAer request for the weight increase on the -10/-12 models if titanium was removed from the engine design, WAGT sent the following data in the adjacent table.[42]

WAGT suggested that the specifications for all engines be modified to indicate the maximum fuel flow for a given engine to simplify the airframe designers' work.[43]

The October progress report now changed the progress to "fair". Some items had still not been released to manufacturing. It still appeared a great deal of work remained to be accomplished on the XJ40-WE-12 to have a prototype ready by March 1951. The turbine discs were expected to be available by December 15 and the engine blades plus spares available by November 20. However, the scale model (0.3 scale) of the turbine was not going to start until December 1. This indicated that some elements involved in the prototype were not scale model tested before the parts were released. Full testing of the 0.3 model was most desirable to confirm the released design. The A/B design for the XJ40-WE-10 would follow very closely to the design of the -8. The latter design was still experiencing difficulty.[44]

As an additional update on the titanium metal issue, WAG stated that the problem had not been supply of the material, but that the forging process of the material had taken time to develop and that problem was now resolved.[45]

Substitution of Alternate Materials for Titanium in XJ40-WE-10 Engine			
Component	Titanium Weight	Substitute Material Weight	Weight Change (lbs)
Compressor Disc – 13th Stage	41.0 (PDS 9466)	67.3 (AMS 5615)	26.3
Compressor Disc – 12th Stage	42.6 (PDS 9466)	64.0 (AMS 5615)	21.4
Compressor Disc – 11th Stage	41.0 (PDS 9466)	64.2 (AMS 5615)	23.2
Compressor Disc – 10th Stage	66.9 (PDS 9466)	66.9 (PDS 6599-2)	28.9
Compressor Disc – 9th Stage	38.0 (AMS 135)*	30.1 (AMS 4135)*	-7.9
Compressor Housing Ring – 13th Stage	4.5 (PDS 9466)	7.6 (AMS 5351)	3.1
Compressor Housing Ring – 12th Stage	3.9 (PDS 9466)	6.7 (AMS 5351)	2.8
Compressor Housing Ring – 11th Stage	4.2 (PDS 9466)	7.2 (AMS 5351)	3.0
Compressor Housing Ring – 10th Stage	5.2 (PDS 9466)	8.9 (AMS 5351)	3.7
Turbine Housing Assembly	23.0 (PDS 9466)	40.0 (AMS 5651)	17.0
Turbine Housing Liner-Front	4.9 (PDS 9466)	8.5 (AMS 5651)	3.6
Turbine Housing Liner-Rear	6.5 (PDS 9466)	11.3 (AMS 5651)	4.8
Total Weight Change			129.9

* A revised aluminum alloy contour disc (heavier) was required to mate with the 9th stage titanium compressor disc.

November's progress report indicated progress had accelerated over the last month. The longest lead time item at that time was the inlet duct casting, not expected until the latter part of January. The compressor housing sample casting delivery date had moved until the end of December. The accessories area was considered to be in good shape, with most of the accessories being common to the -8. The initial -12 prototype engine was not expected to be ready for testing before March. Meeting the 50-hour qualification test completion date of the end of April would require an extreme effort. The XJ40-WE-6 model turbine test was started at the AGT laboratory but was stopped because of imbalance.[46]

In early January, WAGT surveyed the modified side oil filler and scupper as EC 63 for -10 engines and EC 63 for -12 engines. The change was identical to a change on the -6 and -8 engines. The location was outside of the current engine envelope and if not acceptable to the airframe manufacturers, they were asked to give comments on:

1. The feasibility of removing the oil filler and scupper during engine installation and replacing them in their indicated position after installation had been accomplished.
2. The feasibility of WAGT supplying the side oil filler tube only, and requiring the airframe manufacturer to supply the scupper and filler port.
3. The feasibility of retaining the present oil filler and scupper.[47]

December 1950 found the BAR reporting satisfactory performance on the program for both engines. Initial testing of the first prototype of the -12 engine was scheduled to start in May, slipping from the earlier date of April. The supplier dates for the compressor housing had slipped. The combustor-diffuser tests were not yet underway, but should begin during the week of January 22, 1951.[48]

January's monthly progress report summary indicates that work had been accelerated on the -10 and -12. A considerable amount of work needed to be accomplished if the availability date of the first -12 engine in May would be met. That date was considered "conservative" by the BAR, but the report gives the firm impression that overall he meant it to mean "aggressive" and/or "high-risk". The power regulator and combined fuel system had been through some 400 hours of testing, of which 40 hours had been on engine testing. (The latter must have been on a J34 or J40-WE-4 house engine.) Fuel valve instability was still a problem, demonstrating a high-frequency hunt. Some improvement had been attained, but the problem was still unsolved. Minor problems with the test apparatus to be used in the XJ40-WE-12 turbine model tests were noted.

The over-eagerness of the laboratory personnel to get the test started caused a total of four weeks to be lost. A -6 turbine model had been improperly balanced and test data had to be rechecked. The -12 turbine model was only partially completed. The laboratory tests of the -12 combustor with a make-up vaporizer were started with 10 hours completed, but temperature probes in the hook-up drooped over due to the high temperature and had to be redesigned. Combustor development testing was continuing. The BAR noted considerable progress had been made over the last 3 months.[49]

WAGT notified BuAer in February that high temperature wire needed to manufacture relays for the engine (and other WAGT models) was in short supply, but a supply of this wire was at the manufacturer of the relay and assigned to contract Nob(s) 46383. They requested BuShips be asked to release this wire to contract NOa(s) 10067.[50]

A request from WAGT was made in February to be able to use a test cell at AEL for the -10/-12 testing (along with other J34 and J40 models).[51] This request was not approved until a full year later, when BuAer agreed to give WAGT use of a test cell with the proviso that AEL could insert higher priority work into the schedule if necessary and that AEL would perform all the installation/de-installation and testing work on engines in the cell under WAGT supervision. All fuel use at AEL was to be documented and the cost reimbursed by WAGT.[52]

The Installation Manual for the -10 was forwarded to BuAer on February 28, 1951.[53] *(This manual was not found in the files.-Auth.)*

The progress report for February notes that although considerable progress had been made, it was unlikely the -12 would be available for testing before late May or early June. Testing by April 30 did not appear possible. The TLN-11 ignition system had proved erratic during low temperature testing in the altitude chamber and further testing had been stopped. An investigation by Bendix and Scintilla as to the cause of the problem was to be done. Turbine model tests were underway with some testing accomplished on the first stage. The second stage was expected to be available the week of March 19. Failure of the laboratory compressor was going to delay any further testing until rectified. Further, BuShips was going to use the laboratory test facility to test a CO_2 compressor before J40 testing would resume.[54]

In March, WAGT submitted their proposed amendment to the contract to cover the development of the alternate control system. A proposed contract specification was included. The subcontractor would have the following deliverables:

1. Monthly Progress Reports
2. Design Study and Report (9 months from authority to proceed)

3. Specifications (9 months after authority to proceed)

4. Initial Component Development, Bread Board Testing and Engineering Report (19 months after authority to proceed)

5. Development and Evaluation Testing of Prototype Control and Summary Report (28 months after authority to proceed)

6. Final qualification and Delivery of Two Prototype Control Systems (31 months after authority to proceed)

The schedule reflected removal of a phase by phase hard stop approval to proceed requirement before the next phase could begin. The cost was to be $1,332,608 including the fixed fee.[55] This was negotiated, approved, and became Amendment 6 of the contract.

In April, a survey was done of the planned fuel control system of the -8, -10 and -12.[56]

The activity report for March indicated that enough parts (50%) had been constructed to begin assembly of the first -12 engine in May. The power regulator tested on a J34 house engine and the XJ40-WE-8 had shown need for improvement of control during accelerations. Only slight improvement had been noted to date.

Turbine model testing had to be stopped longer than anticipated, the testway having not been available for the prior three weeks. The BuShips CO_2 compressor tests were to be run before the -10/-12 turbine testing could start again. Progress overall was considered good, although the main effort was still on the -12 engine. The -10 A/B design was still in layout. The exhaust nozzle control had shown a need for improvement and another supplier (not named) had been obtained as an alternate source of supply.[57]

The BAR summary was attached to the full WAGT report, which contained interesting details the BAR did not note in the summary. WAGT reported the -6 gearbox was to be used on the first -12 engine. The assembly of the -10/-12 gearbox was 75% complete. (*This could not be the gearbox designed to handle the "fast" starter or a constant speed transmission and was likely to be basically a -8 gearbox, but the report does not offer more detail on this point. –Auth.*) The disc cutting tools for the turbine had not been received and were delaying completion of the machining on the discs. The variable 1st stage nozzle was complete and ready for testing. The fabricated second stage nozzle was to be replaced with cast nozzles after the first engine. More interestingly, two new compressor blade angles and stator blades had been developed and a new, experimental, 12-stage constant hub diameter compressor was being built. Funding for a second experimental compressor had been obtained. This detail gives a different picture from that represented by the summary.

Report A-1096 on the qualification test of the TLN high energy surface gap ignition system was submitted for approval in April. This was designed for the -10 and -12 and ended up replacing the initial high voltage system originally designed for those engines. It offered considerably improved service life and high reliability.[58]

Full assembly of the prototype XJ40-WE-12 engine was expected in the latter part of June with development testing to begin shortly thereafter. The model turbine testing continued to be delayed, the testway still not cleared and the BuShips CO_2 compressor still awaiting test. The -10 turbine was next in line to test after that. The A/B design was still in layout.[59]

The WAGT survey of airframe mounted power regulators was rejected by BuAer, who stated they should stay on the engine unless a method could be shown how a specific installation could improve by mounting it on the airframe. This put the onus back on WAGT.[60]

Amendment 5 was approved which made a change in how the General and Administrative Expense rate would be charged for that fiscal year.[61]

EC 68 for -10 and 67 for -12 engines was surveyed in May. This changed the fuel inlet connection from 6 studs to 8 and rotated the studs to a position different from the mock-up. The face of the fuel inlet pad was lowered ½ inch over the 17 31/32 inch dimension in order to prevent choking of the fuel passage due to the new studs.[62]

In May, WAGT forwarded copies of all the subcontractors' proposals related to the alternate control system for the -10 and -12. These were forwarded with WAGT report A-1100, Preliminary Analysis of Alternate Power Control for the XJ40-WE-10 and -12.[63] The report was not found in the contract files.

The request to be able to use an AEL test cell for engine testing was reviewed with AEL management and it was determined that such testing could be accommodated if it was not of "extremely long duration". WAGT stated they usually knew two weeks to a month in advance of the need for a test cell.[64]

The monthly progress report for May stated the first -12 engine was assembled and ready for test, but a test cell would not be available until June 18. The second -12 was expected to be ready for test by the latter part of August. The first -10 was not expected to be ready until the latter part of 1951 or early 1952. Overall progress was reported as "good".[65]

In June, a survey was done of a revised auxiliary electric control. The changes made were not listed.[66] Another survey a bit later was done on an increased power regulator envelope.[67] A further survey was done on a revised electric control.[68]

A fully revised Specification WAGT-X40E8A-2C was forwarded to BuAer for review and comments on May 14,

1951. WAGT's comments in their cover letter are here merged with BuAer's response dated July 6.[69,70]

1. The requirement for the igniter leads to be able to operate at 350°F temperatures (up from the original 250°F) was deferred until testing was complete to determine if this was possible.

2. The current acceleration times from idle to full thrust were considered unacceptable. (WAGT reported they were in the 20 second range and could not be lowered until considerable testing had been accomplished. They also noted that this applied to the -12 only and that -10 times would be slightly longer due to the need to accelerate into full A/B, taking more time.)

3. BuAer desired a separate set of thermocouples be available to provide control sensing and cockpit indication. WAGT was at that time pursuing a temperature sensing approach that they had stated would eliminate the need for the pilot to have these indicators in the cockpit. A final decision on this was delayed until further proposals on the matter had been received.

4. Radio interference testing would be studied further with the contractor.

5. More study was required to determine if the maximum time of operation at 200°F (of the electrical components) could be increased from 5 to 30 minutes. Continuous operation would remain 165°F.

6. The specification included three different gearbox proposals to convert the -10 gearbox into one for the -12 and BuAer preferred number three. The three alternatives presented were:
   a. Eliminate two generator drives and leave only one generator and one hydraulic pump drive; or
   b. Convert the two generator drives to such high speeds as to make them unusable for aircraft equipment; or
   c. Provide three usable generator drives at a weight cost of 5 lbs and still permit reconverting the gearbox to the constant speed drive configuration by disassembly and replacement of the gears. (For the 50-hour type test, the "#1" gearbox layout would be used.)

7. Power lever torque was still too high, particularly on the -10 when in A/B. WAGT stated that was because the entire range of A/B had to be accommodated in only 20 degrees of throttle movement. It was planned to increase the band for A/B modulation to reduce the torque.

8. Eliminate Appendix D from the Specification, which covered the -18 model. (See Chapter 13)

EC71 for A/B air cooling lines was surveyed in early July. The design was laid out to clear the existing envelopes of airframes planning or contemplating the use of the -10 engine.[71]

The June 1951 progress report continued to reflect good progress. The first -12 had been run and a second stage turbine nozzle failure resulted in distortion of the turbine housing. The engine was awaiting a replacement of the second stage nozzle to be running again, a delay of about 3 weeks. The second engine was expected to be ready for test in late September. The redesign variable nozzle afterburner (still refers to the -12 non-A/B engine – the language is confusing – Auth.) would not be available until late August. The -10 was expected to be ready the latter part of 1951 or early 1952.[72]

Figure 4 XJ40-WE-12 #1 Second stage turbine nozzle failure, WAGT P44895, 7/19/1951. *Courtesy Hagley Museum and Library*

Technical Report A-1170 was transmitted to BuAer in August, covering the *"Interim Technical Report: Design and Development of the J40-WE-10 and -12 Combustors"*.[73] The report was not retained in the files, however the design of the basic combustor can be seen in Figure 6 to have moved to a "walking stick" type vaporizer approach.

August also found WAGT sending out a survey requesting future requirements for bleed air, noting that the increasing flight performance and increasing pressure ratio of engines made it more important to understand the volumes, temperatures and pressures required so that the bleed location on the compressor designs could be situated in the correct place and the blade designs could accommodate the down-stream conditions with the air bleed operating.[74]

The monthly progress report for July showed the 2nd stage turbine nozzle failure had been corrected through redesign and the engine was ready to test again. The new design incorporated a nine segment assembly instead of a one piece assembly. The new part was ready for incorporation by July 17, but faulty heat treatment of the 2nd stage turbine segment "shoes" held up assembly of the engine for two weeks. The second engine would be ready for test about mid-September according to WAGT, but the BAR felt a date in early October was more likely. The qualifying test of the exhaust nozzle control had been changed to a life test to enable the gathering of more data before qualifying the element for service use.[75]

Figure 5 XJ40-WE-10/-12 Fuel spray pipes into diffuser assembly feeding the walking stick evaporators, WAGT P46815, 4/25/1952. *Courtesy Hagley Museum and Library*

Figure 6 Combustion basket inner liner with walking stick evaporators attached to rear of diffuser assembly, WAGT P46813, 4/25/1952. *Courtesy Hagley Museum and Library*

BuAer notified WAGT in early September that use of the simplified control system should only be used on the XJ40-WE-8 engine until testing proved the alternative system was satisfactory. A decision on which control

system to use on the -10/-12 would be deferred without further delaying qualification tests of those engines.[76]

The August progress report showed that testing of the XJ40-WE-12 #1 was approximately 5.5 hours. Total to date on the engine was approximately 11 hours. Two failures occurred in the recent testing. The turbine out straightening vanes failed due to temperature and pressure distortion and had to be removed from the engine so that testing could continue. Two semi-circular flat plates were mounted instead. At 5.5 hours, buckling and tearing of the diffuser inner and outer splitters stopped further testing. Slight engine damage was found (6 first stage turbine blades and one second stage blade). The diffuser of the second engine was being reinforced by additional supports added to the failed section for use on this engine. The first engine was to be back on test by the end of September. The engine had not been up to top speed (7,260 rpm), the maximum achieved to-date being 6,800 rpm. The engine had not achieved its designed airflow of 160 lbs/sec as designed. The top turbine temperature had been achieved (1,500°F). Redesign of the failed elements was being considered and new designs probably used in the second development engine. The completion of the second engine was now not anticipated until November, 1951.[77]

In October, an amendment request was processed to cover cost over-runs on the -10/-12 as well as the -8 and J34 programs. The details for the overruns were not found.[78]

The September 1951 progress report began to reflect serious development problems with the -12 engine. The earlier failures were noted, along with the news of another failure on September 30 of the 2nd stage compressor disc (due, WAGT believed, to operating the compressor for prolonged periods in the stall range). Only changes to alleviate stall problems were being considered for the engine. The redesign of the turbine out straightening vanes would not be incorporated into the engine design until the 3rd or 4th development engine.[79]

WAGT requested permission in November to transmit information on the -10, -12 (and -18) engines to North American. North American was modifying bomber B-45A-5 Serial 47-049 as a test bed for flight testing the engines and it was desired they be supplied with the necessary engine data.[80] The request was approved in December.[81] *(No -10/-12 or -18 was ever flight tested.-Auth.)*

October's progress report showed only 1.95 hours of additional testing up to the date of the report (November 15) on the -12. The reason for the shutdown was not specifically mentioned in the summary, but something (not specified) needed strengthening of its supports. Testing would resume November 19. The second engine would now be available for testing in very late December.[82]

Figure 7 XJ40-WE-12 #1 Compressor failure in September 1951, WAGT P45375, 10/1/1951. *Courtesy Hagley Museum and Library*

The foregoing report raised questions at BuAer, particularly about the statement in the detail report that the high pressure oil pump was to be moved from the engine tailcone to the aft side of the gearbox. BuAer wanted to know the feasibility of installing the solid state starter in the tailcone location. Earlier conversations had indicated the original location of the high pressure oil pump was the only feasible one from the standpoint of lengths of high pressure oil lines required by a gear box location. BuAer specifically requested:

1. The degree of finality of the decision to move the pump from the tailcone and the developments that led to such a decision at such a late date,
2. Availability and accessibility of the space on the engine gearbox for mounting the pump and its required adapter gearbox,
3. Detail WAGT planning and scheduling with regard to qualification testing of the engine with the gearbox installed and with the pump in its final location.[83]

The progress report for November noted that the total engine running time on the XJ40-WE-12 had been 16.73

hours program to date. The engine was ready for testing, but one of the test cells was down for extensive muffler repairs and would not be available for testing until December 10. The current schedule for completion of the qualification test appeared unattainable and the BAR felt the earliest possible dates were September 1952 for the -12 and November 1952 for the -10.[84]

Amendment 6 was signed in December. The change moved the progress report delivery date each month.[85]

A report on the progress of Item 16 of the contract, namely the Alternate Hydraulic Control Development, was filed as report A-1298. This item had been reported as part of the comprehensive project status report, but was broken out separately beginning in December 1951. Engineering review had been completed on 3 reports and they were now being reproduced. They were:

1. A-1264 "Evolution of the Alternate Hydraulic Control"
2. A-1266 "Description of the Alternate Hydraulic Control"
3. A-1267 "Simulation of the Alternate Hydraulic Control"

The three reports were considered to be the design study for the Alternate Hydraulic Control. The basic choice of the design was made due to the excellence of its inherent failure provisions. There was no apparent single failure which would cause the pilot to lose control of the fuel flow to the engine. Only a very elementary emergency fuel system had been provided as a result. The schematic included in Report A-1266 was accurate except for a few minor changes. Opinions of BuAer and the airframe manufacturers were expected to modify the design, so modifying the detailed procurement specifications at that time were not advisable.

The Woodward Governor Company had very nearly finished the design and construction of a breadboard version of the alternative control for sea level testing on a J34. The unit used a Woodward X838 governor fitted with an external throttle valve. By varying the pressure drop across the external valve, the unit could be made to simulate the sea level static performance of the final unit with a high degree of accuracy.

The exhaust nozzle control unit for the breadboard system was designed for use with a fixed displacement pump but was expected to be a close simulation of the final system. A variable displacement pump had been planned, but delivery could not be made until the middle of December. Testing was expected to start in January. Negotiations were underway with Marquardt Aircraft Company to acquire flow nozzles for use in a test to evaluate the effect of temperature on sonic velocity. The nozzles would be endurance tested on house engines.

A report on the evaluation of the Alternate Hydraulic Control would be produced in the spring of 1952. Efforts would be made to press progress on this design so as to provide an alternative to the Simple Hydraulic Control if necessary.[86] Interestingly, report A-1267 indicated that the testing and evaluation process and procedures for engine controls needed to be significantly improved based on the initial testing on a J34 house engine, but the summary does not mention this.

WAGT had stated the -10 A/B fuel control would be ready to run on the first production -8, implying the fuel flows, etc. on both A/B's would be the same. BuAer asked for information on the specified fuel schedule, both transient and steady state for the -8 and -10 engines. The -10 control was developed to provide improved fuel scheduling at high altitudes and BuAer wanted a description of the improvement.[87] WAGT responded in February 1952 that the fuel control was mechanically identical among the -8, -22 and -10, but that the fuel schedules were different. The work underway in the laboratory at that time was to determine the proper fuel schedule for the -8 and -22.[88] Improvement of the high altitude fuel scheduling was accomplished through redesign of the throttle valve spring allowing higher throttle valve controlling pressures. The memo had the calculated fuel flow schedules for the -8 and -22 with the 675 in^2 and 726 in^2 nozzle areas. The -10 schedule was also attached. Transient A/B information for the -10 was not yet available.

EC 74 for the -10 and EC 70 for the -12 were surveyed in January. The change was a relocation of the fuel inlet port for the fuel booster pump on the engine. The change moved the pump to 18 15/32" from the horizontal centerline of the engine and lowered the inlet location vertically ½" from the previous location. Vapor performance of the pump was improved by the change.[89]

A funding issue had arisen in 1951 when the total contract value was projected by WAGT to be expended by August 1951. WAGT estimated in a letter in February of that year that $2,268,387 was required to complete the contract. No reply was received. In March, 1951, Amendment 6 was added to the contract for Lot III increasing the value of the contract by an estimated cost of $1,245,428. Westinghouse now asked if it was OK to charge activities from the earlier lots against Lot 3 funds. The cover sheet notations at BuAer show confusion over this. Contract NOa(s) 10067 was to be cancelled in full during this period (it appears it actually was not at that point), but how funding for follow-on contract activities was to continue is not clear from these notes.[90] Further communication a few days later reminded BuAer that they had been informed in October that an increased estimate of the total brought the

figure to $2,354,856. Material price increases and costing rates had driven the increase. Since then, there had been another increase in the costing rates and BuAer had been notified of that. The February estimate had been based on completion of the XJ40-WE-10 qualification by October 1953 with completion of engine development by December 1952. As of January 11, 1952, this schedule had slipped 5 months.[91]

The December, 1951 monthly progress report showed the -12 was mostly idle during the month. Testing resumed on December 27 but was stopped after 4 hours due to a failure of the turbine outlet straightening vanes. They were removed and testing continued on January 7. Total running time to date of the report was 20 hours. The compressor was troubled with low speed stall. The second engine was expected to be completed in February 1952, using a new compressor design. The estimated qualification test dates for the -10 and -12 engines were now October 1952. Production of the -12, then scheduled for starting in June 1952, would be delayed until November 1952.[92]

EC75 on relocation of the ignition spark plugs was surveyed in February. The change was made to improve ignition at higher altitudes. Four spark plug bosses were to be provided and the airframe manufacturer could use any two and insert plugs (provided) in the other two, depending on airframe accessibility. The ignition leads provided would be long enough to connect to any of the four boss locations.[93]

The January 1952 monthly activity report for the -10 and -12 became available in late February and showed that test time on -12 #1 for the month was 35 hours, total time now up to 56 hours. All compressor configurations tested to date were still troubled with low speed stall. Mechanical problems such as failure of the seal strip in the 2nd stage turbine nozzle, buckling and cracking of the exhaust collector and distortion of the vaporizing plate also occurred. More seriously, cracks were found in the base of the blade roots of the 4th stage compressor disc. It had been sectioned and a deep investigation was underway. The second development engine would be available for test during the first week of March. The qualification test completion of December 1952 still appeared to be attainable.[94]

The control system design was of concern to the airframe contractors and they requested the status, concerned that changes were being made without being surveyed first for their inputs. WAGT responded that the final decision had not been made yet. Only some components of the electronic control, definitely not being used in any new design, were being removed at that time. They were the Power Regulator, Governor Alternator and Compressor Pressure Limiter. Airframe manufacturers were asked not to substitute auxiliaries and/or that equipment not be substituted in place of equipment removed by WAGT during the change-over in design. All changes of major importance would be surveyed by WAGT using the established survey system.[95]

McDonnell sought answers to many engine installation design issues at this time and BuAer found that they were not able to respond from their own knowledge and forwarded many of the items on to WAGT for responses and clarifications. A summary of these items with the WAGT response[96] follows:

1. Could 400 cycle power from the AC bus be fed directly to the engine junction box without controlling it from the master engine switch? If the answer was no, an additional relay would have to be installed.
   WAGT: Yes, 400 cycle power could go to the junction box. It was still recommended that a separate emergency switch be incorporated for the Automatic Temperature Control power circuit.

2. The present 250 VA inverter could not supply the 460 VA required for air starts. Could the engine start with 205 VA? If no, another larger inverter would have to be used resulting "in significant weight penalty.
   WAGT: Airframe AC power was required for the Exhaust Nozzle Control during all operations including starting. Normal operation would require only very low currents but the exact value had not yet been determined. It was expected that 50 VA would be adequate. An emergency switch for the Temperature Control was recommended.

3. Was the pump switch shown on drawing EDS-L-214132 momentary? BuAer added: If it was momentary, *"will the pump return to primary operation when the switch is released?"*
   WAGT: The switch was momentary and returned to primary pump operation when released. All J40 engines would comply with the standard three-position switch specified for fuel pump selection and indicating as follows:
   a. When the switch was in the "Automatic" position, in the event of a failure of either the primary or emergency fuel pump, the remaining operable pump was automatically selected.
   b. If the automatic selection of the emergency fuel pump occurred, the circuit to the cockpit warning light was closed (the light came "on"). On the -10/-12 with parallel primary and emergency pumps operating at low

altitude, the cockpit warning light would come on if either pump failed.
   c. The "check" position on the switch was not there on the -10 engine (unlike the -8/-22), as there was no need for it since failure of either pump closed the cockpit warning light circuit.

4. Were any of the following 2-second surge currents simultaneous? If not, what was the maximum surge at any one time?
   a. 5 amps. For A/B bleed valve
   b. 20 amps. For A/B control valve
   c. 5 amps. For A/B shutoff valve

   WAGT: This did not apply to the -10.
   The present loads required large circuit breakers and heavy wires to every circuit through one pole of the master switch. Did the junction box have a relay that could be used as a power load for the A/B loads?
   WAGT: The junction box had no provision for A/B loads. What type was the torch ignition valve and what current did it require?
   WAGT: The -10 used a "Saval" torch ignition valve, (Not described). Current requirements were 0.2A steady state with 5.0A peak for engine starting only.

5. When will firm dates on the -8/-22 specifications, performance curves and performance correction factor curves be presented to the BAR in Essington?
   WAGT: All curves available had already been sent. The correction factor curves would be available on March 15.

6. What change was affected when the inlet duct contours change was cancelled? BuAer was not aware of the nature of the change that was proposed. Clarification was requested.
   WAGT: The proposed change was to switch to a pure elliptical shape to ease manufacture. This change was proposed by the Mercury-Lincoln Division of Ford Motor Company, but the survey showed all airframe manufacturers were opposed to the change except McDonnell.

7. Please inform BuAer of firm dates when -8, -22 and -10 exhaust nozzle drawings and surveys would be presented to the BAR in Essington.
   WAGT: A letter had been presented to the BAR at Essington that constituted the survey of exhaust nozzle design configurations for all the engines.

8. What is WAGT's response in connection to resolving the deficiency of the lack of a discharge pressure limiter on the simplified control system?

WAGT: This would be handled in separate correspondence.

9. Confirm if the engine control could tolerate the airframe voltage limits. A copy of the McDonnell curve of airplane line voltage vs. response time was forwarded, as it appeared a copy had never been sent to WAGT.

WAGT: The AC power specifications contained in report A-1265 dated 10 Jan. 1952. WAGT would have to test the power control for proper functioning with the power levels described in airframe power limits that had been forwarded, particularly:

    a. Control accuracy is in accordance with applicable model specification,

    b. Extreme limits of power variation permitting minimum acceptable control operation,

    c. Excessive power variations which would cause control malfunction, and

    d. Destructive power variations.

WAGT reminded everyone that any engine using the Simple Hydraulic Control System required AC power at all times for normal operation. Also, on the -10, loss of AC power for any reason caused complete loss of exhaust nozzle control requiring emergency use of the Manual Override Exhaust Nozzle Control. This was an emergency situation requiring the pilot to take immediate action.

The above queries give excellent insight into the challenges all parties were encountering as they attempted to match 3 different models of the engine into the same airframe while the airframe and the engine models were all in development and experiencing rapid and constant design changes for various reasons.[97]

In March, Amendment 8 changed the contract to point to specification WAGT-X40E8A-2C instead of WAGT-X40E8A-2A, and added language regarding the constant speed transmission and fast starter. (Amendment 7 was not located and content is unknown.-Auth.) Other minor language changes were made to bring the rest of the contract into alignment with those main changes. The "50-hour flight substantiation" guarantee clause was removed. Interestingly, a note on the circulation sheet indicates that at one time the XJ40-WE-10/-12 was to be re-designated the "XJ50" but in the time frame that Amendment 8 was being approved the re-designation action was cancelled.[98]

The new designations were to acknowledge that the engine models involved were technically very different from the original XJ40 designs. It is also likely that with the entire program in serious schedule trouble and few of the earlier model engines likely ever to be built, WAGT was attempting to disassociate the later models from the earlier program problems. After review, the change in model numbers was rejected by BuAer because of the administrative implementation burden it would impose on the program.[99]

The WAGT proposed new designations were to be:

Current	Proposed
XJ40-WE-10	XJ50-WE-2
XJ40-WE-12	XJ50-WE-4
XJ40-WE-14	None (Cancelled Model)
XJ40-WE-16	None (Cancelled Model)
XJ40-WE-18	XJ50-WE-6
XJ40-WE-20	XJ50-WE-8

In early March, WAGT had notified BuAer of the need to increase the exhaust nozzle area from 675 in² to 726 in² and BuAer responded with a wire asking for information on this very late change and what effect it would have on performance.[100] See the -8 development for the discussion of this change.

The February progress report showed the XJ40-WE-12 #1 had not accumulated any more test hours than reported the previous month (56.12 hours). The engine had been in teardown and at the time of the report was being tested before delivery to NACA. The second engine would be available March 22. Low speed compressor stall was still a problem, although progress had been made.[101]

WAGT's acceptance of the AEL test cell use conditions was forwarded in March, but the receipt in the Power Plant Division contracts office was delayed by two weeks, as the endorsement letter was erroneously stamped "BRICK-BAT", routing it through the "BRICK-BAT" process office before it went into the correct channel. The BuAer Power Plants Division office immediately filled out the proper paperwork to set up the contract between WAGT and AEL authorizing the use of test cell 4E.[102]

The specifications for the -12 were updated in March to refer to the use of JP-4 (MIL-F-5161B Referee Fuel) throughout the document. This fuel and MIL-L-7808 oil were now specified for all acceptance testing. Avgas and JP-3 remained alternative fuels for service use.[103] The change in the oil was required because of the better high temperature operating characteristics of the MIL-L-7808 oil which was required when JP-4 was used in a liquid to liquid intercooler. WAGT noted also that the performance characteristics of the engine might have to change again once the testing with JP-4 results became known.

Amendment 9 to the contract was processed, removing the requirement for a 50-hour flight substantiation test and modifying the language throughout the contract for the engine to be "in overhauled condition and in assembled form" and instead of "50-hour flight substantiation" stating "modified 150-hour."[104]

In response to the WAGT notification that the size of the engine nozzle needed to be increased, BuAer notified the BAR at Essington that they did not want different ("unique") nozzle sizes for each airframe, but expected all -10 engines for Douglas to be the same, and ultimately all -10's to have one nozzle size, less than 750 in². [105]

WAGT had delivered in February the three previously mentioned reports (A-1260, A-1264 and A-1267) regarding the Alternate Hydraulic Control, stating together they constituted the Design Study deliverable required by contract Lot III, Item 16. They noted that when the deliverable study began, they had been engaged in design of an electronic control system for the -10/-12 and since that time the Simple Hydraulic Control had been developed for use on the -8, -22 and J46-WE-3, -8 and -10 engines. A report was in preparation comparing the Alternate Hydraulic Control to the Simple Hydraulic Control with the expected report to be available by May 15, 1952. Since it appeared that there was great similarity between the expected performance of the subject control and the Simple Hydraulic Control, there appeared to be some question as to the direction which should be taken in the development of the subject control system. They felt justified in delaying issuing the detailed procurement specifications required by Lot II, Item 16 until a thorough review of the entire program had been made. The detailed procurement specifications would not be issued until BuAer and WAGT had reached a "mutually satisfactory agreement on the most advantageous way of meeting the requirements for the subject control in light of recent developments (unspecified – Auth.). Acceptance of this program approach was requested from BuAer. [106]

In a long response following a conference with WAGT on March 18, BuAer accepted the three reports but rejected the proposed approach in developing and release of the detailed specifications for the Alternate Hydraulic Control. Their main reason was that it would further delay development, which was very late in the development cycle for the engine at that point. Further engineering and design reasoning points were made to buttress their feelings that the Alternate Hydraulic Control was needed to supplant the Simple Hydraulic Control:

1. The SHC metered fuel only as a function of compressor pressure rise and power lever position. Lacking flight or estimated engine performance data, the ability of the SHC to provide satisfactory fuel metering over the extremes of pressure and temperature conditions of flight when using a fuel with a broad density range was seriously questioned. The AHC using the Woodward X838 type control under development had the means to independently vary fuel metering schedules as functions of power lever position, air temperature, air pressure and engine RPM, thus allowing much greater flexibility of adjustment in fuel flow scheduling. The control would automatically compensate for fuel density variations. The X838 had been successfully demonstrated in flight. The foregoing factors indicated the developing control would have a much better probability of meeting the as yet unestablished fuel metering requirements for the -10 engine than did the SHC.

2. The AHC eliminated the need for the afterburner blow-out switch required with the SHC, the switch being considered relatively unreliable.

3. The hydraulic exhaust nozzle governing a portion of the AHC was considered inherently more reliable than the electrical nozzle positioning system being used with the SHC.

4. The lack of a satisfactory control system that would be an alternative to the electronic control for the J34-WE-32 and -38 engines had had serious consequences. It also had seriously delayed the completion of the development of the J46-WE-8 and -32 engines. Only the development of the Manning, Maxwell, and Moore temperature amplifier had reduced the delays.

5. The AHC program was put in place to ensure the success of the J40-WE-10 and -12 engines. If the results were not obtained until late in the production schedule, considerable money and effort would have been wasted.

6. The contract called for delivery of the design study report and specification in December 1951. The study report submitted was dated 5 February, 1952 and stated the specification would be submitted 15 May. It was expected that changes of delivery dates would be sent notifying BuAer prior to the date when any item would be delayed and not be notified after the fact.

BuAer accepted the three submitted documents as satisfactorily completing the design study report required under Lot III, Item 16 subject to WAGT's compliance with the requirements that:

1. The comparison of the AHC system with the SHC system should not be conducted at the expense of delay in development of the AHC.

2. Provision for automatic compressor pressure limiting by a single and reliable means was not provided in the AHC design. Pressure limiting by reducing RPM and/or fuel flow with something

less than an optimum relation between these two variables would be acceptable if considerable increase in control simplicity was achieved over a design which retained the optimum relationship.

3. The sea level only incomplete stability study should be extended with AEL facility help to examine flight conditions.

It was considered that the evolving AHC control showed promise of considerable merit and they commended WAGT on the technical results from the program to date.[107]

The March progress report stated the C-19 compressor was assembled into the XJ40-WE-12 #2 and testing began. At three hours the outer shroud of the exhaust collector had numerous cracks, which were repaired and then testing resumed. Part-load stall was still occurring in the first seven stages and changes were made with testing resuming on April 10. After another 2.5 hours of testing the engine was wrecked and completely destroyed. The cause was under investigation. The #1 engine was at NACA for development test work. Total -12 development engine time was approximately 62 hours at that point.[108]

Figure 8 XJ40-WE-12 #2 Wrecked in the test cell after compressor failure, WAGT P46655, 4/11/1952. *Courtesy Hagley Museum and Library*

The discussion of relocating the fast starter in the engine tailcone (as reported in the October 1951 progress report) and movement of the high pressure oil pump to the gearbox from the tail cone housing caused WAGT to write a long memo on the elements affecting a decision:[109]

1. The high-pressure oil pump relocation reasons were:
   a. The availability of a pad on the aft side of the gearbox vacated by the governor alternator which was not required for the SHC system.
   b. Anticipated development problems involved in

obtaining a pump to operate in the high temperature stream immediately downstream from the turbine.
   c. The difficulty of introducing a mechanical linkage to control the pump in the position aft of the turbine, a difficulty that did not exist when the pump was controlled electrically.
   d. The pump when located on the gearbox was in a more accessible location for service.
   e. The problem of long high pressure lines from the pump was considered minor in comparison to the development problems on the pump if located in the rear hub.

2. The installation design change was regarded as minor since the pump would be on the aft side of the gearbox and would not affect the external envelope of the engine. The oil lines would also be well within the envelope. The relocated pump with its adapter gearbox had been mocked up and found to fit properly in the space available.

3. Gearbox related:

   a. Detailed drawings of the adapter gearbox were complete and were to be issued.
   b. The finished gearbox would be available in September 1952.
   c. The high-pressure oil pump would be available in June 1952.
   d. The engine qualification test date would not be affected by the change.

4. The decision to relocate the pump was considered final for the reasons given and also because the design of the rear bearing support was simplified and no significant engine envelope change was involved.

The BAR noted that the above memo did not fully cover the problems. Specifically, it did not treat the subject of the feasibility of locating a solid propellant starter in the position of the turbine vacated by the high pressure pump. The BAR requested of BuAer that any information available on the starter and its specifications be forwarded to his attention.[110]

BuAer responded in a Speed Letter stating the possibility that the governor alternator might be required on the -10 engine and that they no longer desired further consideration be given at that time to locating the solid-propellant starter in the tailcone. A survey on the pump location change of the airframe manufacturers was recommended.[111]

The April progress report indicated the #2 -12 cause of failure had still not been determined. The #3 engine would be available in June. These delays were seriously affecting the test plan. The #1 engine was requested to be returned by NACA and was then in a disassembled state. It would be reassembled with a C-17 compressor, which differed only slightly from the C-16. (*Technical details on differences between the various compressors was not found. Auth.*) Only 2.5 hours of testing had been accomplished in April, with total test time standing at 64 hours. It was estimated by the BAR that an additional 300-400 hours of engine running would be needed prior to qualification testing in December. Based on the rate of progress to date, he felt the qualification test would be considerably behind schedule.[112]

BuAer wrote a response to the April report, particularly focusing on the control system discussion. The Auxiliary section had referred to work on the Auxiliary Electric Control and Manual Temperature Trim Control. Since then, WAGT had assured representatives that the planned control for the -10/-12 would be essentially the same as the configuration planned for the J40-WE-22 engine subsequent to June 30, 1953 production. BuAer requested submission of a single schematic drawing showing all subcomponents of the control system with their proper interconnections. Things such as the fuel pumps, flow divider, fuel manifold, oil pumps or the ignition system needed to be included in the drawing. BuAer suggested the control system be given a type number, as it was cumbersome to keep referring to a specific control system with descriptions such as "Simplified Control System for the J46-WE-8 engine with electrical exhaust nozzle positioning and manual two-position over-ride".[113]

In June, in answer to Mock-up Board Items 20 and 21, WAGT requested approval of their actions. Subsequent to the action taken to comply with Item 20, a new gearbox was designed to use a constant speed drive. The space aft of the gearbox which was used for the DC generator was found adequate for the constant speed drive unit. The DC generator was moved to a new pad on the front of the gearbox where a clearance envelope, in accordance with AND 10305, was provided. The gearbox was moved aft ¼ inch from the position shown on the Mock-up.[114]

WAGT completed report A-1207 (*not located – Auth.*) on research into possible A/B cooling valves requested by Mock-up Board Items, 2,3,4 and 5 and submitted it in early July. Concurrence with the results was requested.[115] BuAer agreed with the recommendation of using pneumatically-actuated valves with a fuel-actuated control. They noted, however, that the report did not describe the installation features of the particular scheme chosen and did not summarize the installation's acceptability to the several affected airframe manufacturers. In particular, the following was requested and until the information was received, the Item was considered not having been fulfilled:

1. Specifications and descriptions, or manufacturer's part numbers, of the particular air shut-off and control valves chosen.
2. A description and appropriate schematic diagram of operation of this cooling-air valve system.[116]

Another Mock-up Board Item, number 51, was closed with a memo from WAGT stating that no correspondence on the subject had been received from BuAer, hence they were using the Bendix 9109-0, which measured main engine fuel flow only. The position of the flowmeter had been relocated slightly and new brackets designed. The flowmeter was now within the firewall diameter. Approval was requested. The notes on the transmittal sheet indicate this was approved.[117] The memo was re-routed to the Power Plant Division and noted the drawing was for a 20,000 lbs/hr flow. The -10 with A/B operating at maximum thrust required 47,500 lbs/hr and BuAer asked when a 50,000 lbs/hr flowmeter transmitter drawing would be available. At the conference they indicated it would be another year. If the larger flowmeter did not become available, they would use the 20,000 lbs/hr meter and not measure A/B fuel flow.[118]

Figure 9 Exhaust nozzle of XJ40-WE-12 #2. Control on right apparently moved the variable 1st stage turbine nozzles, WAGT P46666, 4/11/1952. *Courtesy Hagley Museum and Library*

The May progress report stated the #1 -12 engine was awaiting newly designed stator vanes for stages 1 through 5 (due June 9), the cause of the destruction of the #2 engine was still not determined and the #3 engine not available for work until the latter part of July. Only a total of 64 hours of engine running to date had been accomplished. The BAR now felt the completion of the qualification test by the end of December was not possible.[119]

In June WAGT wrote a memo regarding discussions that were held earlier on design changes under consideration for the elimination of low and high RPM compressor stalls. They were:

A. Variable inlet guide vanes.
B. Variable first and second stage diaphragms.
C. Seventh stage bleed.
D. Controls for provisions A-C above.

They noted that there was a need to increase the length of the compressor outlet diffuser by approximately 4 inches. The change was expected to improve the efficiency of the diffuser and provide room for an outlet mixer if required. This change caused the following associated changes:

A. Downstream relocation of the rear bearing support for both the subject engines by four inches.
B. Increase the distance between the front and rear mounts by four inches on the J40-WE-5 (tentative description) because of (A) above and a comparable increase in the overall length of the engine.
C. A reduction in length of the J40-WE-10 afterburner diffuser in order to retain the present length of the engine.

The memo stated they were in the process of acquainting the applicable airframe contractors with the proposed configuration of Block II engines by way of the Informal Survey system.[120] *(Details on this process were not located.–Auth.)* This memo referenced contract NOa(s) 10385 which was either in error or indicates some additional development work on the -10/-12 was charged to this contract.

WAGT followed up with BuAer in June on Mock-up Board Item 12, asking what type of connection for the customer was required for the thrust measuring device. No action had been taken yet pending a BuAer response. It was decided by BuAer to drop the requirement given that accuracy would depend on knowing how far open the exhaust nozzle was.[121]

WAGT notified BuAer they had taken no action on Mock-up Board Items 48 and 49, awaiting a response from BuAer on what was required in the mounting and envelope requirements.[122]

EC 74 for the -10 and EC 70 for the -12 moving the fuel inlet port for the subject engines was cancelled due to the subcontractor indicating the pump performance could be met with an internal change not requiring the pump inlet port relocation.[123]

Mock-up Board Item 60 had requested that the turbine nozzle actuator control height be reduced. A decision was made to eliminate the variable turbine nozzle from production models. The mock-up was updated.[124] BuAer's response was to support the decision, but recognized that high compression engines would likely have to use other means in supersonic aircraft to avoid compressor stalls, such as variable inlet ducts, variable compressor-inlet guide vanes and so forth. They asked to be advised if Westinghouse had decided to incorporate on the subject engines some other device which was effectively similar to variable turbine nozzles. The weight reduction amount from removing the variable turbine nozzle feature from the engine was also requested. WAGT was instructed to survey the change.[125]

Mock-up Board Item 27 was responded to in regards to using flexible and rigid lines. With BuAer having responded that no known flexible metal hose had proven satisfactory, WAGT had taken the following actions:[126]

A. Bulkhead fittings had been provided at the fire seal on all lines except the electrical and anti-icing lines.
B. The anti-icing lines had been provided with a quick disconnect at the fireseal.
C. All lines aft of the fireseal except the electrical lines had been changed to flexible metal hose and metal lines.
D. Present design used hose with interspersed flexible metal hose and rigid metal lines, rather than flexible metal hose alone.
E. The rigid lines were usually straight lines with the flexible metal hose providing the needed freedom for thermal expansion, with hose design closely controlled as to position and length of the flexible metal sections.
F. All hoses were provided with suitable metal clamps, brackets and restraints.
G. To ensure good sealing coupling, machined hose end couplings rather than flared ends were utilized. High temperature brazing material was generally used where the flexible hose joined the rigid line.
H. High pressure lines used heavy wall rigid metal hose in conjunction with double braid flexible metal hose.

The Power Plant staff notes on the circulation sheet are telling. *"This proposed WAGT design appears to embody the bad features of both rigid and flex lines and gains nothing from doing so. As long as any metal flex is used the system will have the unsatisfactory characteristics associated therewith and the same applies for the rigid parts of the system. In addition, the number of joints is increased, providing more potential sources of trouble. Currently metal flex A/B fuel lines are used on J34-WE-42 engines with no reported troubles. Recommend not approving this design until it is explored and evaluated further."[127]*

Mock-up Board Item 53 was addressed in a WAGT memo stating the fuel and electrical lines on the lower right side of the engine had been rerouted to clear the hydraulic pump envelope and the fuel booster pump envelope. The fuel inlet fittings of the aircraft contacts would not extend beyond the firewall diameter.[128] BuAer refused to approve the changes until WAGT advised BuAer that the revised routing had been coordinated with the airframe contractors and that it was acceptable to them.[129]

At the end of June, WAGT submitted a report (A-1358, May 19, 1952), on the research into the production of titanium compressor blades. The summary of the report states that sound ductile blades were made from commercially pure titanium. Blades made from P.D.S. 9913 (RC130B-4% Mn, 4% Al) were less consistent in soundness and degree of ductility. Blades made from other commercial alloys were very brittle. If the steel blades in the -10/-12 were replaced with titanium ones, a weight savings of 97.9 pounds would be realized. Early attempts at forging indicated many problems associated with the new metal. The problems included oxidation, forging procedure, and subsequent heat treatments.

The report concluded that commercially pure blades are easy to make, and will produce sound ductile blades if made in the following manner:

1. Forge "Bradley Blank" from 1,600°F.
2. Descale blanks in Sodium Hydride bath. Grind out any cracks if necessary and possible.
3. Finish forge from 1,600°F leaving the blade .020" heavy.
4. Descale in Sodium Hydride bath or go directly to machining.

The damping and endurance properties of such blades were in question and tests were to be conducted on the first and second stages of a J46. If higher strength blades turned out to be needed, they would make blades from RC 130B material and test them.[130] The report was accepted by BuAer on July 23.[131]

In July BuAer authorized overtime work on contracts NOa(s) 9670, 10067 and 52-403-c.[132] The request was approved later that month.[133]

The June progress report showed no further running of the -12 had been accomplished in June. The new design compressor stator vanes for stages 1-5 arrived on June 30 and installation began on the number one engine. Test cell availability was holding up restarting tests of this engine. The current compressor changes were designed to improve the part load stall, the main reason for delay on the engine model. The compressor which was scheduled for the #3 engine was being made ready to ship to NACA for test work. It would be delivered along with a test rig at the end of July or early August. Another compressor for the #3 engine would not be available until August. In the first year the engine had been run only 64 hours. The qualification test in December 1952 appeared unlikely and the qualification test on the -10 scheduled for March 1953 would also be delayed.[134]

In July, WAGT asked for a waiver to any clause in the contract which would prevent expenditures to purchase two Jet Temperature Controls Amplifiers from the British government for the purpose of testing. They would be expendable material.[135]

BuAer closed Mock-up Board Items 20, 21 and 23, noting that they had been achieved through the redesign of the gearbox.[136] They also accepted WAGT's action on Item 18, namely they redesigned the forward mounts, removing the two centerline studs and replacing them with tapped holes and inserts to be used with .625-18NF-3 cap screws. A boss was added to the forward face of the mount to provide added thickness for sufficient strength. Both the -10 and -12 were modified to maintain commonality, even though the Mock-up Board item only referenced the -10.[137]

BuAer closed Mock-up Board Item 57 because the control system of the -10/-12 had been changed (the exact configuration yet to be received at BuAer) and the governor alternator and power regulator had been eliminated.[138] A later WAGT memo finally answered BuAer's request for a description of the control system to be used on the -10/-12 and they stated it was to be basically similar to the Simple Hydraulic System being used on the -22. Its main elements were:

1. Hydro-Mechanical exhaust nozzle control.
2. Automatic temperature trim.
3. Provisions for starting and stopping the afterburner without using electrical power.
4. Variable Displacement oil pump as a hydraulic pressure source only to actuate the iris exhaust nozzle.

WAGT stated they were working on a series of type numbers for the various control systems in use and contemplated and would let BuAer know what they were. It was estimated that work would be complete by September 15, 1952.[139]

In response to finding a location for an engine driven generator to supply the engine with its own electrical power, BuAer agreed to drop the requirement and also to accept WAGT's plan to mount the solid propellant starter on the gearbox, as mounting it in the tailcone would require a total redesign of the back of the engine.[140]

EC 79 for the -10 and EC 73 for the -12, adding three holes in a rear bearing support frame were surveyed in late July.[141]

At the end of July, EC 80, which surveyed a 766 in² nozzle on the -10, was sent out. The change shortened the engine length by 1 5/32 inches and the radial position of the center of the lifting eye changed from 18 11/16 inches to 19 5/16 inches as a result of a change in the casing taper angle from 7° 59' 30" to 5° 30".[142]

Mock-up Board Item 15 was responded to by WAGT in June. They noted that in spite of the recommendation of the board that Resistoflex hose couplings be removed from the engine, WAGT recommended they be retained for the following reasons:[143] BuAer agreed with the recommendation on July 31.[144]

1. They offered considerable advantage from the standpoint of envelope clearance.
2. They were now included in AN Standards, AN762, which should remove any objections to special tools required for overhaul or field repairs.

On July 9 WAGT responded to the request for description of the control systems to be used on the various model engines under development. They were summarized as:

1. The J40-WE-8 "semi-production" engines would use the Simple Hydraulic Control only. It contained certain non-fail-safe characteristics associated with the electrical exhaust nozzle positioning control which made it questionable for service engine use without a complex emergency system. The temperature trim was manual.
2. The J40-WE-22 would use the Simple Hydraulic Control using aircraft power for the basic control and automatic temperature trim, while emergency control of the exhaust nozzle actuator operated from the aircraft emergency battery supply.
3. The XJ40-WE-10 would be equipped with the Simple Hydraulic Control of the -22. Future production J40-WE-10's would be equipped with the Simple Hydraulic Control incorporating the Hydro-Mechanical Exhaust Nozzle Positioning System. This would contain an exhaust nozzle actuator control, a mechanical feedback assembly and overrides (including speed signal, afterburner detection signal and temperature trim). It was tentatively planned to mount the temperature trim motor on the control unit. The Hydro-Mechanical

Exhaust Nozzle Positioning System was in a fluid design state.

The -10/-12 development contract was apparently terminated in the early August 1952 time frame, however no specific documentation such as a formal termination letter was found in the files. However, the patent attorney granted permission to settle the contract in August, indicating termination had occurred and the contract was in the final settlement stage to release the final payment to WAGT.[145] It is believed any remaining funds were moved to the -24/-26 production contract, which apparently may also have included continuing development (See Chapter 11).

In late August, the test cell at AEL dedicated to J40-WE-12 testing was re-allocated to J46 test work to be effective as soon as WAGT forwarded a proposed schedule of the testing contemplated.[146]

With the contract now terminated and development activities stopped, publication of the Block II, III, and IV performance projections (Chapter 11) along with the request that they be distributed to the airframe companies was likely an attempt by WAGT to get the airframe designers to put pressure on BuAer to continue development. There is no evidence in the file that the compressor stalling problems being experienced were ever resolved to allow full power development to continue.

Douglas was asked to formally respond to the survey on the 766 in² nozzle, estimating the cost of any airframe changes that would be needed. Why such follow-up was made after contract termination is not clear, unless a larger nozzle was also being considered for the -22 engine, but nothing in the -22 documentation refers to such a change.[147]

J40-WE-10/-12 Description

These engines achieved major upgrades in thrust primarily by increasing the mass flow, using a higher compression ratio and design of a different control system that did not rely in any way on the aircraft electrical system. Assembly basically followed the same approach as the early models. The -10/-12 was to use the Hydro-Mechanical Control System with Automatic Temperature Control, finally providing full fail-safe features in the event of various control system failures. This control never evolved into a definitive design and was in flux at the time the contract ended. The control system was intended to safely allow higher turbine inlet temperatures for short periods at the expense of shortening the overhaul life of critical components. The compressor was originally bid as a 14 stage unit but was designed as a 13 stage unit using the same construction approach as the earlier models. The compressor blades were steel alloy and did not use air cooling. No provisions were originally included for variable stator vanes or blow-off valves, but later blocks might have included such provisions when the compressor was found to be in deep trouble with stalling that seemed to resist solution. One memo indicates WAGT sought to preserve a pad on the -10/-12 gearbox for a hydraulic pump to operate variable stator vanes or blow-off valves. Westinghouse was aware of GE and Rolls-Royce developments along those lines to deal with compressor stalling. The first development -12 engine featured a variable guide vane installation on the first turbine stage. This feature was later dropped from the design, but was in fact used on the NACA test -12 (Block II - 24 engine) and found very useful in running tests to determine the compressor stall limits of the engine. Testing of a -6 with such a variable guide vane installation showed no improved efficiencies over static guide vanes and this test may have led to the abandonment of the approach on the -12/-10. The -10 version Block II was apparently to depart from the eyelid type exhaust nozzle and switch to an iris type design. There is no record that a -10 engine was ever constructed or run on a test bed, although the iris type A/B might have been run on a -8 or -22 house engine. The engines used "walking stick" fuel vaporization instead of the direct injectors of the -6/-8/1/-22.

Research was conducted into the feasibility of using titanium for compressor blades and discs to reduce weight. The former were found to be easy to produce once certain manufacturing procedures were evolved. Apparently no engine tests were run with titanium blades or discs. The discs made of titanium were found to exhibit unacceptable disc growth when made from the alloy available at WAGT and later experiments were planned but no records were found to indicate such tests were conducted by program end.

J40-WE-10/-12 Production

Neither model engine emerged from development or achieved the original specifications. Only four XJ40-WE-12 engines were operated and the first, third and fourth destroyed themselves with compressor failures on the test stand. The exact cause was never determined, but was likely due to the fatigue failure of the blades and/or discs in the compressor while operating in a deeply stalled condition.

Another development engine was partially constructed but was cannibalized for parts to enable the first engine to be returned to NACA for further testing without the variable first stage turbine guide vane assembly.

J40-WE-10 Block I Specifications			
Rating	Thrust lbs	RPM*	SFC
Full Afterburning	13,700	7,260	2.50
Military	9,275	7,260	0.950
Normal	8,330	7,260	0.893
90% Normal	7,500	7,260	0.850
75% Normal	6,250	7,260	0.805
Idle	555	3,000	5.00

J40-WE-12 Block I Specifications			
Rating	Thrust lbs	RPM*	SFC
Maximum	9,500	7,260	0.930
Military	9,500	7,260	0.930
Normal	8,550	7,260	0.870
90% Normal	7,700	7,260	0.830
75% Normal	6,400	7,260	0.767
Idle	560	3,000	

* These engines initially were designed to operate at a constant speed with turbine inlet temperature and speed governors to protect the engine. Block IV engines apparently would have been operated as variable speed.

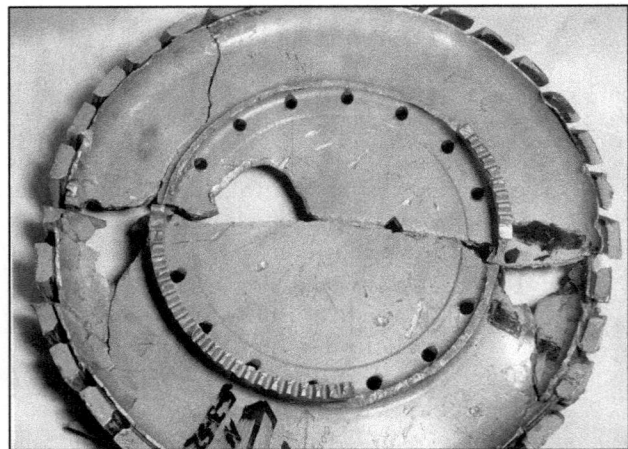

Figure 10 XJ40-WE-12 #1 Seventh stage compressor disc failure which destroyed the engine, WAGT P47722, 9/16/1952. *Courtesy Hagley Museum and Library*

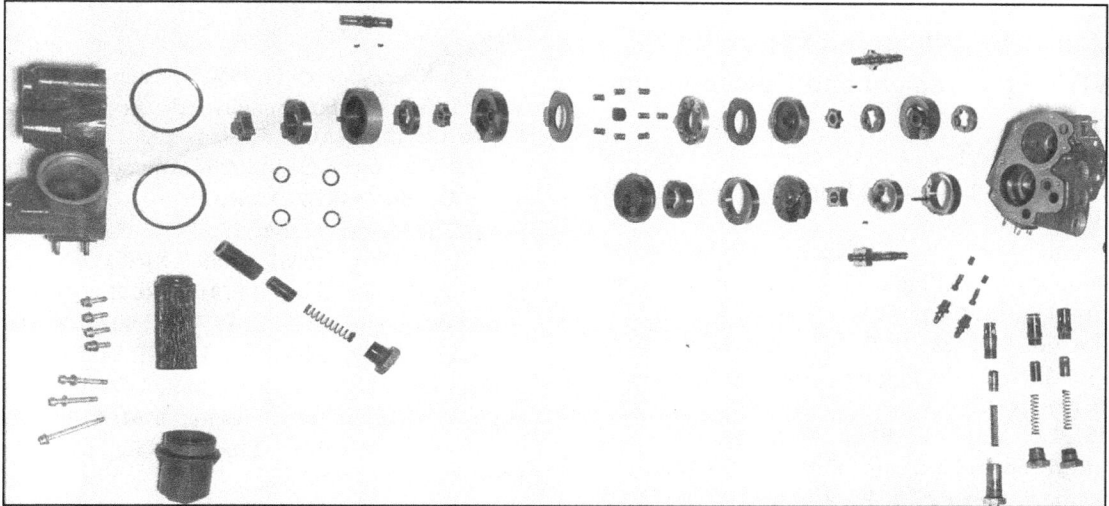

Figure 11 XJ40-WE-10 Exploded parts view of lube pump assembly, WAGT P48162, 11/28/1952. *Courtesy Hagley Museum and Library*

Figure 12 Douglas XA3D-1 Mockup, note flow splitter in engine intake for J40-WE-12. Mock-up Review Report, Douglas ES 76360, 9/12/1949

Figure 13 Douglas XA3D-1 Mockup with Douglas built J40-WE-12 engine mockup installed in nacelle showing engine access, Mock-up Review Report, Douglas ES 76355, 9/12/1949

Chapter 10 - XJ40-WE-10 and XJ40-WE-12 (XJ40-WE-3) Procurement and Development Citations

[1] BuAer Memo, December 29, 1948, "Contract NOa(s) 10067 – XJ40-WE-10 Engine Program; Request for Modification of." **B794VE1RG72.3.2NACP**

[2] WAGT Memo, April 19, 1949, "Contract NOa(s) 10067 – Engine Model XJ40-WE10 and-12 Turbine Nozzle Design Features" **B794VE1RG72.3.2NACP**

[3] BuAer Memo, May 23, 1949, "Contract NOa(s) 10067 XJ40-WE-10, -12 Turbo-Jet Engines; Proposed Changes in." **B794VE1RG72.3.2NACP**

[4] BuAer Memo, May 12, 1949, "Model J40-WE-10, -12 Performance, Proposed Increase in" **B794VE1RG72.3.2NACP**

[5] BuAer Memo, October 26, 1949, "Contract NOa(s) 10067, XJ40-We-10 and -12 engine ratings, changes in" **B794VE1RG72.3.2NACP**

[6] WAGT Specification WAGT-X40E8-2, September 26, 1949, Model X40E8 **B794VE1RG72.3.2NACP**

[7] WAGT Specification WAGT-X40E8-2, September 26, 1949, Model X40E8 **B794VE1RG72.3.2NACP**

[8] WAGT Memo, September 6, 1949, "Plancor 2061, Additional Test Machinery" **B796VE1RG72.3.2NACP**

[9] WAGT Monthly Progress Report A-910, October 1949 **B794VE1RG72.3.2NACP**

[10] WAGT Monthly Progress Report A-918, November 1949 **B794VE1RG72.3.2NACP**

[11] WAGT Monthly Progress Report A-925, December 1949 **B794VE1RG72.3.2NACP**

[12] BuAer Memo, January 31, 1950, "Contract NOa(s) 10067; Westinghouse XJ40-WE-10 engines" **B794VE1RG72.3.2NACP**

[13] WAGT Memo, May 2, 1950, "Proposed Development of an Accessory Gearbox for use with a Solid Propellant Starter for the XJ40-WE-10 Engines being furnished on NOa(s)-10067, Westinghouse WG-62231" **B794VE1RG72.3.2NACP**

[14] Convair Memo, February 20, 1950, "Contracts NOa(s)-9670, 10385, 10114 and 10067. WEC Turbo Jet Engine Models XJ34-WE-32, J34-WE-38, J40-WE-6, XJ40-WE-8, XJ40-WE-10 and X40-WE-12. Removal of Forward Combustion Chamber Drain Valve, Survey of." **B794VE1RG72.3.2NACP**

[15] WAGT Memo, May 2, 1950, "Inclusion of a fuel dump valve as a component of the J40 engines applicable on Contracts NOa(s) 9212, 10114, 10067, 10385, 10943." **B794VE1RG72.3.2NACP**

[16] WAGT Monthly Progress Report for February, 1950 **B794VE1RG72.3.2NACP**

[17] WAGT Monthly Progress Report for March, 1950 **B794VE1RG72.3.2NACP**

[18] WAGT Monthly Progress Report, April, 1950 **B796VE1RG72.3.2NACP**

[19] BuAer Memo, March 21, 1950, "Westinghouse XJ40-WE-10 and -12 engine turbo-jet engines; development program for" **B794VE1RG72.3.2NACP**

[20] BuAer Contract NOa(s) 10067, Amendment 1, Dated June 21, 1950 **B794VA4RG72.3.2NACP**

[21] WAGT Memo, March 23, 1950, "Letter of Intent for Contract NOa(s) 10067 Model XJ40-WE-10 Installation Drawing, Transmittal of." **B794VE1RG72.3.2NACP**

[22] BuAer Memo, June 13, 1950, "Contract NOa(s)-10067, Westinghouse Specification WAGT-X40E8A-2 dated 26 September 1949 covering the Models XJ40-WE-10 and -12 engines; performance curves for" **B794VE1RG72.3.2NACP**

[23] WAGT Monthly Progress Report, May, 1950 **B794VE1RG72.3.2NACP**

[24] BuAer Memo, June 28, 1950, "Contract NOa(s) 10067 XJ40-WE-10 and -12 engines; extension of" **B796VE1RG72.3.2NACP**

[25] WAGT Report A-974, July 25, 1950, Calculation of #1 Bearing Thrust Load on XJ40WE-10 & -12 Engines **B794VE1RG72.3.2NACP**

[26] BuAer Memo, July 28, 1950, "Contract NOa(s)-10067, Westinghouse Specification WAGT-X40E8A-2A of 7 June 1950 covering the Models J40-WE-10, -12 engines; generator drive requirements for" **B794VE1RG72.3.2NACP**

[27] WAGT Monthly Progress Report, June and July, 1950. **B794VE1RG72.3.2NACP**

[28] WAGT Memo, September 1, 1950, "Contract NOa(s)-10067 Our Reference WG-62231 Development Delivery Dates" **B794VE1RG72.3.2NACP**

[29] WAGT Monthly Progress Report, August, 1950 **B794VE1RG72.3.2NACP**

[30] BuAer Memo, September 25, 1950, "Contract NOa(s) 10067; request for additional information on XJ40-WE-10/-12 engine design" **B794VE1RG72.3.2NACP**

[31] BuAer Memo, October 12, 1950, "Contract NOa(s) 10067; Model XJ40-WE-10 engine; flowmeter to measure total engine fuel flow: request for policy concerning" **B794VE1RG72.3.2NACP**

[32] BuAer Memo, October 12, 1950, "Contract NOa(s) 10067; Model XJ40-WE-10, -12 engines; metal flexible hose for; request for data on" **B794VE2RG72.3.2NACP**

[33] BAR Memo, November 1, 1950, "Contract NOa(s) 10067; Model XJ40-WE-10, -12 engines; metal flexible hose for" **B794VE2RG72.3.2NACP**

[34] BuAer Memo, November 14, 1950, "Contract NOa(s) 10067; Model XJ40-WE-10 and -12 engines; information concerning flexible metal hose for" **B794VE2RG72.3.2NACP**

[35] BAR Memo, October 19, 1950, Monthly Progress Report for September, 1950 **B794VE2RG72.3.2NACP**

[36] WAGT Memo, October 26, 1950, "NOa(s)-10067 Alternate Control Our Reference WG-62231" **B796VE1RG72.3.2NACP**

[37] WAGT Memo, November 7, 1950, "Contract NOa(s)-10067 Redesign of the XJ40-WE-10 and -12 Gearbox to Incorporate Constant Speed Drives" **B794VE2RG72.3.2NACP**

[38] BuAer Contract Amendment 2, Contract NOa(s) 10067, November 22, 1950 **B794VA4RG72.3.2NACP**

[39] BuAer Contract Amendment 3, Contract NOa(s) 10067, November 24, 1950 **B794VA4RG72.3.2NACP**

[40] BuAer Memo, November 22, 1950, "Contract NOa(s) 10067; XJ40-WE-10/-12 engine gearbox design" **B794VE2RG72.3.2NACP**

[41] BuAer Memo, October 29, 1950, "Contracts NOa(s) 9670, 10067, and 10114; use of titanium in XJ46 and XJ40 engine designs" **B794VE2RG72.3.2NACP**

[42] WAGT Memo, November 16, 1950, "Contract 10067, XJ40-WE-10 & -12 Engines. Removal of titanium." **B794VE2RG72.3.2NACP**

[43] WAGT Memo, November 17, 1950, "Revision of Engine Model Specifications to show maximum fuel consumption with Primary Control Contracts NOa(s) 10385, 10067, 10943, 11028 and 9670." **B794VE2RG72.3.2NACP**

[44] BAR Memo, November 20, 1950, Monthly Progress Report for October, 1950 **B794VE2RG72.3.2NACP**

[45] WAGT Memo, November 21, 1950, "Contracts NOa(s)-9670, 10825, 10114, 10067 and 10385 – Use of Titanium in J46 and J40 Engines" **B794VE2RG72.3.2NACP**

[46] BAR Memo, December 13, 1950, Monthly Progress Report for November, 1950 **B794VE2RG72.3.2NACP**

[47] WAGT Memo, January 13, 1951, "Contract NOa(s) 10067 WEC Turbo Jet Engine Models XJ40WE-10 and -12. Proposed Side Oil Filler and Scupper, Survey of. Engine Change No. 63 for -10 engines and No. 63 for -12 engines." **B794VE2RG72.3.2NACP**

[48] BAR Memo, January 18, 1951, Monthly Progress Report for December, 1950 **B794VE2RG72.3.2NACP**

[49] WAGT Monthly Progress Report, January, 1950 **B794VE3RG72.3.2NACP**

[50] WAGT Memo, February 21, 1951, "Auxiliary Electrical Control for Westinghouse Turbo Jet Engines" **B796VE1RG72.3.2NACP**

[51] WAGT Memo, February 26, 1951, "Request for Use of a Test Cell at AEL" **B796VE1RG72.3.2NACP**

[52] BuAer Memo, February 19, 1952, "Testing of J40 Engines at Aircraft Engine Laboratory, Philadelphia Navy Yard; Contracts NOa(s) 10067 and 10114" **B796VE1RG72.3.2NACP**

[53] WAGT Memo, February 28, 1951, "Contract NOa(s) 10067, WEC turbo jet engine model J40WE-10, installation manual; transmittal of." **B795VE5RG72.3.2NACP**

[54] WAGT Monthly Progress Status Report for February, 1951 **B794VE3RG72.3.2NACP**

[55] WAGT Memo, March 13, 1951, "Proposed Amendment to NOa(s)-10067 Our Reference Neg. 93637" **B794VE3RG72.3.2NACP**

[56] BAR cover memo, April 2, 1951, "Contracts NOa(s) 10114 and 10067, WEC turbo jet engine models XJ40WE-8, -10 and -12, proposed fuel control configuration, survey of" **B794VE3RG72.3.2NACP**

[57] BAR Memo, Monthly Progress Report, March, 1951 **B794VE3RG72.3.2NACP**

[58] WAGT Repost A-1096, April 25, 1951, "Component Qualification Test of Scintilla Type TLN Low Voltage High t Energy Surface Gap Ignition System for J34 and J40 Turbo-Jet Engines Project 1375, S.O. 4A3310 Contract NOa(s) 10067" **B794VE3RG72.3.2NACP**

[59] WAGT Monthly Progress Report for April, 1951. **B794VE3RG72.3.2NACP**

[60] BuAer Memo, May 17, 1951, "Contract NOa(s) 10385, 10114, and 10067 WEC Turbo Jet Engines Models J40WE-6, -8, -10 and -12. Disapproval of Surveys Concerning Airframe Mounted Power Regulators, Information on." **B794VE3RG72.3.2NACP**

[61] Amendment 5, May 24, 1951 **B794VA4RG72.3.2NACP**

[62] WAGT Memo, May 16, 1951, "Contract NOa(s) 10114 and 10067 WEC Turbo-Jet Engine Models J40WE-8, -10 and -12. Eight Stud Fuel Inlet Connection, Survey of. Engine Change Nos. 82 for -8 engines 68 for -10 engines and 67 for -12 engines." **B794VE3RG72.3.2NACP**

[63] WAGT Memo, May 29, 1951, "Item 16, Lot III, Contract NOa(s) 10067, Development of XJ40-WE-10 and -12 Engine" **B794VE3RG72.3.2NACP**

[64] WAGT Memo, June 5, 1951, "Request for Use of Test Cell at AEL Contracts Nos. NOa(s) 9670, 10067 and 10014" **B796VE1RG72.3.2NACP**

[65] BAR Memo, June 15, 1951, Monthly Progress Report for May, 1951. **B794VE3RG72.3.2NACP**

[66] BAR Memo, June 22, 1951, Contracts NOa(s) 11028, 9670 and 10067, WEC turbo jet engine models XJ46WE-1, -2, -4 and XJ40WE-10; revised auxiliary electric control, survey of." **B794VE3RG72.3.2NACP**

[67] BAR Memo, July 5, 1951, "Contracts NOa(s) 11028, 9670, 10114 and 10067, WEC turbo jet engine models XJ46WE-1, -2, -4, XJ40WE-8, -10 and -12, increased power regulator envelope; survey of." **B794VE3RG72.3.2NACP**

[68] BAR Memo, July 5, 1951, "Contracts NOa(s) 11028, 9670 and 10067, WEC turbo jet engine models XJ46WE-1, -2, -4 and XJ40WE-10; survey concerning a revised electric control; information on." **B794VE3RG72.3.2NACP**

[69] BuAer Memo, July 6, 1951, "Contract NOa(s) 10067, Westinghouse Specification WAGT-X40E8A-2C covering the models XJ40-WE-10 and -12 engines." **B794VE3RG72.3.2NACP**

[70] WAGT Memo, May 14, 1951, "Contract NOa(s) 10067 Westinghouse Model Specification WAGT-X40E8A-2B dated 20 December 1950" **B794VE3RG72.3.2NACP**

[71] WAGT Memo, July 5, 1951, "Contract NOa(s) 10067 WEC Turbo Jet Engine Model XJ40WE-10. Afterburner Cooling

Chapter 10 - XJ40-WE-10 and XJ40-WE-12 (XJ40-WE-3) Procurement and Development

Air Line Configuration, Survey of. Engine Change No. 71." **B794VE3RG72.3.2NACP**

[72] BAR Memo, July 13, 1951, "Contract NOa(s) 10067, items 7a and 14a; Monthly Progress Report A-1169 (June 1951); forwarding of." **B794VE3RG72.3.2NACP**

[73] BAR Memo, August 3, 1951, "Contract NOa(s) 10067; technical report; transmittal of." **B794VE3RG72.3.2NACP**

[74] WAGT Memo, August 10, 1951, "Contract NOa(s) 10067 WEC Turbo Jet Engine Models XJ40WE-10 and XJ40-WE-12. Present and Future Design Considerations for Compressor Air Bleed, Survey of." **B794VE3RG72.3.2NACP**

[75] BAR Memo, August 15, 1951, Monthly Progress Report A-1180 for July 1951 **B794VE3RG72.3.2NACP**

[76] BuAer Memo, September 7, 1951, "Contract NOa(s) 10067; type of control system for XJ40-WE-10/-12 engines" **B794VE3RG72.3.2NACP**

[77] BAR Memo, September 14, 1951, Monthly Progress Report A-1215 for August 1951 **B754VE4RG72.3.2NACP**

[78] BuAer Memo, October 4, 1951, "Contract NOa(s) 10067; status of preparation of Amendment Request covering cost over-run" **B795VE4RG72.3.2NACP**

[79] BAR Memo, October 16, 1951, Monthly Progress Report A-1232 for September, 1951 **B795VE4RG72.3.2NACP**

[80] WAGT Memo, November 16, 1951, "Engine Information to North American Inglewood, Calif., Request for distribution of." **B795VE4RG72.3.2NACP**

[81] BuAer Naval Speedletter, December 4, 1951, "Request for distribution of engine information to North American, Inglewood, Calif.; approval of" **B795VE4RG72.3.2NACP**

[82] BAR Memo, November 19, 1951, Monthly Progress Report A-1255 for October 1951 **B795VE4RG72.3.2NACP**

[83] BuAer Memo, December 10, 1951, "Contract NOa(s) 10067; location of high-pressure oil pump" **B795VE4RG72.3.2NACP**

[84] BAR Memo, December 12, 1951, Monthly Progress Report A-1280 for November 1951 **B795VE4RG72.3.2NACP**

[85] WAGT Memo, December 29, 1951, "Amendment No. 6 (dated March 31, 1951) to Contract NOa(s)-10067 Our Reference WG-62231" **B795VE4RG72.3.2NACP**

[86] WAGT Report A-1298, January 5, 1952, "Monthly Report, Month of December 1951, Alternate Hydraulic Control Development, Contract NOa(s) Item 16" **B795VE4RG72.3.2NACP**

[87] BuAer Memo, January 8, 1952, "Contract NOa(s) 10067; afterburner fuel control for XJ40-WE-10 engine" **B795VE4RG72.3.2NACP**

[88] WAGT Memo, February 25, 1952, "Contract NOa(s) 10067; Afterburner Fuel Control for XJ40-WE-10 Engine." **B795VE5RG72.3.2NACP**

[89] WAGT Memo, January 7, 1952, "Contract NOa(s) 10067 WEC Turbo Jet Engine Models XJ40WE-10 and XJ40WE-12. Proposed Relocation of fuel inlet port of the fuel booster pump, Survey of. Engine Change nos. 74 for -10 engines and 70 for -12 engines." **B795VE4RG72.3.2NACP**

[90] WAGT Memo, January 11, 1952, "Contract NOa(s)-10067 Development of XJ40-WE-10 and -12 Engines Our Reference WG-62231 Compensation" **B795VE4RG72.3.2NACP**

[91] WAGT Memo, January 14, 1952, "Contract NOa(s)-10067 Additional Funds Required Our Reference WG-62231" **B795VE4RG72.3.2NACP**

[92] BAR Memo, January 14, 1952, Monthly Progress Report A-1298 for December 1951 **B795VE4RG72.3.2NACP**

[93] WAGT Memo, February 13, 1952, "Contract NOa(s) 10067 WEC Turbo-Jet Engine Models XJ40-WE-10 and XJ40-WE-12. Spark Plug Relocation, Survey of. Engine Change NOa(s) for -10 engines." **B795VE4RG72.3.2NACP**

[94] BAR Memo, February 18, 1952, "Contract NOa(s) 10067; Monthly Progress Report A-1309 for Month of January 1952 – Items 7(a) and 14(a); Forwarding of" **B795VE4RG72.3.2NACP**

[95] WAGT Memo, February 18, 1952, "Contract NOa(s) 10067 WEC Turbo-Jet Engines XJ40WE-10 and XJ40WE-12. A new Westinghouse Power Control System on the XJ40We-10 and XJ40WE-12, Incorporation of." **B795VE4RG72.3.2NACP**

[96] WAGT Memo, March 7, 1952, "Contract NOa(s) 10067 and 10385; WEC Turbo Jet Engine Models J40-WE-22 and J40-WE-10; Engine Installation and Performance Information for" **B795VE5RG72.3.2NACP**

[97] Naval Speedletter, February 20, 1952, "Contracts NOa(s) 10067, 10385, and NOa(s) 52-503c (Notice of Award); shortage of engine installation and performance information" **B795VE5RG72.3.2NACP**

[98] Amendment 8, March 8, 1952. **B794VA8RG72.3.2NACP**

[99] BuAer Memo, March 5, 1952, "Westinghouse XJ40-WE-Series Engines: re-designation of" **B3VJ40RG72.4.3NACP**

[100] BuAer Telegram, March 4, 1952, Request for Reasons for Late change in Nozzle Size **B796VE2RG72.3.2NACP**

[101] BAR Memo, March 14, 1952, Monthly Progress Report for February 1952 **B795VE6RG72.3.2NACP**

[102] WAGT Memo, March 13, 1952, "J40 Engine Testing at AEL" (Received at BuAer PP Division April 7) **B796VE4RG72.3.2NACP**

[103] WAGT Memo, March 25, 1952, "Contracts NOa(s) 10067 and NOa(s) 8670 Model Specification changes to provide for the use of JP-4 in J40 and J46 engines" **B797V3RG72.3.2NACP**

[104] Amendment 9, March 31, 1952 – Removal of and adjust of contract language relating to the 50-hour flight substantiation test. **B794VA4RG72.3.2NACP**

[105] BuAer Naval Speed Letter, April 5, 1952, "Contract NOa(s 10067; XJ40-WE-10 afterburning exhaust nozzle area" **B795VE5RG72.3.2NACP**

[106] WAGT Memo, February 5, 1952, "Contract NOa(s) 10067, Lot III, Item 16" **B795VE6RG72.3.2NACP**

[107] BuAer Memo, April 5, 1952, "Contract NOa(s) 10067, Lot III, Item 16, alternate hydraulic control development" **B795VE6RG72.3.2NACP**

[108] BAR Memo, April 15, 1952, Monthly Progress Report A-1352 for March 1952 **B795VE6RG72.3.2NACP**

[109] WAGT Memo, April 4, 1952, "Contract NOa(s) 10067; Location of High-Pressure Oil Pump" **B795VE5RG72.3.2NACP**

[110] BAR Memo, April 9, 1952, "Contract NOa(s) 10067; Location of High Pressure Oil Pump" **B795VE5RG72.3.2NACP**

[111] BuAer Naval Speed Letter, April 25, 1952, "Contract NOa(s) 10067, location of high-pressure oil pump" **B795VE5RG72.3.2NACP**

[112] BAR Memo, May 14, 1952, Monthly Progress Report A-1377 for April 1952 **B795VE8RG72.3.2NACP**

[113] BuAer Memo, June 16, 1952, "Contract NOa(s) 10067, XJ40-WE-10, -12 engines control system for" **B795VE6RG72.3.2NACP**

[114] WAGT Memo, June 3, 1952, "Contract NOa(s) 10067, XJ40-WE-10 and 12 Engines – clearance envelop for DC generator" **B795VE7RG72.3.2NACP**

[115] WAGT Memo, June 3, 1952, "Contract NOa(s) 10067, XJ40-WE-10 Engine – Study of Afterburner Cooling." **B795VE7RG72.3.2NACP**

[116] BuAer Memo, July 25, 1952, "Contract NOa(s) 10067; XJ40-WE-10 engine, study of afterburner cooling" **B795VE7RG72.3.2NACP**

[117] WAGT Memo, June 3, 1952, "Contract NOa(s) 10067, XJ40-WE-10 and 12 Engines – Design of flowmeter transmitter to stay within firewall diameter." **B795VE6RG72.3.2NACP**

[118] BuAer notes attached to WAGT Memo of June 17 asking for flowmeter data. **B795VE6RG72.3.2NACP**

[119] BAR Memo, June 12, 1952, Monthly Progress Report A-1319 for May 1952 **B795VE8RG72.3.2NACP**

[120] WAGT Memo, June 9, 1952, "Contract NOa(s) 10385 Westinghouse Turbojet engine models Block II J40-WE-10 and J40-WE-(-5 tentative) installation configuration of." **B937VE11RG72.3.2NACP**

[121] WAGT Memo, June 12, 1952, "Contract NOa(s) 10067, XJ40-WE-10 Engine – Provide customer connection to give thrust indication" **B797V3RG72.3.2NACP**

[122] WAGT Memo, June 17, 1952, "Contract NOa(s) 10067, XJ40-WE-10 Engine – Flowmeter mounting." **B796VE6RG72.3.2NACP**

[123] WAGT Memo, June 20, 1952, "Contract NOa(s) 10067, !0385; 52-048 WEC Turbojet Engine Models XJ40-WE-10 and -12. Proposed Relocation of Fuel Inlet Port of the Fuel Booster Pump, Cancellation of. Engine Change nos. 74 for -10 engines & 70 for -12 engines." **B795VE7RG72.3.2NACP**

[124] WAGT Memo, June 12, 1952, "Contract Noa(s) 10067, XJ40-WE-10 and 12 Engines – Reducing height of proposed turbine nozzle actuator control to stay within engine firewall diameter." **B795VE7RG72.3.2NACP**

[125] BuAer Memo, June 30, 1952; XJ40-WE-10/-12 engines, proposed turbine nozzle actuator control" **B795VE7RG72.3.2NACP**

[126] WAGT Memo, June 27, 1952, "Contract NOa(s) 10067, XJ40-WE-10 and 12 Engines – Bulkhead fittings at fire seal, stainless steel or metal flexible hose back of fire seal." **B795VE7RG72.3.2NACP**

[127] Power Plant Circulation Notes on WAGT Memo of June 27 regarding rigid and flexible metal lines. **B795VE7RG72.3.2NACP**

[128] WAGT Memo, June 12, 1952, "Contract NOa(s) 10067, XJ40-We-10 and 12 Engines – Rerouting of electrical and fuel lines." **B795VE7RG72.3.2NACP**

[129] BuAer Memo, July 2, 1952, "Contract NOa(s) 10067; XJ40-We-10/-12 engines, re-routing of electrical and fuel lines" **B795VE7RG72.3.2NACP**

[130] WAGT Report A-1358, May 19, 1952, "Development of Titanium Compressor Blades" **B795VE7RG72.3.2NACP**

[131] BuAer Memo, July 23, 1952, "Contract NOa(s) 10067; approval of interim technical report" **B795VE7RG72.3.2NACP**

[132] BuAer Memo, July 11, 1952, "Contracts NOa(s) 9670, 10067, and 52-403-c – Request for Overtime Approval" **B795VE7RG72.3.2NACP**

[133] BuAer Memo, July 11, 1952, "Contracts NOa(s) 9670, 10067, and 52-403-c – Request for Overtime Approval" **B795VE7RG72.3.2NACP**

[134] BAR Memo, July 16, 1952, Monthly Progress Report A-1418 for June 1952 **B795VE8RG72.3.2NACP**

[135] WAGT Memo, July 16, 1952, "Necessary Waiver of Clauses Under Contract NOa(s) 10067" **B795VE8RG72.3.2NACP**

[136] BuAer Memo, July 17, 1952, "Contract NOa(s) 10067; J40-WE-10/-12 engine gearbox design" **B795VE7RG72.3.2NACP**

[137] BuAer Memo, July 18, 1952, "Contract NOa(s) 10067; XJ40-WE-10/-12 engines, revision of fore and aft centerline studs on forward mount" **B795VE7RG72.3.2NACP**

[138] BuAer Memo, July 22, 1952, "Contract NOa(s) 10067; XJ40-WE-10/-12 engines, pressure differential across walls of control components" **B795VE7RG72.3.2NACP**

[139] WAGT Memo, July 28, 1952, "Contract NOa(s) 10067, XJ40-WE-10, -12 engines, control system for." **B795VE7RG72.3.2NACP**

[140] WAGT Memo, July 25, 1952, "Contract NOa(s) 10067. Location of High Pressure Oil Pump." **B797V1RG72.3.2NACP**

[141] WAGT Memo, July 28, 1952, "Contract NOa(s) 10067, 10385; 52-048. WEC Turbojet Engine Models J40-WE-10, J40-WE-12. Addition of Three (3) Holes to Rear Bearing Support Frame, Survey of. Engine Change Number 79 for J40-WE-10; Engine Change Number 73 for J40-WE-12." **B795VE7RG72.3.2NACP**

[142] WAGT Memo, July 28, 1952, "Contract NOa(s) 10067; 52-048. WEC Turbojet Engine Model J40WE-10. Seven Hundred and Sixty Six Square Inch Exhaust Nozzle Area, Survey of. Engine Change No. 80." **B795VE7RG72.3.2NACP**

[143] WAGT Memo, June 3, 1952, "Contract NOa(s) 10067, XJ40-WE-10 and 12 Engines – Elimination of Resistoflex Hose elbow couplings." **B795VE8RG72.3.2NACP**

[144] BuAer Memo, July 31, 1952, "Contract NOa(s) 10067; XJ40-WE-10/-12 engines, elimination of Resistoflex Hose elbow couplings" **B795VE8RG72.3.2NACP**

[145] BuAer Patent Counsel, August 19, 1952, "Contract NOa(s) 10067 – Westinghouse Electric Corporation" **B795VE8RG72.3.2NACP**

[146] BuAer Naval Speed Letter, August 28, 1952, "Static sea level testing of XJ40 and XJ46 engines at AEL" **B796VE6RG72.3.2NACP**

[147] BuAer Memo, September 29, 1952, "Westinghouse engine installation survey; request for reply to" **B796VE6RG72.3.2NACP**

Chapter 11

XJ40-WE-24 and XJ40-WE-26 (XJ40-WE-5)
Procurement and Development

As stated in the previous chapter, in early June, 1952, with progress slow on the -10 and -12 development and the delivery schedule slipping steadily, WAGT proposed to BuAer that to obtain usable production of the -10 and -12 engines as quickly as possible, BuAer should accept intermediate de-rated production models. The initial production block was designated "Block II" for both models which were later named the XJ40-WE-24 for the de-rated -10 and XJ40-WE-26 for the de-rated -12. Discussion of -10 and -12 design and development shifted to the -24/-26 engines, although the development documentation is not always clear on which model was now being discussed within the -10/-12 program.

The Block II engines, as XJ40-WE-24 and XJ40-WE-26 were placed under production contract NOa(s) 52-1171. WAGT specifications assigned were WAGT-X40E8A-6A for the -24 and WAGT-X40E8A-8 for the -26.

Calculations and graphs showing the expected performance of the -10 Block II production engines were released to BuAer in early July. They were marked "preliminary". The information was forwarded to the airframe contractors with the warning noted on each copy. The numbers (extrapolated from the graphs) were:[1]

J40-WE-10 Block II Preliminary Performance Projections		
Setting	Thrust (lbs)	SFC
Maximum	13,600	2.425
Military	8,750	
Normal	7,750	

The contract covering the -24 and -26 is not in the National Archives. Only fragmentary documentation has been found in other locations that in any way relate to the -24/-26 engines and later blocks. It has not been possible to establish an actual build specification for each block based on that information.

In May, WAGT had surveyed the airframe companies regarding the maximum size exhaust nozzle they could accommodate in their current airframe design without major redesign. The driving force behind this change was a decision to accept a six month schedule delay on -10 engines in lieu of jumping directly to the J40-WE-10 Block II (J40-WE-24) model with an iris type exhaust nozzle of larger size. The survey responses were:[2]

A. McDonnell – A size of 750 in² still applied without major redesign, but an increase to 766 in² could be tolerated if absolutely necessary by reducing clearance of the jet wake with fuselage after structure.

B. Douglas – Preferred the maximum nozzle area possible.

C. Grumman – The largest nozzle they could take without airframe delay over and above six months was 750 in² due to jet wake problems. However, going to the 766 in² nozzle would only entail a ¼ inch larger diameter change and it was very difficult to state definitely that the larger nozzle would be the difference between an acceptable and unacceptable design.

BuAer approved the use of the 766 in² nozzle on J40-WE-24 engines in July.[3]

WAGT submitted a memo covering the control system design state at the time of the memo.[4]

Flight test requirements similar to those required for the J46 model engines were added for the -24 engine in February. The same memo added the requirement that the engine be acceptance tested under the MIL-E-5010 specifications except for certain modifications:[5]

1. Alternate fuel testing with MIL-F-5572 avgas had to be demonstrated

2. The A/B control had to be cycled for 6 hours, simulating operation.

3. A hot test 15 hours long at +/-200 °F hot soak for the exhaust nozzle control

4. After the hot soak in (3), maintain an ambient temperature of 375°F +/-5 degrees and with a fluid temperature of 250°F, cycle the control once a minute for 5 hours.

5. After Item 4, increase the fluid temperature to 300°F +/- 5 degrees and continue cycling at once per minute for an additional 5 hours.

6. Upon completion of Item 5, reduce the operating fluid temperature to 200°F +/- five degrees and continue to cycle at once per minute for 40 hours.

7. The ambient temperature during the execution of Items 5 and 6 should remain 350°F +/- five degrees.

8. The A/B nozzle and actuator should be cycled for 6 hours with the operating fluid maintained at or below minus 65°F for 72 hours.

9. While at the cold soak of minus 65°F, five simulated starts should be made. After the simulated starts, the control should be cycled for 10 hours at the cold soak temperature.

In May, BuAer stated that their operational requirement for engine electrical components was 200°F continuous with a five minute rating at 250°F. WAGT had stated that only a five minute rating at 200°F could be met, with a normal continuous rating of 165°F of the earlier engines.[6]

In June, BuAer added a requirement that the engine fuel system must be run for 150 hours with 8 grams of foreign matter contamination per 1,000 gallons in the fuel and for the last 2 hours of the test, 50 grams per 1,000 gallons of fuel. The fuel must be contaminated with salt water to demonstrate the ability of all components to continue to function without damage from corrosion during and after contamination.[7]

The same memo stated that the proposed acceleration times for the engine were considered poor, particularly for the variable area engine (which Block or model they are referring to is not specified). Eight seconds from 40% military to maximum thrust and 20 seconds from idle to maximum thrust were considered excessive. The maximum time to start the A/B should not exceed 1.5 seconds.

The weight of 4,218 pounds would be acceptable for the -24 engine. The oil should be changed to a synthetic MIL-L-7008 compliant type with additives for preservation of the engine in lieu of the current MIL-O-6081 mineral oil. Engines should use the preservative synthetic oils during acceptance tests to eliminate the need for special preservation runs.

Figure 1 XJ40-WE-12 #3 Showing damaged combustion basket inner liner shape for -10/-12 engines and walking stick evaporators. Basket damaged in testing. WAGT P49296, 4/29/1953. *Courtesy Hagley Museum and Library*

Figure 2 XJ40-WE-10 and -24 No. 2 Bearing assembly holder and A/B connection duct. Note the change to only two fuel spray rings. WAGT P49511, 6/10/1953. *Courtesy Hagley Museum and Library*

Figure 3 Afterburner design using a multi-petal nozzle for -10 type engines. Note that the use of a yoke with single activator was retained, in spite of the mock-ups earlier showing multiple hydraulic rams, WAGT P49509, 6/10/1953. *Courtesy Hagley Museum and Library*

In early August, BuAer submitted specific comments on the model specifications. They now required an acceleration time of less than 10 seconds from idle to Maximum thrust at an airspeed of at least 150 knots at 20,000 feet up to the operational altitude.[8] In mid-August, WAGT forwarded updated performance estimates in Report A-1430 for the -10/-12 Block II (J40-WE-10 and J40-WE-12 / YJ40-WE-5) and for the first time, the J40-WE-10/-12 Block III engines. For the Block III, IV and V engines, new model designations were never mentioned or apparently assigned. WAGT's block designation numbers have been used for these later follow-on models. No weights or dimensions of these projected later production block engines were included in the report. (However, charts (consolidated below) were found that show the expected thrust, SFC and weight achievements for Blocks II-V.) Complete model specification curves were scheduled for completion on January 2, 1953 for Block II and March 16, 1953 for Block III. Briefly, they projected:[9]

Block II (Projected August 11, 1952)						
	XJ40-WE-10 (XJ40-WE-24)			XJ40-WE-12 (XJ40-WE-26) XJ40-WE-5		
Power	Thrust lbs	RPM	SFC	Thrust lbs	RPM	SFC
Maximum	12,050	7,260	2.50	9,500	7,260	1.020
Military	9,275	7,260	1.045	9,500	7,260	1.020
Normal	8,330	7,260	1.028	8,550	7,260	1.005
90% Normal	7,500	7,260	1.027	7,700	7,260	1.003
75% Normal	6,250	7,260	1.068	6,400	7,260	1.047
Idle	555	3,000	5.000	560	3,000	5.000

During June when the de-rated approach was proposed by WAGT, they included the following charts:

The two graphs that follow were used by WAGT to show BuAer where the program was and how the different Blocks would step by step achieve BuAer's performance objectives. It is not known exactly when these were presented to BuAer, but the charts are dated in June of 1952, just about the time the proposal to move to de-rated engines was undergoing review by BuAer.

J40-WE-12
PERFORMANCE GROWTH
SEA LEVEL STATIC

THRUST (LBS) — SFC W_f/F_N — WEIGHT (LBS)

AFTER WAGT P47194, JUNE 25, 1952

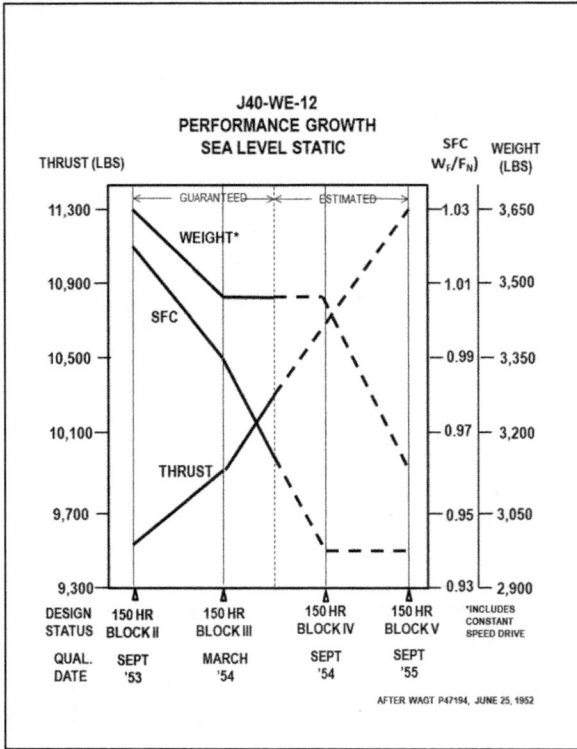

Figure 4 Detail of multi-petal afterburner nozzle, size unknown. WAGT P49513, 6/10/1953. *Courtesy Hagley Museum and Library*

Block III (Projected August 11, 1952)					
	XJ40-WE-10			XJ40-WE-12 / XJ40-WE-5		
Power	Thrust lbs	RPM	SFC	Thrust lbs	RPM	SFC
Maximum	13,100	7,260	2.50	9,900	7,260	0.990
Military	9,640	7,260	1.013	9,900	7,260	0.990
Normal	8,610	7,260	0.980	8,820	7,260	0.958
90% Normal	7,750	7,260	0.970	7,940	7,260	0.950
75% Normal	6,450	7,260	0.993	6,610	7,260	0.975
Idle	590	3,000	4.500	580	3,000	4.500

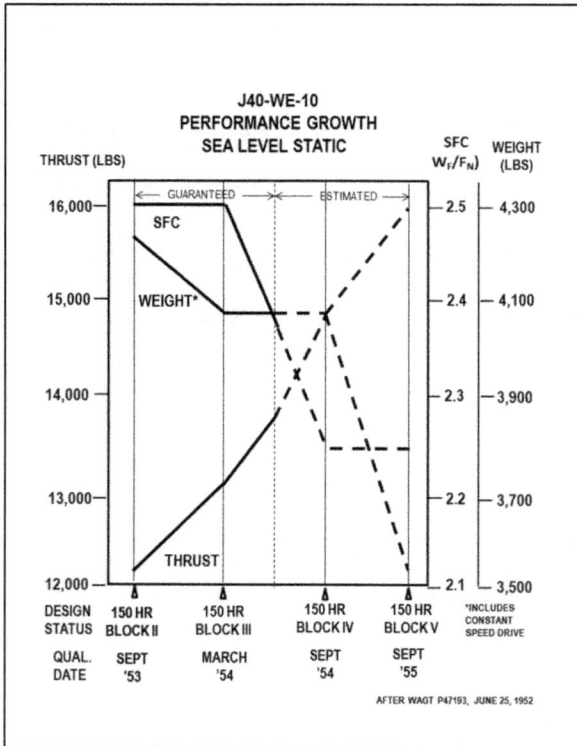

J40-WE-10
PERFORMANCE GROWTH
SEA LEVEL STATIC

THRUST (LBS) — SFC W_f/F_N — WEIGHT (LBS)

AFTER WAGT P47193, JUNE 25, 1952

Figure 5 XJ40-WE-10 and -24 Afterburner flameholder assembly and duct leading aft to the nozzle section shown above. Note new design. WAGT P49515, 6/10/1953. *Courtesy Hagley Museum and Library*

The contract for the -24 and -26 was cancelled in mid-September 1953.[10] In spite of contract cancellation, WAGT submitted their performance estimates for -10/-12 Block IV production engines and requested they be distributed to the airframe manufacturers. The charts were marked preliminary.[11] Note the XJ40-WE-5 no longer appears on the charts from this point.

Block IV (Projected August 28, 1952)						
	XJ40-WE-10			XJ40-WE-12		
Power	Thrust lbs	RPM	SFC	Thrust lbs	RPM	SFC
Maximum	14,200	7,260	2.45	10,250	7,260	0.962
Military	10,050	7,260	0.980	10,250	7,260	0.962
Normal	9,050	7,020	0.958	9,230	7,020	0.940
90% Normal	8,150	6,800	0.948	8,300	6,800	0.932
75% Normal	6,790	6,460	0.960	6,920	6,460	0.948
Idle	600	3,000	4.500	620	3,000	4.500

As can be seen from this chart, the engine would no longer be operated across the normal operating range as a "constant speed" engine, but would vary the RPM in that range.

A set of graphs dated June 25, 1952 and titled "J40-WE-10 (and -12) Performance Growth Sea Level Static" show a Block V projection to be achieved in the September 1955 timeframe. The performance projections were:

	XJ40-WE-10 Block V	XJ40-WE-12 Block V
Thrust	16,000 lbs (A/B)	11,300 lbs
SFC	2.25 (A/B)	0.94
Weight	3,600 lbs	3,125 lbs

Even though the -10/-12 and -24/-26 contracts were now terminated and development activities stopped, WAGT's publication of the Block II, III, and IV performance projections with the request that they be distributed to the airframe companies was likely an attempt by WAGT to get the airframe designers to put pressure on BuAer to continue development. There is no evidence in the files that the compressor stalling problems being experienced were ever resolved to allow full power development to continue.

No testing data was found relative to the new A/B design and if tested, it had to have been at the lower powers of the earlier -8/-22 models, since no -10 engine is believed to have been assembled.

After contract termination, Douglas was asked to formally respond to the survey on the 766 in² nozzle, estimating the cost of any airframe changes that would be needed. Why such follow-up was made after contract termination is not clear, unless a larger nozzle was also being considered for the -22 engine, but nothing in those records was found to indicate such a change.[12]

Figure 6 XJ40-WE-24 TIN-11 Ignition unit. Container was about a foot long and connected to the aircraft power system. WAGT P49676, 6/30/1953. *Courtesy Hagley Museum and Library*

NACA Testing of the Block II (XJ40-WE-24) Compressor

NACA tested a development compressor ("C" *number unknown*) for the Block II engines (the report calling the engine the J40-WE-24) and completed their report on April 30, 1953. The report summary stated:

"A 13-stage development compressor for the J40-WE-24 engine was investigated over a range of speeds from 30 to 112 percent of design speed. It was found that design equivalent flow and pressure ratio could not be attained at design speed. The design-speed peak pressure ratio of 5.58 was obtained at a flow of 145.8 pounds per second. The peak efficiency at design speed was 0.783. The maximum efficiency of 0.830 was obtained at 85 percent of design equivalent speed. The surge pressure ratio at 112 percent of design speed was 6.02 and was obtained at a flow of 156.1 pounds per second. The maximum efficiency at 112-percent speed was 0.730.

"An analysis was conducted to determine the reasons for the poor performance obtained at the design and over-design speeds. The results indicated that excessively high relative Mach numbers combined with negative angles of attack at the tip of the first-stage rotor limited the attainable performance of the compressor. Therefore, the failure to meet design performance specifications may be mainly attributed to the fact that first stage was over compromised for low-speed operation. Analytical investigation of the effect of reducing the turning angle of the inlet guide vanes indicated that even if design flow could be attained by such a modification, the efficiency of the compressor would be lower than that presented herein."[13]

The report states the compressor design was aimed at allowing efficient part-speed operation. The key design targets were: Airflow: 164 lbs/sec; Pressure Ratio – 6.0:1; Efficiency – 0.838, Max design rpm – 7,260. The report went on to explain the design approach of the compressor being tested:

"Simple radial equilibrium, correct for the mutual interference of adjacent blade rows, was used in the calculation of the original J40-WE-24 design, referred to in the present discussion as the "equilibrium" design compressor. Compressor tests made by the Westinghouse Electric Company on an engine having a compressor similar to the equilibrium design, except for higher rotor blade-setting angles as measured from the axial direction at the hub sections, indicated severe part-speed acceleration problems. On the basis of these tests, the manufacturer decided to modify the equilibrium compressor by twisting rotor and stator blade angles in an attempt to improve the part-speed operation. The resulting compressor, whose performance is presented herein, was modified from the equilibrium compressor by twisting the first seven rotors closed and the first five stators open."

Chapter 11 - XJ40-WE-24 and XJ40-WE-26 (XJ40-WE-5) Procurement and Development Citations

[1] WAGT Memo, June 17, 1952, "XJ40-WE-10 Block II Preliminary Estimated Performance" **B795VE7RG72.3.2NACP**

[2] WAGT Memo, May 6, 1952, "Contract NOa(s) 10067 WEC Turbo-Jet Engine Model J40WE-10. Survey Results Concerning Enlarged Exhaust Nozzle Area, Summation of." **B796VE6RG72.3.2NACP**

[3] BuAer Memo, July 22, 1952, "Contract NOa(s) 10067; maximum exhaust nozzle area for J40-WE-24 engines" **B796VE6RG72.3.2NACP**

[4] WAGT Memo, August 4, 1952, "Contract NOa(s) 10385, 10067; 52-403C, 52-048. WEC Turbojet Engine Models

J40WE-8, -22, -10; XJ40WE-10. Interim Exhaust Nozzle Control System and Hydro-Mechanical Exhaust Nozzle Control System, Description of." **B737VE11RG72.3.2NACP**

[5] BuAer Memo, February 27, 1953, "Contract NOa(s) 52-1171, Westinghouse Specification WAGT-X40E8A-6 dtd 17 Oct 52 covering the Model XJ40-We-24 engine" **B938VE12RG72.3.2NACP**

[6] BuAer Memo, May 11, 1953, Westinghouse J46-WE- and J40-WE- Series engines; temperature requirements for electrical and electronic components" **B3VJ40RG72.3.4NACP**

[7] BuAer Memo, June 11, 1953, "Contract NOa(s) 52-1171, Westinghouse Specification WAGT-40E8A-6A of 1 May 53 covering the Model J40-WE-24 engine" **B3VJ40RG72.3.4NACP**

[8] BuAer Memo, August 7, 1953, Contract NOa(s) 52-1171, Westinghouse Specification WAGT-50E8A-6A of 1 May 53 covering the Model J40-WE-24 engine" **B3VJ40RG72.3.4NACP**

[9] WAGT Report A-1430, (not dated), submitted to BuAer August 11, 1952 **B948VE16RG72.3.2NACP**

[10] BuAer Memo, September 18, 1953, "Contract NOa(s) 52-1171, Westinghouse Specifications WAGT-40E8A-6A and WAGT-40E6A-8 covering Models J40-WE-24 and -26 engines, respectively, status of" **B3VJ40RG72.3.4NACP**

[11] WAGT Memo, September 5, 1952, "Transmittal of Westinghouse Confidential Report A-1438 entitled "Preliminary Altitude Performance Estimates for the XJ40-WE-10 and -12 Block IV" **B796VE6RG72.3.4NACP**

[12] BuAer Memo, September 29, 1952, "Westinghouse engine installation survey; request for reply to" **B796VE8RG72.3.4NACP**

[13] NACA Research Memorandum, RM SE58D17, April 30, 1953, "Performance of a 13-Stage Development Compressor for the J40-WE-24 at Equivalent Speeds from 30 to 112 Percent of Design." **NACA Internet Report Archive Database**

Chapter 12

XJ40-WE-14 and XJ40-WE-16

◆━━━◆

Procurement

These engines were to follow the 40 inch diameter of the engines already under contract combined with non-A/B fuel consumption lower than the -6 model. The engines would use two spool compressors of 16:1 compression ratio. The -14 would be approximately 11 inches longer than the current -6. The Navy expressed their thinking on design targets in a memo of August 11, 1949 to WAGT. They stated WAGT should try to raise the individual pressure rise in the high pressure compressor spool to eliminate stages and save weight. They also asked for at least a 40% thrust rise on the A/B model. In summarizing BuAer's analysis and recommendations, they stated:

"In July of 1949, WAGT and BuAer signed a Letter of Intent contract NOa(s) 10943 for engines that stayed close to the sponsorship. It appears mandatory that aircraft gas turbine power plants to be manufactured under BuAer should reflect the best current American information regarding analytical, design, test and performance data. It is suggested that current British practice and attainment be properly evaluated and applied to future engines. It would also be advisable to leave some room in the design for possible future improvements. This type of program will obviate the possibility of obtaining a scaled up version of existing engines or engines based upon the design characteristics of obsolescent machines. It is believed that this type of thinking will result in a genuine research and development program, completely worthy of being sustained with the "hard to get" funds obligated for such endeavors."[1]

This statement shows that as far back as July 1949 BuAer considered the single spool designs of the earlier models, including the -10 and -12, to be limited and approaching obsolescence even while they were in development.

The detailed model specification (WAGT-X40E9A-2A for -14 with Appendix A covering the -16) was to be delivered to BuAer not later than January 31, 1950 (later moved to May 25, 1950). The justification supporting documentation states these would be "vastly superior" engines of 12,000 and 17,400 pounds thrust.[2] The development cost was estimated to be $9,875,000 at the

time the Letter of Intent was signed. The XJ40-WE-14 was WAGT model X40E9 and the XJ40-WE-16 was WAGT model X40E10.

No Air Force developments at that point were in direct conflict with the -14/-16 aimed at fighter aircraft missions. The only designs mentioned in the discussions were the XJ57-P-1 being developed for the XB52 program and the General Electric XJ53-GE-1. The former was seen as a bomber engine that was excessively heavy and too big for any known or projected Navy application. The latter was far too big for any projected Navy projects.

Allison and General Electric were offering future versions of the J35 and J47 at 9,700 lbs and 9,100 lbs thrust respectively. Specific performance and fuel consumption were inferior to the -12 under contract and appreciably inferior to the -14 being proposed.

Specifications for the XJ57-P-1 included in the justification memo:

Maximum Dry S.L. Thrust*	9,000 lbs
*projected to grow to 9,250 lbs with a variable area exhaust	
Normal S.L. Thrust	7,800 lbs
Dry Weight	4,380 lbs
Maximum S.L Mach Number	0.69M
Two spool axial compressor	12:1 CR

Specifications for the XJ53-GE-1 (at time of justification memo):

Maximum Dry S.L. Thrust	18,000 lbs
Normal S.L. Thrust	16,200 lbs
Dry Weight	5,500 lbs
Maximum S.L Mach Number	0.95M
Two spool axial compressor	6:1 CR

At the time of the -14/-16 procurement, BuAer considered that their engine development program over time would support the airframe developments planned as shown in this BuAer presentation slide attached to the -14/-16 procurement justification:

J40 Model Application Program As of November 19, 1949		
1950-51	**1951-1952**	**1953-1955**
J40-6 (7,500 lbs) XA3D-1 (2 engines)	J40-12 (9,500 lbs) A3D-1 (2 engines)	J40-14 (12,000 lbs) XA3D-2 (2-engines) A3D-2 (2-engines)
J40-8 (A/B 10,900 lbs) XF3H-1 XF4D-1 XF10F-1	J40-10 (A/B 13,700 lbs) F3H-1 F4D-1 F10F-1	J40-16 (A/B 17,400 lbs) a) Experimental and Production Version of high altitude interceptor fighter to replace the F4D-1 & F3H-1. b) Probable engine for single engine night fighter to replace the F3D-1 (J46). Also probable engine for attack or fighter seaplane under design study.

The proposal, dated September 23, 1949, stated that the proposed engines would be constructed to MIL-E-5007 and tested to MIL-E-5009 but that environmental testing was not part of the proposal because WAGT did not have existing test facilities to accomplish the work. If such testing was requested, WAGT would need at least three periods in the designated facilities to prepare the engines for the tests.

In November, BuAer asked WAGT to give them revised weight estimates on the -14 if it was built to lower maneuver load specifications as a bomber engine. Apparently BuAer did not give WAGT formal maneuver load specifications and WAGT initially used their own:

1. 6.0 g acting vertically up or down
2. 2.0 g acting fore or aft
3. Gyroscopic moments of inertia due to engine rotation of 2 radians/sec at maximum engine speed for 30 seconds
4. Maximum Sea Level speed 0.7M

Proposal Engine Specifications as of September 23, 1949		
	J40-WE-14 (X40E9)	**J40-WE-16 (X40E10)**
Maximum thrust	12,000 lbs	17,400 lbs
Military thrust	12,000 lbs	11,550 lbs
Military SFC (S/L)	0.781	2.20/0.810
Normal SFC (S/L)	0.766	0.800
Minimum in Flight (.85M/35,000 ft.)	0.905	0.929
Weight	4,850 lbs	5,723 lbs
Specific Wt.	.404	.329
Width	43.0 in	44.0 in
Height	47.9 in	47.9 in
Length	205.0 in	304.6 in
Compression ratio	16:1	16:1

Given the WAGT assumed flight loads, a study resulted in the following weight reductions possible:

Study of Weight Reduction of the J40-WE-14 as a Bomber Engine	
Initial bid specification weight	4,850.0 lbs
Removal of stiffening provisions to mount A/B	-30.0
Removal of lubrication system provisions for inverted flight	-20.0
Lowered "g" loading requirements for bomber engine	-80.0
Lowered sea level flight speed limit to 0.7M	-406.0
Bomber Engine Weight	3,414.0 lbs

Upon receipt of a subsequent BuAer letter[3] specifying slightly higher load limits than used in the above study, which also added the requirement that the -14/-16 be interchangeable, and asking that the -16 also be evaluated for weight reduction with those flight loads, WAGT responded with the following:

Revised Study of J40-WE-14 and J40-WE-16 as a Bomber Engine		
	J40-WE-14	**J40-WE-16**
Initial Bid Specification Weight	4,850 lbs	4,850 lbs
Afterburner (1.0M) weight	0	873
Lower maximum loadings	-35	-35
Reduction in S.L Speed (0.7M)	-406	-406
Removal of A/B Stiffening	-27	0
A/B weight reduction (0.7M)	0	-75
Est. Bomber Engine Weight	4,382 lbs	5,207 lbs

Design analysis raised the weight estimates of the above reduced factors engines to 4,409/5,207 pounds respectively if the interchangeability requirement was restored.[4] The new weights were approved by BuAer on March 30, 1950.[5]

Minimal design work beyond the work in Report A-1006 (*Not found- Auth.*) was done. The Progress Report for October, 1950 stated that matching the components was complete and only minor changes in design thinking for the study in Report A-1006 had resulted.[6]

No actual development work beyond that stated above was accomplished on either engine. The engineering resources were re-allocated to the J46 and XJ40-WE-10/-12 engine contracts. The contract was terminated in its entirety on July 16, 1951. It was anticipated that engines in the categories of the contract would be needed in the future and WAGT was encouraged to keep the data from their studies to be used in the future.

The specification was submitted by WAGT but a copy was not found in the files.

Chapter 12 - XJ40-WE-14 and XJ40-WE-16 Procurement Citations

[1] BuAer Memo, August 11, 1949, "Westinghouse Compound Axial Flow Compressor Design" **B1116VARG72.3.2NACP**

[2] BuAer contract justification document, December 5, 1949. **B1116VARG72.3.2NACP**

[3] BuAer Memo, January 13, 1950, "Contract NOa(s) 10943 – Westinghouse XJ40-WE-14/-16 turbo-jet engine design limitations, requested revisions to" **B1116VERG72.3.2NACP**

[4] WAGT Memo, February 8, 1950, "Contract NOa(s) 10943 Westinghouse XJ40-WE-14/16 Turbo Jet Engine Design Limitations, Revisions to" **B1116VERG72.3.2NACP**

[5] BuAer Memo, March 30, 1950, "Contract NOa(s) 10943 XJ40-WE-14/-16 engine weight and design limitations, request for revisions to engine model specifications" **B1116VERG72.3.2NACP**

[6] WAGT Memo, November 6, 1950, "Contract NOa(s) 10943, Progress Report" **B1116VERG72.3.2NACP**

Chapter 13

XJ40-WE-18 and XJ40-WE-20

The XJ40-WE-18 engine was proposed to support the "Minelayer" aircraft competition. Ultimately, this program produced the Convair P6M-1 SeaMaster flying boat powered by Allison J71-A-4 engines. The WAGT proposal was submitted as WAGT-X40E8A-2C Appendix D under Contract NOa(s) 10067 on April 28, 1951 as a follow-up to an earlier communication in February. The proposed J40 engine variant would be a modified J40-WE-10 with a simplified afterburner giving modest performance increases from sea level to six thousand feet at low airspeeds. As such it would be lighter and give the sea plane the needed thrust for a heavy high speed takeoff. The development schedule was to be 27 months from receipt of authorization to proceed.[1]

The U.S. Navy reviewed the proposal and rejected it as unacceptable on May 18, 1951. The reasons stated were:[2]

a. The estimated cost was significantly in excess of the cost of anticipated engine development work of that nature. A comparison had been made with the proposals for development of the high performance J40 and J46 afterburners for operation at high altitude that had confirmed that the proposed cost in this case was excessive.

b. The estimated development time of 27 months through qualification test was considered excessive. The project appeared to be of relatively moderate technical difficulty, particularly compared to the high altitude afterburner work.

c. It was believed that availability of "take-off" afterburners would prove excessive for such aircraft as attack and bomber types. Such afterburners were already under development for other engine types competitive with the J40. After further development work by Westinghouse on the current afterburners, it was felt Westinghouse would find themselves in a better position to make an acceptable proposal.

Even though rejected as is, the proposed engine remained as a part of the specification for the -10. In July, as part of a review of the updated specification WAGT-X40E8A-2C, WAGT was requested to remove Appendix D

covering the -18, stating that if the -18 proposal was accepted, a separate proposal would be requested.[3] This seems to indicate that consideration of the -18 in some form was still underway.

WAGT responded with a modified proposal on July 30, improving the schedule and lowering the cost. If authority to proceed was received by September 15, 1951, all work could now be completed by December, 1952 (16 months). The estimated cost was $645,482.

The shorter schedule and lower costs were arrived at by considering only the refinement and simplification of the XJ40-WE-10 afterburner for the purpose of providing take-off augmentation for the XJ40-WE-12 engine. The proposal was based on the 150-hour verification of only the component parts of the -18 engine which differed from those of the -10 engine in order to eliminate the need for a full qualification test which was originally part of the earlier proposal. Some materials and equipment derived from the development of the XJ40-WE-10, -12 and -20 would need to be made available as Government Furnished Equipment for the verification of the -18 parts.[4]

BuAer responded on November 16, 1951 acknowledging the receipt of the proposal. Possible airframe applications were under investigation. A final decision would not be possible until approximately May 1, 1952. The proposal was held in abeyance.[5]

No further memos on the proposal were found in the files and the "Mine Layer" project selected another engine. There the matter rested.

Specifications (Initial)

The engine would basically be a -10 engine with a different "short stack" non-modulated afterburner modified to operate between sea level and 6,000 feet at airspeeds below 155 knots. The exhaust nozzle was to be of the variable area type and was to be comprised of a fixed area outer cone and an axially moveable central plug. The plug would be actuated by a lever extending radially through one of the struts. The outer end of the lever was to be connected to a hydraulic actuator mounted externally on the engine. The A/B was designed to be used only for take-offs and it would appear no provision for air starts was to be provided. No other details were found.

Specifications (Revised)

The construction details submitted with the revised schedule and price are vague. They stated only that *"our change in approach to the development of this engine by considering only the refinement and simplification of the XJ40-WE-10 afterburner for the XJ40-WE-12 engine"*. This memo refers to unspecified equipment derived from the development of the XJ40-WE-10, -12 and -20 engines.

No final rejection memo regarding the -18 was found in the files. It likely died more as a result of lack of progress with the -10/-12 than anything else.

XJ40-WE-20

This was to be a modified -10 engine able to produce power to higher altitudes when operating in A/B. It would have a maximum A/B operating altitude of 55,000 feet and an A/B relight capability up to 50,000 feet. The Air Force in 1951 was notified of this capability, so it can be assumed the Air Force had some project in mind for the engine. No other details of this engine were found in the records.[6]

Chapter 13 - XJ40-WE-18 and XJ40-WE-20 Citations

[1] WAGT Memo, April 28, 1951, "Proposed Amendment to Contract NOa(s) – 10067 Our Reference Neg. 93856" **B794VE3RG72.3.2NACP**

[2] BuAer Memo, May 16, 1951, "Proposal for development of J40-WE-12 type engine with limited thrust augmentation" **B794VE3RG72.3.2NACP**

[3] BuAer Memo, July 9, 1951, "Contract NOa(s) 10067, Westinghouse Specification WAGT-X40E8A-2C covering the models XJ40-WE-10 and -12 engines" **B794VE3RG72.3.2NACP**

[4] WAGT Memo, July 30, 1951, "Proposed Amendment to Contract NOa(s)-10067 Our Reference Neg. 93856" **B794VE3RG72.3.2NACP**

[5] BuAer Memo, November 16, 1951, "Proposal for development of XJ40-WE-18 engine" **B795VE43RG72.3.2NACP**

[6] BuAer Memo, May 28, 1951, "Westinghouse XJ40-WE series engines; AN designations for" **B3VJ40RG72.4.3NACP**

Chapter 14

J40 Flight Test Program

Well into the development of the various J40 engine models, the lack of altitude testing facilities at Westinghouse and the limited availability of such facilities elsewhere indicated that a flight test program needed to be initiated. Early flight test results from flying the -6 in the XF3H-1, XF4D-1, and XA3D-1 experimental airframes further emphasized the need for extensive development outside of the experimental airframe programs.

Bailment contract NOa(s) 52-020-b was signed September 21, 1951 to establish the program. Airframes would be bailed to Westinghouse for flight testing of the J34, J40 and J46 engines. The J34 would be tested in a Douglas F3D-2 SkyKnight. J46 versions would also be tested using an F3D-2, a North American B-45 Tornado, and much later, a Chance Vought F7U-3 Cutlass. Of interest here is the testing to be carried out on various J40 models using the B-45. A project identification of TEC-WEC-PP-401 was assigned.

B-45A-5 (AF serial 47-049) was surveyed and selected for modification by North American under contract NOa(s) 52-621. The aircraft had been superseded in Air Force use by the B-45C model and this particular airframe was in poor condition. Extensive repairs were required to deal with corrosion. Spare parts were hard to obtain in a timely manner. This problem would plague the program as it moved ahead.

The airframe was modified by building and installing an engine nacelle with engine intakes on a pylon that could be lowered below the bomber in flight and raised back up closely under the bomb bay for landing. Instrumentation, fuel tanks and fuel line hook-ups were added to enable the engine to be operated in flight. Engines were to be monitored by testing instrumentation that recorded the operation for evaluation on the ground. The ultimate cost of $707,075 for repairing and modifying the airframe ended up being almost double the original estimate.

The establishment of a bailment program involving flight vehicles being flown and managed by a non-airframe company introduced many unanticipated problems. Airframe companies typically have all of their own flight related personnel, equipment, and an airfield to maintain and operate a bailed airframe. In Westinghouse's case, all of these had to be procured either as GFE or through purchase contracts. For one of the F3D programs involving J34

engines, an airfield had to be identified and approved (New Castle, Delaware in that case.) Establishment of a radio broadcasting station posed a problem, as the U.S. Navy did not have the authority to license such a station. Ultimately this was solved by BuAer arranging through the local BAR to use dedicated U.S. Navy frequencies and broadcasting equipment at each base for the duration of testing.

The B-45 presented major challenges. The Korean War was in progress and severe manpower shortages moved the modification work to a low priority. Corrosion had to be remediated, most of the landing gear seals had to be replaced, and the cockpit canopy panels needed repair or replacement. The airframe was polished and finally repainted in Navy blue lacquer to stop further erosion. The rear fuselage had to be strengthened to allow two 550 gallon fuel tanks for the test engine to be installed, with suitable vents and piping to the test engine. Minor modifications to the bomb bay doors and bomb bay itself had to be made. The press of other work forced North American to delete items from their work schedule.

All through the readiness-for-flight work-up period and during the actual test flying phase, the aircraft had continuing problems with fuel systems and tanks; landing gear malfunctions, cabin pressurization; canopy cracking and crazing; electrical wiring; control surfaces and the J47 engines. Much of the repair had to be done at locations other than the flight test location, involving much lost time and many ferry flights.

The initial testing location was to be Edwards AFB but was moved to the Chance Vought's Dallas facilities, requiring all the materials and personnel to move. At the beginning of 1955, the U.S. Navy consolidated and moved the Westinghouse Air Test program to the Naval Air Station in Olathe, Kansas, the contract becoming NOa(s) 55-198b. Again, everything had to be moved and a pit designed and built there to allow test engines to be installed and removed from the nacelle. During this period through the end of the J40 program, little test flying was accomplished as the B-45 aircraft continued to experience long groundings for inspections and repairs to both the airframe and J47 engines.

A constant source of dispute between BuAer and WAGT was the timeliness of reports and the frequency of reporting. Some reports were not submitted until over a month after the reporting period ended. The contract had conflicting clauses calling for bi-weekly and quarterly reporting. Eventually, reporting became monthly by contract modification.

Flight data transmissions also suffered from problems from time to time, adding delays as the data then had to be compiled and reduced manually before sending it to Kansas City.

It is difficult to determine the actual contribution the B-45 air testing made to the J40 program itself, if any. Actual air testing started a full year and a half after the beginning of the contract. By the end of 1955, the nacelle was being modified for J46 testing and the J40 components, engines and personnel were all returned to Kansas City, the J40 program having ended.[1]

Figure 1 B-45A-5 Serial 47-049 in flight with nacelle extended, *WAGT Photo in 12/1/1954 WAGT monthly report*

Figure 2 Test aircraft over the pit in Olathe, KS, *WAGT Photo in 12/1/1954 WAGT monthly report*

Summary of J40 Engine Air Testing Program Activity

Engine	Start	End	No. Flights	Testing Time Air (hrs)	Ground (hrs)
XJ40-WE-8 #4 Modified as a -22A	3/6/54	4/28/54	4	12.75	0.00
XJ40-WE-22A #7	5/19/54	7/19/54	10	21.23	12.54
XJ40-WE-22A #6 with the C-23, 12 Stage Compressor	7/29/54	8/8/54	5	4.9	3.65
XJ40-WE-22 #5	9/18/54	4/8/55	6	13.01	2.33
XJ40-WE-22A #7	Re-installed in June/July of 1955, but never air tested again. Some ground testing may have occurred, but details are not given in the reports.				

One interesting note to conclude this chapter is that the flight test reports covering the flights with the XJ40-WE-22A #6 engine show that the engine was never operated above 51,000 feet, that being attained in a zoom climb above the B-45's service ceiling of 45,000 feet. In their proposal to BuAer for the C-23 compressor modification, WAGT stated the engine had operated up to 60,000 feet without stalling. No reference or test report was found to support that claim.

Chapter 14 - J40 Flight Test Program Citation

[1] Contracts NOa(s) 52-020-b and NOa(s) 55-198-b, Bi-Weekly Reports 1-91. Some reports are missing, particularly in the early months of the first contract when no flight activity of the B-45 occurred. Activity from later missing reports was reconstructed from comments in prior and follow-on reporting. The last two reports reverted to monthly reporting at the U.S. Navy's request. Reports continue after No. 91 and those cover other testing for which the B-45 was used after the J40 program ended. **B2453-2454V1-12RG72NACP; B3685V1-6RG72NACP**

Chapter 15

Concluding Observations

Based on the events and details preserved in the remaining project documentation, can we draw any conclusions about the J40 technical program? The House of Representatives subcommittee review's scope only covered the last stages of the program specific to the F3H-1N and only those aircraft flying with the J40-WE-22A engines. Based on the last grounding order for the F3H-1N with J40WE-22A engines, it is not likely that any F3H-1N ever flew with the -22 version with its different control system. While McDonnell Aircraft Corporation submitted information to the Navy covering their total J40 program experience (See Appendix B) with all the versions of the engine they actually operated in flight, Douglas stayed out of view, having switched their two airframes (the F4D and A3D) to the new Pratt and Whitney J57. Grumman's F10F proved to have major aerodynamic control issues and was cancelled after the first two prototypes flew. In any event, the prime issues discussed were that the F3H-1N airframe suffered from engine unreliability and was underpowered using the J40-WE-22A. It was true that weight growth of the airframe in converting it from a pure day interceptor to a missile-armed all-weather interceptor was significant. All of this discussion failed to point to any specific technical deficiency of the engines. The Navy testified that all of the -22A/-22 engines in current use had passed their acceptance tests and met the specifications.

Westinghouse (W.W. Smith, Director of the Gas Turbine Division) in testimony during the subcommittee hearings, testified that they recognized five major areas where they (WAGT) could have done a better job on the contracts:[1]

1. *"A lack of technical manpower on jet-engine work. We did not expand our engineering staff fast enough.*
2. *We lacked experimental parts development facilities and manpower which we know now are vitally important to successful development in jet engines.*
3. *We placed inadequate emphasis on the basic aerodynamic aspects of research, no doubt because of some complacency resulting from our almost immediate success with the J30 and J34 engines.*

4. *We had too few "house", or experimental engines for house work.*
5. *Our operations were too widely scattered for most effective performance. Our development facilities were at South Philadelphia, Pa., and our production facilities at Kansas City, Mo. Our flight test facilities were in three wide-apart locations-Delaware, Texas and California."*

The technical record certainly supports these observations on the part of Westinghouse. They hide many other interrelated factors that also led to the ultimate failure.

Smith's examples of how the deficiencies had been or were being addressed were focused on the present, not the past situation, seemingly an argument that was intended to encourage BuAer to continue placing contracts with Westinghouse for future engines. Smith specifically mentioned the new 6,500 pound thrust PD-33, which Westinghouse stated was developed completely with $8M of Westinghouse resources and was proving to be highly successful in initial testing. The argument that Westinghouse had now caught up and was capable of development of advanced engines had no relevance to the subject then being reviewed and is of no help in understanding why the program failed.

Many other factors could have been mentioned as possible contributing causes for failure of the project. The late delivery of the first service cleared -6 and -8 engines meant that the engine program was now running in direct competition with viable alternatives from other first tier engine companies such as Pratt and Whitney and Rolls Royce. The aerodynamic efficiencies, materials advances, and growth potential of those engines were leap-frogging the Westinghouse design, which offered no growth/advancement path beyond the basic engines under contract. The engine design was now properly seen by BuAer as a technological dead end. With airframes ready and capable of utilizing more powerful and efficient engines than Westinghouse could deliver, BuAer was being pressured to move to other engines and to their credit, began to do so. With the important -10 and -12 engines not making progress on the test beds, by the end of 1951 the J40 program was in full retreat, its only value being to supply marginally acceptable lower powered engines to allow airframe testing. In such an atmosphere, recruiting top flight engineers and convincing senior management to make major investments was difficult. Very little factual information relative to the -10 and -12 engines has come to light over the years, a situation this volume has sought to rectify. Their importance to the airframe programs was crucial and BuAer's decisions to abandon the J40 completely were tightly tied to the failure of the -10/-12 engines to emerge from testing.

Certification testing was intended to establish a confidence in an engine prior to its use in flight. Here the testing program repeatedly failed to accomplish this. Early flight engines quickly manifested problems and flight safety risks that immediately undermined confidence in the engines. The lack of high altitude ground testing prior to flight demonstrated that such testing was critical in any well planned acceptance process. The compressor stalls experienced by all J40 engines ultimately proved resistant to complete elimination. Combined with frequent failures of accessories and the control system upon introduction to flight, the engine was soon perceived as both fragile and unreliable. In comparison, while on introduction the Pratt and Whitney J57 engine was also found to be prone to compressor stalling, it was a heavy, strong engine and absorbed repeated stalls without damage and soon had a great reputation for toughness.

The persistent leaks of both fuel and oil were not simply maintenance issues. They occurred frequently in flight and caused fires. Fires in flight are terribly dangerous and contributed to pilot wariness relative to the engine's operations. It is disturbing that none of the existing testing reports show any reference to leaks being experienced in test cell testing until the -22 started to be produced. It is unlikely that could be true. When the engines began to be used, such leaks were frequently experienced and a multitude of patches applied. Even as the -22A/-22's were being used, leaks were again much in evidence.

Lack of sufficient quality and inventory controls enabled both defective and/or down-level parts to be installed in engines and for entire engines to be constructed out of compliance with the accepted build standard for that model, tested, and then shipped to the field. While this was also a BuAer responsibility as part of engine acceptance, this was never adequately addressed and the groundings that resulted whenever each situation occurred must have contributed to the ever growing lack of confidence.

Lastly, with most of the J40 program already cancelled, and only a small number of true production engines being produced, the build quality of those engines began to decline just as the engine was finally being used in a production aircraft. It must be recognized that the morale of everyone left on the program had to be lower than ideal at that point and having an effect on individual performance. The complaints filed about engine condition on delivery seem to support this. This situation was not a primary cause of program failure, but repeated the continuing stream of failure in the field. Ultimately, severe unreliability was the final impression of anyone in contact with the engine at that point. Very few facts to the contrary emerged subsequently to change that impression. It is a wonder that WAGT testified that the Navy had never told them the engine was unreliable. Based on four years of engine issues in the field, it begs the question of just what Westinghouse's working definition of an "unreliable" engine was? We will never know.

Both the Navy and Westinghouse had no desire to do long term battle over the J40 failure and were undoubtedly eager to let it sink out of sight immediately after the hearings. Despite the J40 program failure, the Navy still required J46 and J34 engines and had very large contracts with the Westinghouse electronics divisions. Both parties had a continuing need for each other, at least for a few years. Westinghouse had the PD-33 on test (BuAer designation J54-WE-1) and hoped BuAer would buy it.

In one last vain attempt to gain some acknowledgement from the Navy of limited success, WAGT stated that the fact that the engines had produced several world records (both official and un-official) had demonstrated that the engine was not a failure. In such a narrow context of only a few limited situations, the best that can be said of those records is that if you put a powerful enough engine in an airplane it will go fast. True program success required the ability to do this day after day in the hands of everyday servicemen with reasonable safety. Here the J40 did not remotely measure up.

Chapter 15 - Concluding Observations Citations
[1] W.W. Smith testimony at House of Representatives Committee on Government Operations, October 24-27, 1955. **BY41603(aka.Y5123),Compartment 21,Shelf 3,RG287, NAB**

Appendix A

Ratings and Specifications

The compiled ratings and specifications for models of the J40 are presented for those models where sufficient data are available. The last known specification version for a model has been used, supplemented from various data in correspondence and reports. In some cases, actual performance might have been a bit better than the table would indicate, especially for the J40-WE-22/-22A, for which specific production acceptance reports were available. They reported consistent improvements better than the given specifications during almost all acceptance tests.

J40-WE-1 Ratings and Specifications (WAGT-X40E3A-2B)

Rating	Thrust (lbs)	RPM*	SFC Specification
Afterburner	10,900	7,260	2.70
Military	7,250	7,260	1.078
Normal	6,500	7,260	1.035
90% Normal	5,850	7,260	1.020
75% Normal	4,880	7,260	1.040
Idle	450	3,000	N/A

* Engine operated as a constant speed engine at and above 75% thrust levels.

Physical Details

Compressor	Axial, Single Spool, 11 Stages
Pressure Ratio	Design 4.6:1, Actual 5.1:1
Airflow	140.9 lbs/sec
Turbine	Reaction, 2 Stages
Exhaust Nozzle	Eyelid Type, 726 in², (First 2 Engines 675 in²)
Fuel	JP-3 (MIL-F-5624A), Avgas 94 Octane
Power Control	Single Lever, Simple Hydraulic Control, Manual Trim
Ignition	Low Tension, 2 Spark Plugs, A/B Hot Streak
Oil Consumption	Less than 0.7 lbs/hr
Oil Type	MIL-O-6081 Grade 1010, 5 Gal Max, Normal 4.5 Gal
Length	287 in Room Temp / 288.8 in Max Operating Temp
Width	41.5 in Room Temp / 41.8 in Max Operating Temp
Height	43.0 in Room Temp / 43.3 in Max Operating Temp
Diameter	40 7/8 in
Weight	Guaranteed 3,580 pounds

Operating Limits

Turbine Inlet Temperature, Idle	1,350°F
Accelerate from Idle to Military	5 seconds above 1,400°F, Never Exceed 1,700°F
Maximum Altitude	60,000 ft
Maximum Relight Altitude	40,000 ft/ Ram Pressure Ratio 1.2 or Higher
Maximum Altitude in A/B	50,000 ft / Ram Pressure 1.5 or Higher
Maximum A/B Relight Altitude	40,000 ft / Ram Pressure 1.2 or Higher
Continuous A/B Operation Limit	1 Hour
Inverted Negative "G" Flight	No Limit
Overhauls	Initial engines 50 hours, raised to 150 hrs later
Operating Environment Limits	SL Ram 1.38 at -65°F; Ram 1.89 at +100°F / 165°F Maximum

Production and Survivors

Total Production: 30 - Serials WE030101 to WE030130

Survivors: Two. Serials WE030102 and WE030107 are in the X-10 exhibit aircraft (Serial 19307) at the United States Air Force Museum. A serial number match between the engine logs and the actual engines has not been done due to difficulty of accessing the engines in the missile.

J40-WE-6 Ratings and Specifications (WAGT-X40E2-2E)

Rating	Thrust (lbs)	RPM*	SFC Specification	SFC Actual
Military	7,500	7,260	1.04	0.975
Normal	6,800	7,260	0.99	0.948
90% Normal	6,120	7,260	0.96	0.935
75% Normal	5,100	7,050	0.94	0.937
Idle	450	3,000	4.00	3.170

* Engine operated as a constant speed engine above 75% thrust levels.

Physical Details

Compressor	Axial, Single Spool, 11 Stages
Pressure Ratio	Design 4.61, Actual 5.1:1
Airflow	140.9 lbs/sec
Turbine	Reaction, 2 Stages
Exhaust Nozzle	Eyelid Type
Fuel	JP-4 (MIL-O-5624A), Alt. JP-3, Avgas 94 Octane
Power Control	Single Lever modulated by electronic control to prevent over-speeds and over-temps
Ignition	Low Tension, 2 Spark Plugs
Oil Consumption	Less than 0.7 lbs/hr
Oil Type	AN-O-9 type, 1010, 5 Gal Max, Normal 4.5 Gal
Length	186 in Room Temp / 187 in Max Operating Temp
Width	42.25 in Room Temp / 42.50 in Max Operating Temp
Height	45.5 in Room Temp / 45.75 in Max Operating Temp
Diameter	40 7/8 in
Weight	Guaranteed 2,974 pounds

Operating Limits

Turbine Inlet Temperature, Idle	1,400°F
Accelerate from Idle to Military	5 seconds above 1,400°F, Never Exceed 1,700°F
Maximum Altitude	60,000 ft
Maximum Relight Altitude	40,000 ft/ Ram Pressure Ratio 1.2 or Higher
Inverted Negative "G" Flight	30 Seconds Maximum
Overhauls	Initial engines 50 hours, raised to 100 hrs later
Operating Environment Limits	SL Ram 1.6 at -65°F; Ram 1.89 at +100°F / 165°F Maximum

Production and Survivors

Total Production: 23 engines built. Serials WE030001 to WE030023
Survivors: None. All engines were scrapped at Essington at the end of the program.

J40-WE-8 Ratings and Specifications (WAGT-X40E2-4B)

Rating	Thrust (lbs)	RPM*	SFC Specification	Actual**
Afterburner	10,900	7,260	2.48	2.18
Military	7,250	7,260	0.98	1.026
Normal	6,500	7,260	0.93	0.993
90% Normal	5,850	7,260	0.90	0.992
75% Normal	4,880	7,260	0.88	1.052
Idle	450	3,000	4.00	4.000

* Engine operated as a constant speed engine at and above 75% thrust levels. ** WE031002 Acceptance

Physical Details

Compressor	Axial, Single Spool, 11 Stages
Pressure Ratio	Design 4.6:1, Actual 5.1:1
Airflow	140.9 lbs/sec
Turbine	Reaction, 2 Stages
Exhaust Nozzle	Eyelid Type, 726 in²
Fuel	JP-4 (MIL-F-5624A), Alt. JP-3, Avgas 94 Octane
Power Control	Single Lever, Simple Hydraulic Control, Manual Trim
Ignition	Low Tension, 2 Spark Plugs, A/B Hot Streak
Oil Consumption	Less than 0.7 lbs/hr
Oil Type	MIL-L-7808 type, 1010, 5 Gal Max, Normal 4.5 Gal
Length	287 in Room Temp / 288.8 in Max Operating Temp
Width	41.5 in Room Temp / 41.8 in Max Operating Temp
Height	43.0 in Room Temp / 43.3 in Max Operating Temp
Diameter	40 7/8 in
Weight	Guaranteed 3,615 pounds

Operating Limits

Turbine Inlet Temperature, Idle	1350°F
Accelerate from Idle to Military	5 seconds above 1,400°F, Never Exceed 1,700°F
Maximum Altitude	60,000 ft
Maximum Relight Altitude	40,000 ft/ Ram Pressure Ratio 1.2 or Higher
Maximum Altitude in A/B	50,000 ft / Ram Pressure 1.5 or Higher
Maximum A/B Relight Altitude	40,000 ft / Ram Pressure 1.2 or Higher
Continuous A/B Operation Limit	1 Hour
Inverted Negative "G" Flight	No Limit
Overhauls	Initial engines 50 hours, raised to 150 hrs later
Operating Environment Limits	SL Ram 1.38 at -65°F; Ram 1.89 at +100°F / 165°F Maximum

Production and Survivors

Total Production: 15 engines built. Serials WE031001 to WE0310015

Survivors: Serial WE031006 is in the Smithsonian Air and Space Museum engine collection. This was the engine used in the speed record and circular course records set by the XF4D-1 Skyray. Also later used by Grumman in the XF10F-1 for flight testing.

J40-WE-10 Ratings and Specifications (WAGT-X40E2-8A)

(See Chapter 11 for Planned De-rated Block II, III, IV and V)

Rating	Thrust (lbs)	RPM*	SFC
Afterburner**	13,700	7,260	2.48
Military**	9,275	7,260	0.950
Normal	8,330	7,260	0.893
90% Normal	7,500	7,260	0.850
75% Normal	6,250	7,260	0.805
Idle	555	3,000	5.00

* Engine operated as a constant speed engine at and above 75% thrust levels. ** Max 5 min/30 min total before rejection to overhaul. Essentially "sprint" ratings trading fatigue life for thrust.

Physical Details

Compressor	Axial, Single Spool, 13 Stages
Pressure Ratio	6.0:1
Airflow	164.0 lbs/sec
Turbine	Reaction, 2 Stages
Exhaust Nozzle	Eyelid Type, 726 in²
Fuel	JP-4 (MIL-F-5624A), Alt. JP-3, Avgas 94 Octane
Power Control	Single Lever, Hydro-Mechanical
Ignition	Low Tension, 2 Spark Plugs, A/B Hot Streak
Oil Consumption	Less than 0.7 lbs/hr
Oil Type	MIL-L-8188 type, 1010, 5 Gal Max, Normal 4.5 Gal
Length	287 in Room Temp / 289 in Max Operating Temp
Width	41.5 in Room Temp / 41.8 in Max Operating Temp
Height	43.0 in Room Temp / 43.3 in Max Operating Temp
Diameter	42.5 in Room Temp / 42.7 in Max Operating Temp
Weight	Guaranteed 3,968 pounds

Operating Limits

Turbine Inlet Temperature, Idle	1,310°F
Accelerate from Idle to Military	5 seconds above 1,400°F, Never Exceed 1,700°F
Maximum Altitude	60,000 ft
Maximum Relight Altitude	45,000 ft/ Ram Pressure Ratio 1.2 or Higher
Maximum Altitude in A/B	NLT 45,000 ft / Ram Pressure 1.5 or Higher
Maximum A/B Relight Altitude	40,000 ft / Ram Pressure 1.2 or Higher
Continuous A/B Operation Limit	1 Hour
Inverted Negative "G" Flight	No Limit
Overhauls	150 Hours
Operating Environment Limits	Ram 1.38 at -65°F; Ram 1.89 at +100°F / 165°F Maximum, Sea Level – No Limit

Production and Survivors

Total Production: No Complete -10 engines are known to have been built.
Survivors: None. All components scrapped at program end.

J40-WE-12 Ratings and Specifications (WAGT-E40E2-8A)
(See Chapter 11 for Planned De-rated Block II, III, IV and V)

Rating	Thrust (lbs)	RPM*	SFC
Maximum**	9,500	7,260	0.930
Military**	9,500	7,260	0.930
Normal	8,550	7,260	0.870
90% Normal	7,700	7,260	0.830
75% Normal	6,400	7,260	0.767
Idle	560	3,000	5.00

* Engine operated as a constant speed engine at and above 75% thrust levels. ** Max 5 min/30 min total before rejection to overhaul. Essentially "sprint" ratings trading fatigue life for thrust.

Physical Details

Compressor	Axial, Single Spool, 13 Stages
Pressure Ratio	6.0:1
Airflow	164.0 lbs/sec
Turbine	Reaction, 2 Stages (1st Engine Var. 1st Stage Nozzle)
Exhaust Nozzle	Cone Type, 726 in² Max
Fuel	JP-4 (MIL-F-5624A), Alt. JP-3, Avgas 94 Octane
Power Control	Single Lever, Hydro-Mechanical
Ignition	Low Tension, 2 Spark Plugs
Oil Consumption	Less than 0.7 lbs/hr
Oil Type	MIL-L-8188 type, 1010, 5 Gal Max, Normal 4.5 Gal
Length	197 in Room Temp / 198 in Max Operating Temp
Width	42.5 in Room Temp / 41.8 in Max Operating Temp
Height	43.0 in Room Temp / 43.3 in Max Operating Temp
Diameter	42.25 in Room Temp / 42.5 in Max Operating Temp
Weight	Guaranteed 3,396 pounds

Operating Limits

Turbine Inlet Temperature, Idle	1310°F
Accelerate from Idle to Military	5 seconds above 1,400°F, Never Exceed Temp 1700°F
Maximum Altitude	60,000 ft
Maximum Relight Altitude	45,000 ft/ Ram Pressure Ratio 1.2 or Higher
Maximum Altitude in A/B	N/A
Maximum A/B Relight Altitude	N/A
Continuous A/B Operation Limit	N/A
Inverted Negative "G" Flight	No Limit
Overhauls	150 Hours
Operating Environment Limits	Ram 1.38 at -65°F; Ram 1.89 at +100°F / 165°F Maximum, Sea Level – No Limit

Production and Survivors

Total Production: At least 4 house engines built, 3 destroyed by compressor failures.
Survivors: None. All components scrapped at program end.

J40-WE-14 Ratings and Specifications (WAGT X40E9)

Rating	Thrust (lbs)	RPM	SFC
Maximum	12,000	N/A	0.781
Military	12,000	N/A	0.781
Normal	8,550	N/A	0.766
90% Normal	N/A	N/A	N/A
75% Normal	N/A	N/A	0.905
Idle	N/A	N/A	N/A

Physical Details

Compressor	Axial, Dual Spool, 8 Low/13 High Stages
Pressure Ratio	16.0:1
Airflow	N/A
Turbine	Reaction, 2 Stages
Exhaust Nozzle	N/A
Fuel	JP-4 (MIL-F-5624A)
Power Control	New – never developed
Ignition	N/A
Oil Consumption	Less than 0.7 lbs/hr
Oil Type	N/A
Length	205 in
Width	43.0 in
Height	47.9 in
Diameter	N/A
Weight	Proposed 4,850 pounds Bomber Engine - WAGT Assumptions: 3,414 lbs Bomber Engine – BuAer Assumptions: 4,382 lbs

Operating Limits
(Not Published – Would likely have been the same as the -12)

Turbine Inlet Temperature, Idle	N/A
Accelerate from Idle to Military	N/A
Maximum Altitude	N/A
Maximum Relight Altitude	N/A
Maximum Altitude in A/B	N/A
Maximum A/B Relight Altitude	N/A
Continuous A/B Operation Limit	N/A
Inverted Negative "G" Flight	No Limit
Overhauls	150 Hours
Operating Environment Limits	Ram 1.38 at -65°F; Ram 1.89 at +100°F / 165°F Maximum, Sea Level – Mach 1 Ram

Production and Survivors

Total Production: No hardware built. Contract cancelled.
Survivors: None

J40-WE-16 Ratings and Specifications (WAGT-X40E10)

Rating	Thrust (lbs)	RPM	SFC
Maximum A/B	17,400	N/A	2.20
Military	11,550	N/A	0.810
Normal	N/A	N/A	0.800
90% Normal	N/A	N/A	N/A
75% Normal	N/A	N/A	0.929
Idle	N/A	N/A	N/A

Physical Details

Compressor	Axial, Dual Spool, 8 Low/13 High Stages
Pressure Ratio	16.0:1
Airflow	N/A
Turbine	Reaction, 2 Stages
Exhaust Nozzle	Iris, Multi-petal Type
Fuel	JP-4 (MIL-F-5624A)
Power Control	New – never developed
Ignition	N/A
Oil Consumption	Less than 0.7 lbs/hr
Oil Type	N/A
Length	304.6 in
Width	44.0 in
Height	47.9 in
Diameter	N/A
Weight	Proposed 5,723 pounds

Operating Limits
(Not Published – Would likely have been the same as the -12)

Turbine Inlet Temperature, Idle	N/A
Accelerate from Idle to Military	N/A
Maximum Altitude	N/A
Maximum Relight Altitude	N/A
Maximum Altitude in A/B	N/A
Maximum A/B Relight Altitude	N/A
Continuous A/B Operation Limit	N/A
Inverted Negative "G" Flight	No Limit
Overhauls	150 Hours
Operating Environment Limits	Ram 1.38 at -65°F; Ram 1.89 at +100°F / 165°F Maximum, Sea Level – Mach 1 Ram

Production and Survivors

Total Production: No hardware built. Contract cancelled.	
Survivors: None	

J40-WE-18 Ratings and Specifications (Initial Proposed)

Rating	Thrust (lbs)	RPM*	SFC
Afterburner (to 6K ft)	11,400	7,260	1.700
Military (90 Mins)	9,530	7,260	0.945
Normal	8,380	7,260	0.887
90% Normal	7,540	7,260	0.841
75% Normal	6,290	7,260	0.798
Idle	557	3,000	4.000

* Engine operated as a constant speed engine at and above 75% thrust levels.

Physical Details

Compressor	Axial, Single Spool, 13 Stages
Pressure Ratio	Design 6.0:1
Airflow	160.0 lbs/sec
Turbine	Reaction, 2 Stages
Exhaust Nozzle	Two Position, Cone Type
Fuel	JP-4 (MIL-F-5624A), Alt. JP-3, Avgas 94 Octane
Power Control	Single Lever, Simple Hydraulic Control, Manual Trim
Ignition	Low Tension, 2 Spark Plugs, A/B Hot Streak
Oil Consumption	Less than 0.7 lbs/hr
Oil Type	MIL-L-7808 type, 1010, 5 Gal Max, Normal 4.5 Gal
Length	217.4 in Room Temp / 218.5 in Max Operating Temp
Width	42.4 in Room Temp / 42.6 in Max Operating Temp
Height	44.1 in Room Temp / 44.3 in Max Operating Temp
Diameter	39.0/39.2 in
Weight	Guaranteed 3,681 pounds*

Operating Limits

Turbine Inlet Temperature, Idle	1350°F
Accelerate from Idle to Military	5 seconds above 1,400°F, Never Exceed 1,700°F
Maximum Altitude	60,000 ft.
Maximum Relight Altitude	40,000 ft/ Ram Pressure Ratio 1.2 or Higher
Maximum Altitude in A/B	6,000 ft / Ram Pressure 1.5 or Higher*
Maximum A/B Relight Altitude	Take off only – no relight capability*
Continuous A/B Operation Limit	1.5 Hours
Inverted Negative "G" Flight	No Limit
Overhauls	150 Hours
Operating Environment Limits	SL Ram 1.38 at -65°F – Ram 1.89 at +100°F / 165°F Maximum

Production and Survivors

Total Production: None – Proposals for all variations not accepted by BuAer or the AF.

Survivors: No development or production engines were manufactured.

*See Chapter 12 for variations depending on mission assumptions. XJ40-WE-20 version would have had high altitude A/B functioning ability.

J40-WE-22, -22A, -22B,-22A-G Ratings and Specifications
(WAGT-X40E2A-4D)

Rating	Thrust (lbs)	RPM*	SFC Specification
Afterburner	10,900	7,260	2.70
Military	7,250	7,260	1.078
Normal	6,500	7,260	1.035
90% Normal	5,850	7,260	1.020
75% Normal	4,880	7,260	1.040
Idle	450	3,000	N/A

* Engine operated as a constant speed engine at and above 75% thrust levels.

Physical Details

Compressor	Axial, Single Spool, 11 Stages
Pressure Ratio	Design 4.6:1, Actual 5.1:1
Airflow	140.9 lbs/sec
Turbine	Reaction, 2 Stages
Exhaust Nozzle	-22/-22A Eyelid Type, 726 in^2, -22B Iris Type 726 in^2
Fuel	JP-4 (MIL-F-5161B), JP-3, Avgas 94 Octane
Power Control	-22A Simple Hydraulic Control w/Auto Trim -22 Simple Hydraulic Control w/Hydro-Mechanical Trim
Ignition	Low Tension, 2 Spark Plugs, A/B Hot Streak
Oil Consumption	Less than 0.7 lbs/hr
Oil Type	MIL-O-6081 Grade 1010, 5 Gal Max, Normal 4.5 Gal
Length	287 in Room Temp / 288.8 in Max Operating Temp
Width	41.5 in Room Temp / 41.8 in Max Operating Temp
Height	43.0 in Room Temp / 43.3 in Max Operating Temp
Diameter	40 7/8 in
Weight	Guaranteed 3,615 pounds

Operating Limits

Turbine Inlet Temperature, Idle	1350ºF
Accelerate from Idle to Military	5 seconds above 1,400ºF, Never Exceed 1,700ºF
Maximum Altitude	60,000 ft
Maximum Relight Altitude	40,000 ft / Ram Pressure Ratio 1.2 or Higher
Maximum Altitude in A/B	50,000 ft / Ram Pressure 1.5 or Higher
Maximum A/B Relight Altitude	40,000 ft / Ram Pressure 1.2 or Higher
Continuous A/B Operation Limit	1 Hour
Inverted Negative "G" Flight	No Limit
Overhauls	Initial engines 100 hours, raised to 150-180 hrs later
Operating Environment Limits	SL Ram 1.38 at -65ºF; Ram 1.89 at +100ºF / 165ºF Max

Production and Survivors

Total Production: -22 Model: 77; Some not acceptance tested, 12 went to the Air Force, 5 as "-22A-G",
Serials WE302551-WE302627
-22A Model: 50; Serials WE302501 – WE302550

Survivors: One. J40-WE-22A is a cutaway exhibit at the National Naval Aviation Museum.
Original serial number was likely WE032516, but not verifiable.

Appendix B

Engine Operating Limits Charts

The charts presented are the best available copies after being cleaned up. Charts were not found for some models and it is likely they were never produced. Drafting such a chart in the late 1940's and 1950's era was a significant effort and the originals were in a 24 inch by 24 inch layout. The remaining copies in the BuAer contract files had been reduced to fit on 8.5x11 inch paper with one inch margins, the readability suffering considerably.

It is important to remember that all of the data points on such charts were calculated using formulas and do not represent actual data taken during altitude test chamber running or actual flight testing. An enormous amount of manual calculation using slide rules and mechanical calculator machines was required to determine the various data points. When actual flight and tunnel tests returned different values from those calculated, it brought the formulas being used into question. The use of such charts, however, represented the state of the art available at the time to aircraft designers in predicting airframe performance capabilities.

XJ40-WE-1

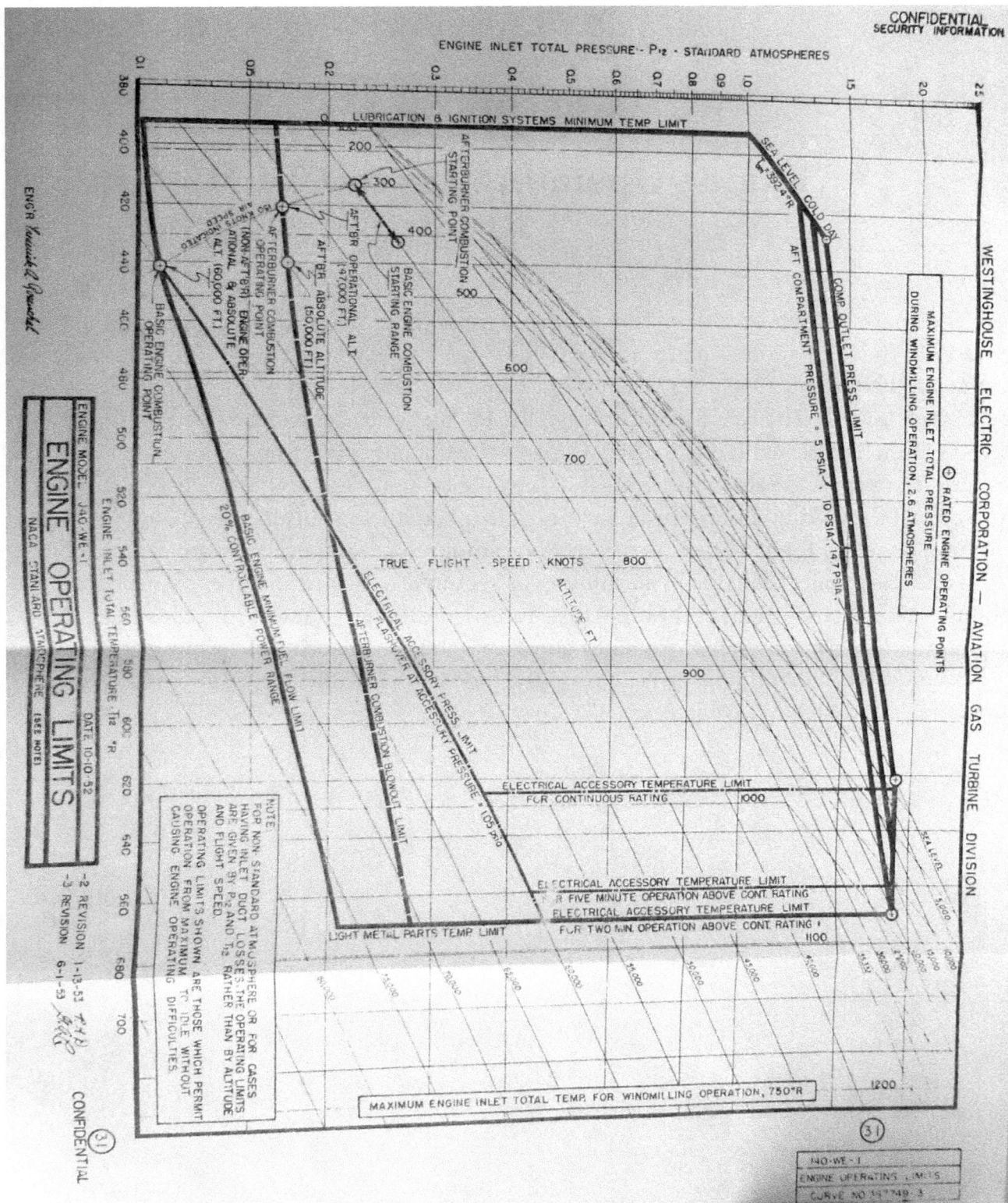

XJ40-WE-10 and XJ40-WE-12

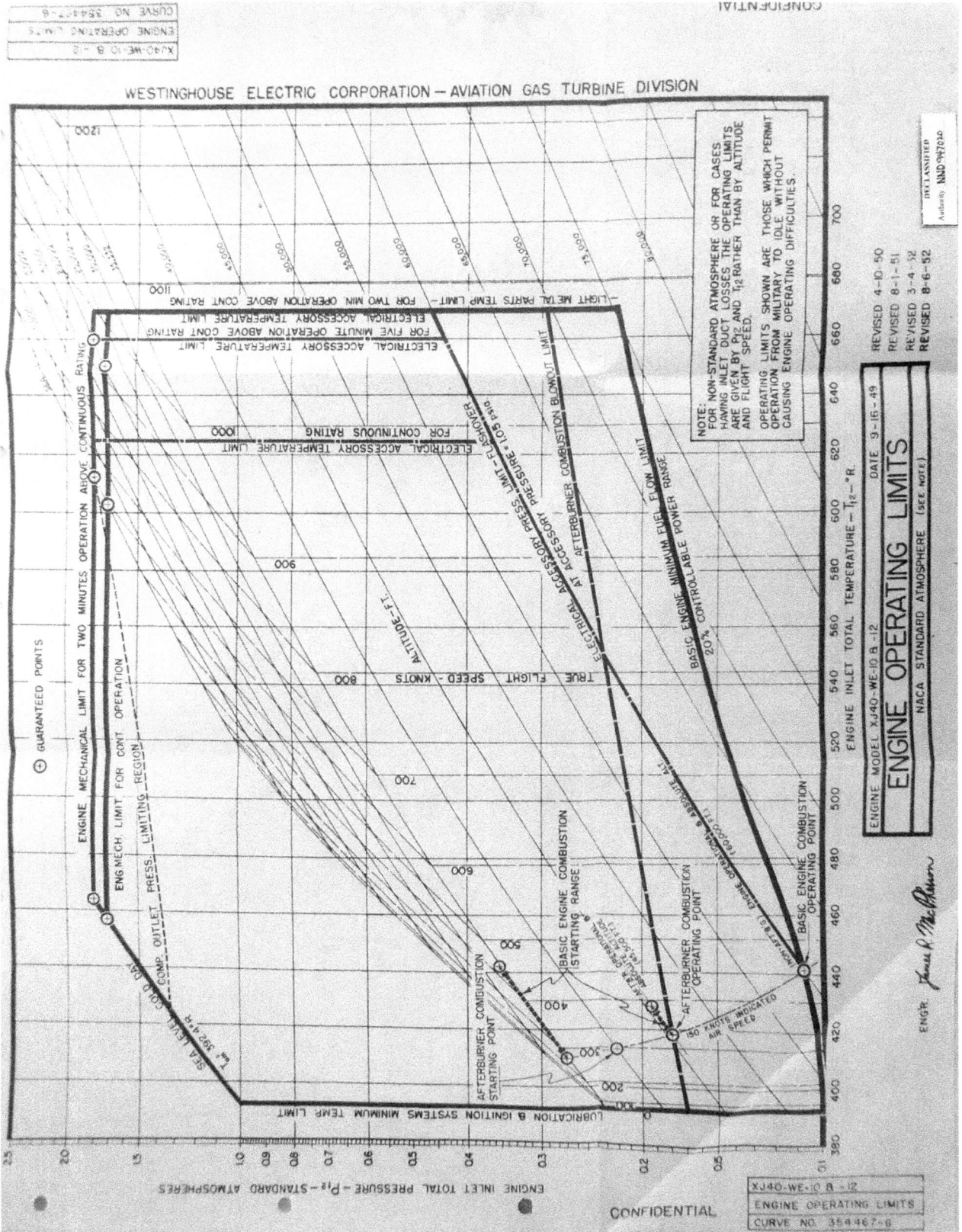

J40-WE-24

XJ40-WE- (Block IV) – Production version of -10

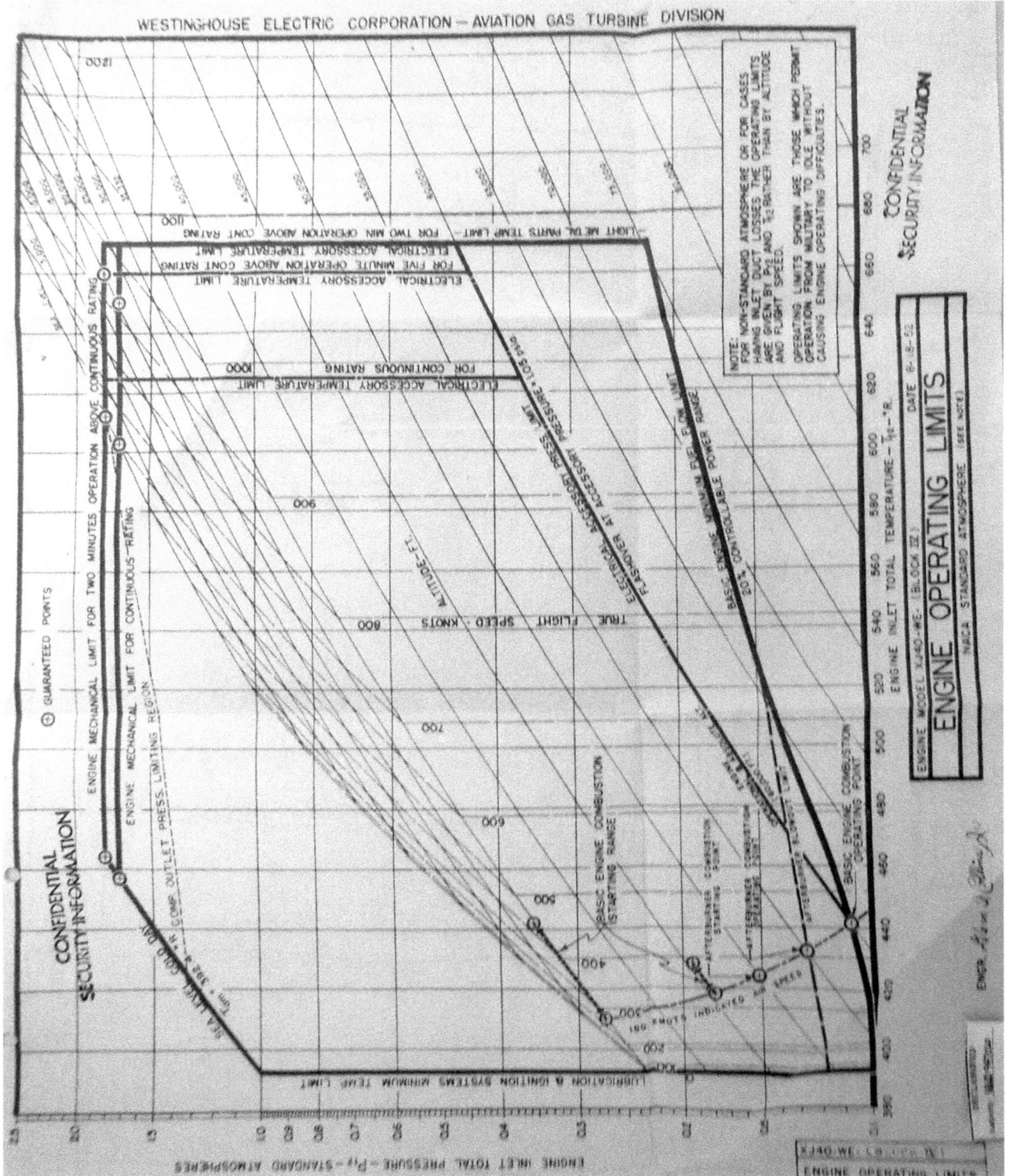

XJ40-WE-8 and -22

XJ40-WE-6 and XJ40-WE-8

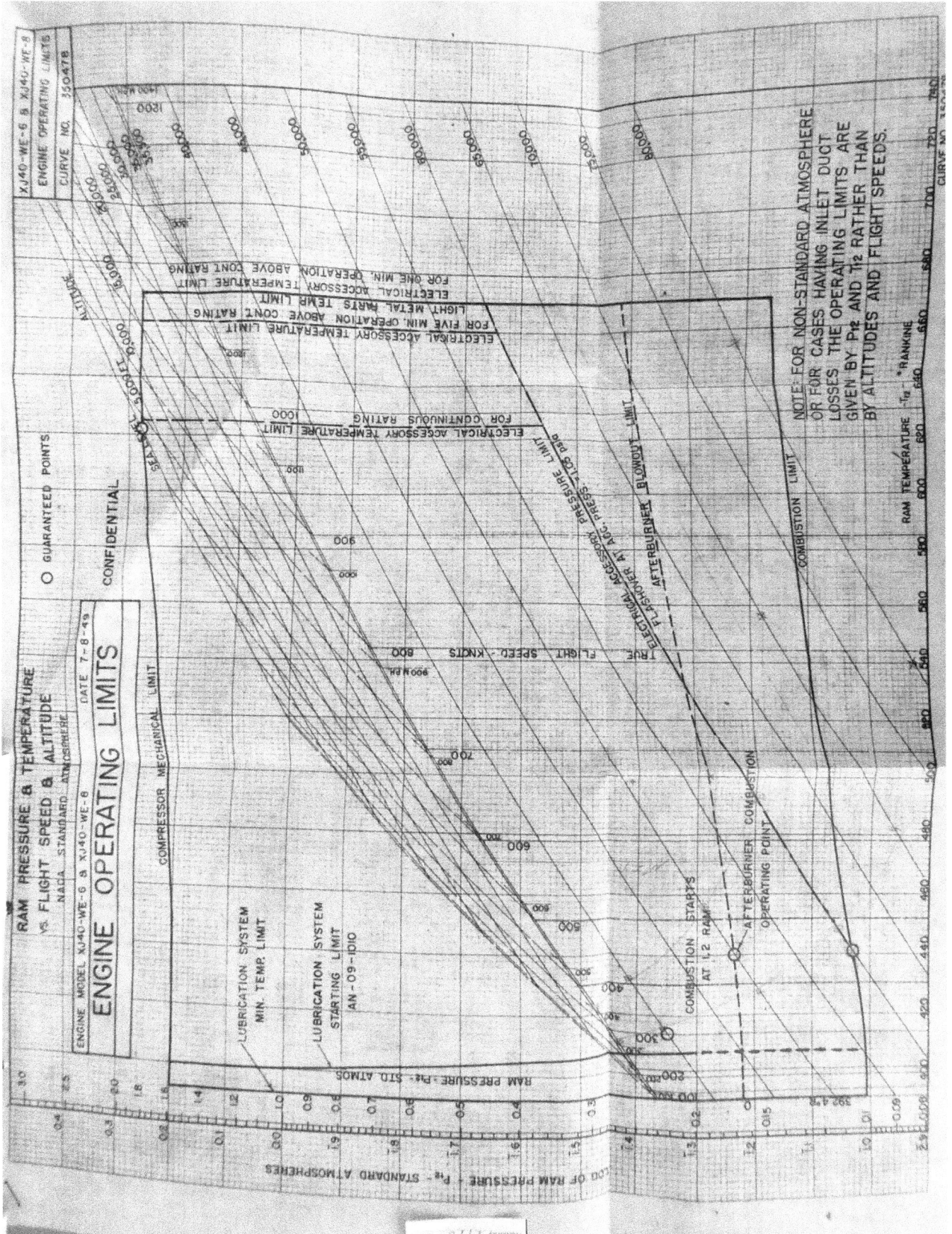

Appendix C

Government Hearings

From October 24-27, 1955, a subcommittee of the Government Operations Committee of the United States House of Representatives held hearings on the U.S. Navy Jet Aircraft Procurement Program. The stated purpose was to hold a public hearing on certain aspects of the Navy jet plane procurement program, particularly those related to the procurement of J40 jet engines manufactured by the Westinghouse Electric Corporation, and the F3H aircraft manufactured by McDonnell Aircraft Corporation. The public hearing was triggered by reports of unflightworthy aircraft sitting on the ramp at McDonnell's factory in St. Louis, having been found unsuitable by the Navy Department when powered with J40-WE-22 Westinghouse engines. It was known that a more powerful engine, the Allison J-71, was being retrofitted to some of the airframes at considerable cost. Airframes found unsuitable for conversion were grounded permanently and were to be used for ground instruction or other testing uses that could be found. Any excess J40-WE-22/-22A engines in inventory after contract settlement with Westinghouse would be used for other purposes but not flown.

The hearing record reflects the difficulties of trying to review highly technical matters in regards to decision-making and collection of data in a comprehensive and understandable form. The members of the committee were not technically trained and also lacked a basic understanding of contemporary government and military procurement policy and procedures. These obstacles were further compounded by the fact that much of the information was classified at the time and many technical details could only be discussed in general terms in the open hearings. The committee held executive closed sessions to cover classified subjects, but none of the details of those sessions were ever made public.[1]

Other issues crept into the hearing, particularly the practice of industry hiring of retired military personnel in work related to their prior government assignments, the procurement use of "termination for default" vs. "termination at government convenience" policies of the Navy, and excessive foreign object damage to engines being incurred at the McDonnell plant and airfield. Being outside the scope of this book, these will not be reviewed here except to say that no pattern of FOD damage to Westinghouse J40 engines by any airframe manufacturer emerged from the correspondence in the contract files.

The Navy's explanation of the situation was that decisions relating to airframe and engine production and the plans to use the F3H-1N aircraft with interim J40-WE-22 engines had been influenced by the delays and final failure of the -10 engine to emerge from development and the late availability of a flight qualified Allison J71 engine as the replacement. (It took two years to get a reliable afterburner for the J71.) This resulted in a decision to proceed with limited production of a knowingly underpowered airframe using the lower thrust -22 engine, with the J71 to be used in the last 30 (of 60 airframes ordered) and the -22 on the first 30. The logic for flying the Demon with the lower powered engine was that pilot and ground crew training on the new Demon fighter were of value and could continue while the transition was made. The problem that then emerged was that the F3H-1N had assumed the -10 engine would be available and the weight of the airframe (up from 22,000 lbs. of the XF3H-1 to 29,000 lbs. in the F3H-1N) reflected this assumption. With the development of the -10 cancelled because of inability to solve the compressor stall issues, the lower power -22 was the only timely available and easily installed alternative. The resulting engine/airframe mismatch apparently had a more serious impact on safe flight operations than expected. This plus the continuing problems with general J40 reliability and pilot deaths when flying the F3H-1N (none attributed to the J40 other than by inference) resulted in the Navy grounding the airframe with J40's installed and the end of J40-WE-22 production.

The Navy defended the use of the J40 even in the face of problems during its development years, citing knowledge gained in airframe development and the various records achieved by aircraft using some models of engine. (These were: XF4D-1 World Speed Record October 3, 1953, 752.9 mph; XF4D-1 World Record 100 km Closed Course, October 16, 1953, 728.12 mph; both using a -8.) McDonnell added that the F3H-1N (using a -22A) had recently set an (unofficial) time to climb record (but did not give the statistics, those being classified) and had used the -22A engine to deck qualify the F3H-1N. (Not part of the hearing, but Douglas could have added that the XF4D-1 was supersonic with a -8 installed (marginally) and they had also run deck trials of the XF4D-1 using a -8. They had moved on to other manufacturers' engines by the time of the hearing and so did not participate.) For completeness here, in 1956 (exact date not known) the Air Force recorded an X-10 (-1 engines) flight speed of 2.04M on Flight 19 of the program flown from the Air Force Missile Test Center in Florida.[5]

Six months after the hearings, a report was submitted by the sub-committee to the House of Representatives and published in March, 1956. Other than doing a reasonable job of summarizing in a non-technical way the substance of the hearings and presenting some recommendations, it does not add anything of a technical nature.[2]

The hearings did capture many details of the J40 program never recorded anywhere else. In particular, the costs, uses and troubles encountered when operating the engine in aircraft. McDonnell sited 101 engine related problems encountered when test flying the aircraft with J40's (-6, -8 and -22A) installed. (See chart at the end of the Appendix.) They also stated they felt there were ten instances where the engine performance was the cause of or a contributing factor in an accident or serious flight emergency. The charts and data to support McDonnell's claims were located in a separate report not included in the hearing records that were published.[3] Westinghouse stated they were unaware of any instance where a pilot was injured or killed as a result of engine performance problems but admitted they might never have been informed.

From the hearing and McDonnell's records, the flight hours of the J40 (all models) was given as:

Aircraft Type	Hours Flown	J40 Model
XF3H-1	161.25	-6
XF3H-1	168.05	-8
F3H-1	659.28	-22A
XF4D-1	357	-6, -8
XA3D-1	153	-6
XF10F-1	35-40	-6, -8

While the Air Force experience with J40 engines was not a discussion point at the hearings, the records for the North American X-10 program would add to the Navy total.[4,5] Since each airframe used two engines, the engine time would be double that shown in the table.

Aircraft Type	Hours Flown	Engine
X-10	23 (approx.)	J40-WE-1

The lack of a consistent process for recording, determination (root cause analysis) and communication of the causes of flight problems encountered was an area the hearing brought into focus. It showed that the three parties did not agree to causes of accidents, demonstrate awareness in some cases that events had occurred or that their equipment was found to be the cause of a problem. McDonnell stated over and over the engine was unreliable and Westinghouse stated the Navy never told them the engine was unreliable. The McDonnell report includes a list of the 101 times and methods engine reliability issues were reported in some form to the Navy. Unfortunately, data of this type was not found for Douglas or Grumman. The hearing discussion is quite confused regarding engine unreliability – the airframe/engine mismatch and the engine or airframe reliability issues are never clarified by the parties to a point any conclusions can be made, other than the obvious one that flying underpowered airplanes using a new engine type will increase the hazards of operation, and in this case the hazards were demonstrated to be too great.

The Navy took full responsibility for their decisions, kept nothing from the committee and stated, in response to a direct question, that given the circumstances as they unfolded at the time, they would make the same decisions again.

Total Government outlays on the J40 program (does not include airframe costs) were estimated in the hearings as: Westinghouse development $27M, Production $41.5M, Termination $10M; and Ford - $15M. The Westinghouse corporate expenditures were not given.

The chart on the following page is taken from McDonnell Aircraft Corporation Report 4451, Page 2.1. *Copyright, Boeing – used under license*

CHRONOLOGICAL SUMMARY OF J40 ENGINE TROUBLES ENCOUNTERED BY MᶜDONNELL

CONFIDENTIAL

	XF3H-I	J40-WE-6 (176 FLTS)	TOTAL FLIGHTS
		J40-WE-8 (304 FLTS)	1,153
	F3H-IN	J40-WE-22A (673 FLTS)	

NO. OF A/c ON FLIGHT STATUS

J40-WE-22A ENGINE GROUNDING PERIODS

OPERATIONAL PROBLEMS

1. COMPRESSOR STALLS, MIL. TRANSIENT
2. COMPRESSOR STALLS, A/B IGNITION
3. COMPRESSOR STALLS, A/B STEADY
4. A/B SQUEAL PLACARD
5. MECHANICAL LIMITATIONS
6. LOW THRUST, MILITARY POWER
7. LOW THRUST, A/B POWER
8. A/B DUD IGNITION
9. A/B INSTABILITY AT HIGH ALTITUDE
10. THROTTLE SENSITIVITY
11. ENGINE OVERSPEED
12. ENGINE FLAME-OUT
13. ENGINE SHUTDOWN FIRES
14. ACCELERATION BY-PASS (STALL CONTROL)

COMPONENT PROBLEMS

15. PRIMARY FUEL CONTROL
16. TURBINE OUT TEMP. CONTROL
17. OIL LEAKS & FIRES
18. FUEL LEAKS & FIRES
19. NOZZLE ACTUATOR & NOZZLE
20. TURBINE, GENERAL STRUCTURE
21. HIGH PRESSURE DUAL FUEL PUMPS
22. TURBINE BLADE FAILURE
23. A/B AEROTEC SWITCH
24. A/B FUEL REGULATOR
25. COMPRESSOR. STRUCTURE
26. AUXILIARY ELECTRIC CONTROL (AEC)
27. A/B FUEL PUMP
28. EMERGENCY NOZZLE SYSTEM
29. HIGH PRESSURE OIL PUMP
30. HIGH PRESSURE OIL LINE FAILURE
31. ENGINE OIL SYSTEM "O" RING FAILURE
32. EXCESSIVE OIL CONSUMPTION
33. REAR BEARING FAILURE
34. A/B STRUCTURE

Years across top: 1951, 1952, 1953, 1954, 1955

FLIGHTS: 176, 32 FLTS, 272 FLTS, 59 FLTS, 90 FLTS, 159 FLTS, 171 FLTS, 144 FLTS

F3H-IN GROUNDINGS

A. 18 MAR. TO 20 JUNE 1954 — GROUNDED FOR TURBINE & GENERAL ENGINE INTEGRITY. MAR. SOUND AFTER A.D.L. SERIAL NO. 4080—19 MAR. 1954.
B. 23 SEPT. TO 30 SEPT. 1954 — GROUNDED FOR SEIZING OF HIGH PRESSURE OIL LINE AT REAR BEARING STRUT.
C. 2 OCT. TO 9 OCT. 1954 — GROUNDED FOR INSPECTION OF ENGINE GEAR BOX SPLINE RETAINER RING.
D. 22 OCT. TO 29 OCT. 1954 — GROUNDED FOR INSPECTION & REPLACEMENT OF HIGH PRESSURE OIL LINES AS A RESULT OF OIL LINE FAILURES.
E. 2 DEC. TO 9 DEC. 1954 — GROUNDED FOR INSPECTION OF DUNBAR OIL LINE FITTING. POSSIBLE FAILURE DUE TO OVER TORQUING.
F. 4 DEC. TO 11 DEC. 1954 — GROUNDED FOR INSPECTION OF A.B. HIGH PRESSURE FUEL LINE RETRIAL.
G. 2 FEB. TO 1 APR. 1955 — GROUNDED FOR REPLACEMENT OF DUNBAR FITTED ON HIGH PRESSURE OIL LINES & INSPECTION OF HIGH PRESSURE OIL FILTERS FOR CRACKS, HIGH PRESSURE OIL PUMP DISCHARGE FILTER FAILURE AND HIGH PRESSURE OIL LINE FAILURE.
H. 7 JULY 1955 — FATAL ACCIDENT. FLIGHT TESTING WITH J40 DISCONTINUED.

INDICATES ROUGHLY TIME PERIOD DURING WHICH TROUBLE EXISTED

EACH NUMBER INDICATES A DOCUMENT REGARDING ENGINE TROUBLES

Appendix C Government Hearings Citations

[1] Navy Jet Aircraft Procurement Program, Hearings Before a Subcommittee of the Committee on Government Operations, House of Representatives, Eighty-Fourth Congress, First Session, October 24-27, 1955., Archives Ref: Y4.G74/7:J51

[2] Navy Jet Aircraft Procurement Program, Tenth Intermediate Report of the Committee on Government Operations, HR Report 1891, Union Calendar No. 666, March 15, 1956

[3] McDonnell Aircraft Corporation Report 4451, 19 October 1955, Summary of J40 Engine Troubles Encountered at McDonnell –Statistics. The numbers in the grid referred to specific communication vehicles where the particular issue was communicated to the Navy.

[4] North American Summary Report, X-10 Missile Flight Tests at Edwards Air Force Base, Project WE-104A, 5 September 1956

[5] The X-Planes, X-1 to X-31, Jay Miller, 1988, Orion Books, Crown Publishers, Inc, New York, New York

Appendix D

J40-WE-6 Components and Supplying Vendors

FUEL SYSTEM

<u>Fuel Booster Pump</u> – Nash Engineering Co., Norwalk, CN – Single-stage, centrifugal with throttling type pressure regulating valve.

<u>Dual Fuel Pump</u> – Pesco Products Div., Borg-Warner Corp., Cleveland, OH – Dual positive-displacement pumps, with valves for automatic switching from one pump to the other in case of failure of one pump, and with solenoid for pilot selection of emergency pumping element.

<u>Fuel Filter</u> – Aero Supply Mfg. Co, Corry, PA – Dual, plated-paper, 10-micron, replaceable elements.

<u>Primary Fuel Regulator</u> – Aviation Gas Turbine Div., Westinghouse Electric Corp., Philadelphia, PA – Hydraulic and electrical.

<u>Emergency Fuel Regulator</u> - Aviation Gas Turbine Div., Westinghouse Electric Corp., Philadelphia, PA – Hydraulic but incorporating a solenoid valve for switching from primary to emergency regulator.

<u>Overspeed Relay</u> – Woodward Governor Co., Rockford, IL – Fly-weight governor limiting top rpm hydraulically when the engine is on emergency system.

<u>Dump Valve</u> - Aviation Gas Turbine Div., Westinghouse Electric Corp., Philadelphia, PA- Spring-loaded poppet.

<u>Fuel Distributor</u> - Aviation Gas Turbine Div., Westinghouse Electric Corp., Philadelphia, PA – Rotary, variable-orifice valve operated by fuel-pressure-sensitive actuator.

<u>Valve & Nozzle Assembly</u> - Aviation Gas Turbine Div., Westinghouse Electric Corp., Philadelphia, PA – Diaphragm-operated metering valve and spring-loaded variable-area nozzle.

<u>Combustion Chamber Drain Valve</u> - Aviation Gas Turbine Div., Westinghouse Electric Corp., Philadelphia, PA – Spring-loaded poppet.

OIL SYSTEM

<u>Oil Reservoir</u> - Aviation Gas Turbine Div., Westinghouse Electric Corp., Philadelphia, PA – Five gallon capacity and fixed pick-ups, with air separator.

<u>Side Oil Filler</u> - Aviation Gas Turbine Div., Westinghouse Electric Corp., Philadelphia, PA – Scupper.

<u>Main Oil Pump</u> – W.H. Nichols & Sons, Waltham, MA – Two pump elements and three scavenge elements, self-priming, with replaceable AN filter.

Rear Oil Pump – Eclipse Pioneer Div., Bendix Aviation Corp., Teterboro, NJ – One high pressure pump element and one scavenge element.

Oil Cooler – Clifford Manufacturing Co. (Division of Standard Thomson Corp.), Boston, MA – Coolant (engine fuel) through tubes, oil over tubes.

Exhaust Nozzle Control - Aviation Gas Turbine Div., Westinghouse Electric Corp., Philadelphia, PA – Electrical and hydraulic.

ELECTRICAL SYSTEM

Power Regulator - Aviation Gas Turbine Div., Westinghouse Electric Corp., Baltimore, MD – Electrical (consists of frequency meter, modulators, and amplifiers)

Power Scheduler – Atlas Instrument Co., Haddonfield, NJ – Two Potentiometers

Governor Alternator – Westinghouse Electric Corp., Lima, OH – High speed, inductor, three-phase alternator with no brushes.

Voltage Regulator - Westinghouse Electric Corp., Lima, OH – Carbon pile

Turbine Outlet Temperature Thermocouples –Aviation Gas Turbine Div., Westinghouse Electric Corp., Philadelphia, PA – Chromel-Alumel

Electric Control Relay – Price Electric Corp., Frederick, MD – Three Relays

Ignition Coil – Scintilla Magneto Div., Bendix Aviation Corp., Sidney, NY – Motor-operated breaker.

Sparkplugs – Auburn Sparkplug Co., Inc., Auburn, NY – Double electrode type

Electric Control Cable – Westinghouse Electric Corp., Baltimore, MD – Radio-shielded

ANT-ICING SYSTEM

Anti-icing Valve - Aviation Gas Turbine Div., Westinghouse Electric Corp., Philadelphia, PA-Solenoid-operated air valve and helical bimetallic thermostatic coil

Bibliography

◆━━━━━━━━━━━━━━━━━━━━━━━━━━━━━━━◆

Jane's All the Worlds Aircraft, 1950-1951, 1951-1952, 1052-1953, 1953-1954, 1956-1957; McGraw-Hill Book Company, Inc, New York, 1950-1957

National Advisory Committee for Aeronautics (NACA) Reports, Internet database:

Altitude Investigation of Gas Temperature Distribution at Turbine of Three Similar Axial Flow Turbojet Engines, W.R. Prince and F.W. Schulze, RM E52H06, August 18, 1952

An Evaluation of Turbojet Engine Thrust Control by Exhaust-Nozzle-Area Modulation and Compressor-Inlet Throttling, James L. Harp, Jr., Wallace W. Velie, and William E. Mallett, RM E54F21, August 10, 1954

Analysis of Stage Matching and Off-Design Performance of Multistage Axial-Flow Compressors, Harold B. Finger and James F. Dugan, Jr., RM E52D07, June 27, 1952

Effect of Blade-Root Fit and Lubrication on Vibration Characteristics of Ball-Root-Type Axial-Flow-Compressor Blades, Morgan P. Hanson, RM E50C17, June 15, 1950

Effect of Mach Number on Over-All Performance of Single-Stage Axial-Flow Compressor Designed for High Pressure Ratio, Charles H. Voit, Donald C. Guentert and James F. Dugan, RM E50D26, July 14, 1950

Effect of Rotor- and Stator-Blade Modifications on Surge Performance of an 11-State Axial-Flow Compressor II – Redesigned Compressor for XJ40-WE-6 Engine, E. William Conrad, Harold B. Finger, and Robert H. Essig, RM E52I10, May 25, 1953

Effects of Obstructions in Compressor Inlet on Blade Vibration in 10-Stage Axial-Flow Compressor, Andre J. Meyer, Jr., Howard F. Calvert and C. Robert Morse, RM E9L05, February 13, 1950

Investigation of Performance of Typical Inlet Stage of Multistage Axial-Flow Compressor, Jack R. Burtt, RM E9E13, July 18, 1949

Comparison of Performance of AN-F-58 Fuel and Gasoline in J34-WE-22 Turbojet Engine, Harry W. Dowman and George G. Younger, RM E8L10a, April 7, 1949

Performance of a 13-Stage Development Compressor for the J40-WE-24 Engine at Equivalent Speeds from 30 to 112 Percent of Design, James E. Hatch, James G. Lucas, and Harold B. Finger, RM SE53D17, April 30, 1953

Preliminary Performance Data on Westinghouse Electronic Power Regulator Operating on J34-WE-32 Turbojet Engine in Altitude Wind Tunnel, James R. Ketchum, Darnold Blivas, and George J. Pack, RM SE50J11, October 27, 1950

Preliminary Transient Performance Data for Afterburner Operation of Westinghouse Electronic Power Regulator on XJ34-W-32 Turbojet Engine in Altitude Wind Tunnel, George Vasu, Glennon V. Schwent, and James R. Ketchum, RM SE50L29, February 14, 1951

North American Aviation, Inc., Missile Development Division, Summary Report, X-10 Missile Flight Tests at Edwards Air Force Base, Project W3-104A, Contract No. AF 33(600)-28469, Task No. 448-45 and 55, G.O. 9505, September 5, 1956. National Museum of the United States Air Force Archives.

McDonnell Aircraft Corporation Publications Summary of J40 Engine Troubles Encountered at McDonnell-Statistics, Report 4451, October 19, 1955. Boeing Aircraft Corporation, www.boeingimages.com.

Paul D. Lagasse, Master of Arts Thesis: The Westinghouse Aviation Gas Turbine Division 1950-1960: A Case Study in the Role of Failure in Technology and Business (unpublished). Aircraft Engine Historical Society, www.enginehistory.org

U.S. Air Force, Air Force Research Laboratory, Propulsion Sciences and Advanced Concepts Division, "History of Aviation Fuel Development in the United States", (date unknown)

United States House of Representatives Committee Reports, Reports for 1955-1956, National Archives Building, Washington, DC

84th Congress, 1st Session, Hearings Before A Subcommittee of the Committee on Government Operations House of Representatives, Navy Jet Aircraft Procurement Program, October 24-27, 1955.

84th Congress, 2nd Session, House Report No. 1891, Navy Jet Aircraft Procurement Program, March 15, 1956

United States House of Representatives Committee Reports, Reports for 1955-1956, National Archives Building, Washington, DC

84th Congress, 1st Session, Hearings Before A Subcommittee of the Committee on Government Operations House of Representatives, Navy Jet Aircraft Procurement Program, October 24-27, 1955.

84th Congress, 2nd Session, House Report No. 1891, Navy Jet Aircraft Procurement Program, March 15, 1956

United States Navy, Turbo-Jet Aviation Engine Log Books, United States Air Force Museum Collection, North American X-10 Documentation:

Westinghouse Model J40-WE-1, Serial No. WE030102, Total Operating Hours: 88.79

Westinghouse Model J40-WE-1, Serial No. WE030108, Total Operating Hours: 154.34

United States Navy, Bureau of Aeronautics Contract Correspondence, Record Group 72.3.2, National Archives at College Park, Contracts:

NOa(s) 9212, Containers 303-310
NOa(s) 10067, Containers 794-797
NOa(s) 10114, Containers 822-825
NOa(s) 10123, Container 828
NOa(s) 10385, Containers 934-945
NOa(s) 10943, Containers 1116-1117
NOa(s) 51-966, Container 2347
NOa(s) 52-020b, Containers 2353-2354
NOa(s) 52-048, Container 2463
NOa(s) 52-117, Container 2498
NOa(s) 52-403c, Container 2561
NOa(s) 54-640, Container 3487
NOa(s) 54-695, Container 222
NOa(s) 55-198b, Containers 3684-3686

United States Navy, Bureau of Aeronautics, Record, Power Plant Division Project Correspondence, J40, 1948-1953, Group 72.4.3, National Archives College Park, Containers 2-3

Westinghouse Electric Corporation

Operating Recommendations for Westinghouse Turbo-Jet Engine Model J40-WE-6, February 12, 1951

Operating Recommendations for Model J40-WE-1, -8 and -22A Westinghouse Turbojet Engines, December 1, 1952

Proposal WAGT-X40E3 for XJ40-WE-8, September 16, 1948

Proposal WAGT-X40E8A for XJ40-WE-10 and -12, September 25, 1949

Report A-496, XJ40E2 Preliminary Design Considerations, April 18, 1947

Report A-695, Anti-Icing Design Study, May 13, 1948

Report A-696, Supplemental Oil Supply Design Study, May 12, 1948

Report A-720, Performance Curve Corrections for Losses of Turbojet Engines, June 15, 1948

Report A-721, Sample Calculations to be Used with XJ40-WE-6 Performance Curves, June 18, 1948

Report A-749, Structural Requirements for Cantilevering an Afterburner and Extension, September 13, 1948

Report A-754, Sample Calculations for Use with the XJ40E3 Performance Curves, September 21, 1948

Report A-810 Hydraulic Starter System, May 2, 1949

Report A-840 Sample Calculations to Use with J40-WE-6 and J40-WE-8 Performance Curves, April 28, 1949

Report A-942 Information on Ignition Systems, April 19, 1950

Report A-969 Power Lever Torque Measurements, September 1, 1950

Report A-974 Calculation of No. 1 Bearing Thrust Load on J40-WE-10 and J40-WE-12, July 18, 1950

Report A-999 Stress and Vibration Analysis of the XJ40-WE-6 Jet Engine, June 19, 1950

Report X-1058 Analysis of XJ40-WE-10 and -12 Lubrication System, February 3, 1951

Report A-1081, Description of the Control System for the J40-WE-6, February 9, 1951

Report A-1092 (Revised), Engine Windmilling Characteristics, March 30, 1951

Report A-1092 (Revised), Windmilling Under Flight Conditions, April 10, 1951

Report A-1093 150 Hour Qualification Test Model XJ40-WE-6 Serial No. WE003003 Navy Contract NOa(s) 9212, February 10, 1951

Report A-1096 Component Qualification Test of Scintilla Type TLN Low Voltage High Energy Surface Gap Ignition System for J34 and J40 Turbo-Jet Engines, Project 1375, S.O. 4A3310, Contract NOa(s) 10067, April 25, 1951

Report A-1123 Interim Report on the Design of the Fuel and Control System, October 31, 1951

Report A-1125 Airflow Testing of Power Regulator, November 9, 1951

Report A-1167 Determination of Blade Clearance on -6 and -8 Engines, June 6, 1951

Report A-1174 Design and Test of Accessory Gear Box, August 31, 1951

Report A-1208 J40-WE-6 Primary and Emergency Fuel Regulator Component Qualifications, August 25, 1951

Report A-1237 Summary Evaluation Report on the XJ40-WE-6 Jet Engine Model, May 20, 1952

Report A-1248 Component Testing of Fuel Pumping System, April 16, 1952

Report A-1264 Evolution of the Alternate Hydraulic Control System for the -10 and -12, January 8, 1952

Report A-1266 Description of the Alternate Hydraulic Control System for the XJ40-WE-10 and -12, December 27, 1951

Report A-1267 Simulation of the Alternate Hydraulic Control System for the XJ40-WE-10 and -12, December 27, 1951

Report A-1283 Qualification Test of Fuel Pump System for the J40-WE-6, December 18, 1951

Report A-1286 Rough Cutover Analysis, January 23, 1952

Report A-1309 Alternate Hydraulic Control, February 5, 1952

Report A-1358 Report on Titanium Compressor Blades, May 19,.1952

Report A-1364 Re-run of Low Temperature of Exhaust Nozzle Actuator, May 19, 1952

Report A-1419 WE030014 Damage Report, September 10, 1952

Report A-1420 Transmittal of Report on Gearbox Failure on WE030006, November 10, 1952

Report A-1430 Preliminary Altitude Performance Estimates for XJ40-WE-10 and -12 Block II and Block III, (Not dated, graphs dated June 14, 1952 for -10 and August 6, 1952 for -12)

Report A-1438 Preliminary Altitude Performance Estimates for the XJ40-WE-10 and -12 Block IV, August 28, 1952

Report A-1456 Trans Response Report of Exhaust Nozzle Using Automatic Trim Control, October 2, 1952

Report A-1500 Overhaul Condition of J40-WE-6 WE030015 After Compressor Failure, August 17, 1953

Report A-1514 Report on Flight Test Analysis, January 5, 1953

Report A-1626 Mechanical Limits Report on the J40, July 10, 1953

Report A-1636 Revised Testing of Constant Speed Transmission, December 31, 1953

Report A-1656 Constant Speed Transmission Qualification Test Report, November 11, 1953

Report A-1669 Qualification Test Report for Accessory Gearbox, October 29, 1953

Report A-1753 Removal of Hydraulic Locks from Hydro-Mechanical Exhaust Nozzle Control System, March 5, 1954

Report A-1776 Analysis on Use of Hydraulic Fuses, May 17, 1954

Report A-1802 150 Hr. Full Load Parts Verification Test of J40-WE-22 Gearbox, June 10, 1954

Report A-1806 Submission of 150 Hr. Test on Lycoming-Spencer Gearbox, August 3, 1955

Service Instruction for Westinghouse Turbojet Engine Model J40-WE-6, December 15, 1951

Service Instructions for J40-WE-1 and -8 Turbojet Engines, January 20, 1954

Specification WAGT-X40E2-2 for XJ40-WE-2, -4 and -6, April 15, 1947

Specification WAGT-X40E8A-2B, Appendix D for the XJ40-WE-18, April 24, 1951

Specification WAGT-X40E3A-1B, for J40-WE-1, March 13, 1953

Specification WAGT-X40E3A-1D for J40-WE-1, April 2, 1954

Specification WAGT-X40E2-2E for J40-WE-6, October 18, 1949

Specification WAGT-X40E2-50 for J40-WE-22 Gearbox, October 31, 1952

Specification WAGT-X40E2-4B for J40-WE-8 and J40-WE-22, March 3, 1953

Illustrations

United States Navy, Bureau of Aeronautics Contract Correspondence, Contract NOa(s) 10414, AER-AC-38, Mock-up Board Report for Model XA3D-1
    Douglas ES 76355, 09/12/1949
    Douglas ES 76360, 09/12/1949

Westinghouse Electric Corporation Steam Division, (Accession 1969.170) Hagley Museum and Library, Wilmington, DE 19807

Photographs by Westinghouse "P" reference number/ Date/Box Number
    P36997 / 04/11/1947 / 243
    P39237 / 07/01/1948 / 244
    P39239 / 07/01/1948 / 244
    P39240 / 07/01/1948 / 244
    P39607 / 09/24/1948 / 244
    P39842 / 10/22/1948 / 244
    P39878 / 10/26/1948 / 244
    P39879 / 10/26/1948 / 244

P39894 / 11/01/1948 / 244
P39895 / 11/01/1948 / 244
P40631 / 03/04/1949 / 244
P40768 / 04/07/1949 / 244
P40770 / 04/07/1949 / 244
P40771 / 04/07/1949 / 244
P40772 / 04/07/1949 / 244
P40779 / 04/08/1949 / 244
P41351 / 08/13/1949 / 244
P41588 / 10/03/1949 / 244
P41589 / 10/03/1949 / 244
P41590 / 10/03/1949 / 244
P41952 / 12/12/1949 / 244
P42038 / 01/12/1950 / 244
P42095 / 01/23/1950 / 244
P43138 / 09/04/1950 / 245
P43161 / 09/04/1950 / 245
P43566 / 11/20/1950 / 245
P43786 / 01/09/1951 / 245
P43793 / 01/10/1951 / 245
P43797 / 01/10/1951 / 246
P43799 / 01/10/1951 / 246
P43802 / 01/10/1051 / 246
P43803A / 07/03/1952 / 246
P43806 / 01/10/1951 / 246
P43830A / 01/12/1951 / 246
P43832A / 07/30/1953 / 246
P43868 / 01/17/1951 / 246
P43884 / 01/19/1951 / 246
P43895 / 01/19/1951 / 246
P43946 / 02/01/1951 / 246
P44137 / 02/28/1951 / 247
P44895 / 07/19/1951 / 248
P45375 / 10/01/1951 / 248
P45610 / 11/01/1951 / 249
P45712 / 11/21/1951 / 249
P45765 / 11/28/1951 / 249
P45766 / 11/28/1951 / 249
P45807 / 12/05/1951 / 249
P46226A / 12/15/1952 / 249
P46550A / 07/03/1952 / 250
P46561A / 07/03/1952 / 250
P46565 / 04/02/1952 / 250
P46655 / 04/11/1952 / 250
P46666 / 04/11/1952 / 250
P46813 / 04/25/1952 / 251
P46814 / 04/25/1952 / 251
P46815 / 04/25/1952 / 251
P47191 / 07/21/1952 / 252
P47227A / 07/03/1952 / 252
P47228A / 12/15/1952 / 252
P47604 / 09/03/1952 / 252
P47722 / 09/16/1952 / 252

P47734 / 09/18/1952 / 253
P47912 / 10/17/1952 / 254
P47913 / 10/17/1952 / 254
P47979 / 10/29/1952 / 254
P48162 / 11/28/1952 / 254
P48171 / 12/01/1952 / 254
P48173 / 12/01/1952 / 254
P48207 / 12/03/1952 / 254
P48298 / 12/12/1952 / 254
P48641 / 01/30/1953 / 255
P48913 / 03/10/1953 / 256
P49160 / 04/13/1953 / 256
P49191 / 04/15/1953 / 256
P49296 / 04/29/1953 / 257
P49509 / 06/10/1953 / 257
P49511 / 06/10/1953 / 257
P49513 / 06/10/1953 / 257
P49515 / 06/10/1953 / 257
P49675 / 06/29/1953 / 258
P49676 / 06/30/1953 / 258
P49763 / 07/16/1953 / 258
P50020 / 08/26/1953 / 258